NEW GCSE SCIENCE

Separate Sciences B

For Specification Modules B5, B6, C5, C6, P5 and P6

OCR

Gateway Science

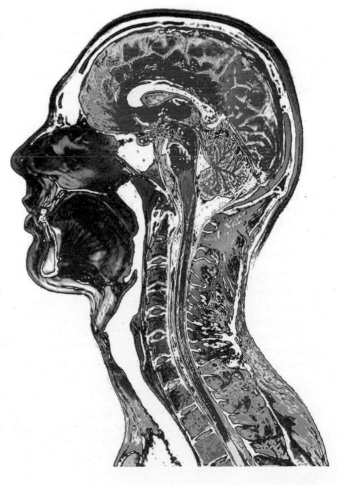

Series Editor: Chris Sherry

**Authors: Colin Bell,
Brian Cowie, Ann Daniels,
Sandra Mitchell
and Louise Smiles**

Student Book

William Collins' dream of knowledge for all began with the publication of his first book in 1819. A self-educated mill worker, he not only enriched millions of lives, but also founded a flourishing publishing house. Today, staying true to this spirit, Collins books are packed with inspiration, innovation and practical expertise. They place you at the centre of a world of possibility and give you exactly what you need to explore it.

Collins. Freedom to teach

Published by Collins
An imprint of HarperCollinsPublishers
77–85 Fulham Palace Road
Hammersmith
London
W6 8JB

Browse the complete Collins catalogue at
www.collinseducation.com

© HarperCollinsPublishers Limited 2011

10 9 8 7 6 5 4 3 2 1

ISBN-13-978-0-00-741534-2

British Library Cataloguing in Publication Data
A Catalogue record for this publication is available from the British Library

Commissioned by Letitia Luff
Project managed by Tammy Poggo and Hart McLeod
Production by Kerry Howie

Edited, proofread, indexed and designed by
Hart McLeod
Proofread by Grace Glendinning
New illustrations by Simon Tegg
Picture research by Caroline Green
Concept design by Anna Plucinska
Cover design by Julie Martin

Printed and bound by L.E.G.O. S.p.A. Italy

Acknowledgements – see pages 318–319

Contents

Chemistry

C5 How much (quantitative analysis)

C6 Chemistry out there

Physics

P5 Space for reflection

P6 Electricity for gadgets

How to use this book

Welcome to Collins New OCR Gateway Separate Sciences B

The main content

Each two-page lesson has three sections corresponding to the Gateway specification:

> For Foundation tier you should understand the work in the first and second sections.

> For Higher tier you need to understand the first section and then concentrate on the second and third sections.

Each section contains a set of level-appropriate questions that allow you to check and apply your knowledge.

Look for:

> 'You will find out' boxes

> Internet search terms (at the bottom of every page)

> 'Did you know' and 'Remember' boxes

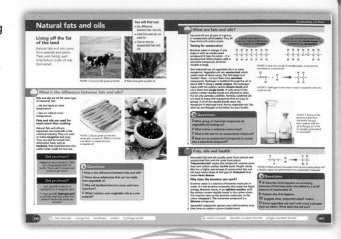

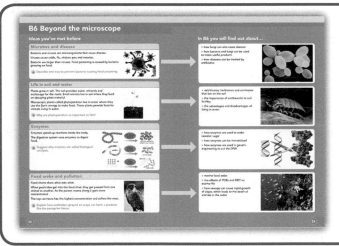

Module introductions

Each Module has a two-page introduction.

Link the science you will learn with your existing scientific knowledge to give you an overview before starting each Module.

Remember!

To cover all the content of the OCR Gateway Science specification you should study the text and attempt the exam-style questions.

Checklists

Each Module contains a checklist.

Summarise the key ideas that you have learned so far and look across the three columns to see how you can progress.

Refer back to the relevant pages in this book if you find any points you're not sure about.

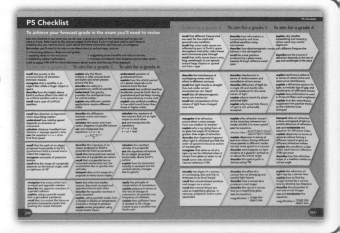

Exam-style questions

Every Module contains practice exam-style questions for both Foundation and Higher tier, labelled with the Assessment Objectives that it addresses.

Familiarise yourself with all the types of question that you might be asked.

Worked examples

Detailed worked examples with examiner comments show you how you can raise your grade. Here you will find tips on how to use accurate scientific vocabulary, avoid common exam errors, improve your Quality of Written Communication (QWC), and more.

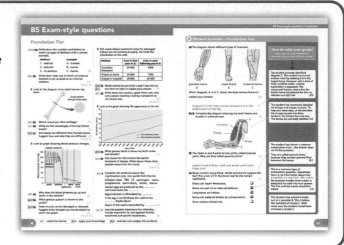

Preparing for assessment

Each Module contains preparing for assessment activities. These will help build the essential skills that you will need to succeed in your practical investigations and Controlled Assessment, and tackle the Assessment Objectives in your written exam.

Each type of preparing for assessment activity builds different skills.

> Applying your knowledge: Look at a familiar scientific concept in a new context.

> Research and collecting secondary data: Plan and carry out research using handy tips to guide you along the way.

> Planning and collecting primary data: Plan and carry out an investigation with support to keep you on the right track.

> Analysis and evaluation: Process data and draw conclusions from evidence. Use the hints to help you to achieve top marks.

Assessment skills

A section at the end of the book guides you through your practical work, your Controlled Assessment tasks and your exam, with advice on: planning, carrying out and evaluating an experiment; using maths to analyse data; the language used in exam questions; and how best to approach your written exam.

B5 The living body

Ideas you've met before

Organisation

Cells are organised into tissues, organs and body systems.

Animal cells make up tissues, for example blood tissue.

Different tissues form an organ, for example the heart.

Different organs work together in a system, for example the circulatory system.

 Name three systems in the human body.

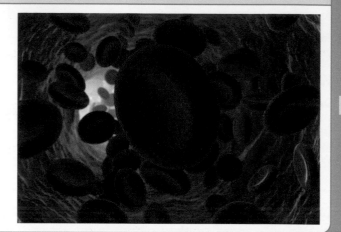

Reproduction

The human reproductive cycle includes adolescence, fertilisation and foetal development.

Living organisms can reproduce sexually.

In humans, the male produces sperm, the female produces eggs.

An egg can be fertilised by a sperm.

In humans (mammals), the fertilised egg develops inside the mother's body.

 Why is a human egg much bigger than a human sperm?

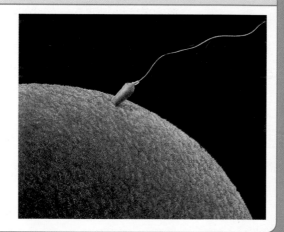

Precautions

Conception, growth and development can be affected by diet, drugs and disease.

Since the unborn baby develops inside the mother, she should be aware of the possible effects of diet, drugs and disease.

Different people grow and develop at different rates.

 What special precautions should be taken by a pregnant mother?

In B5 you will find out about...

> the difference between external and internal skeletons

> fractures and ossification of bone

> the difference between single and double circulatory systems

> pacemaker cells and blood transfusions

> advantages and disadvantages of gills and lungs

> information on lung diseases

> the importance of physical and chemical digestion

> adaptations of the small intestine for food absorption

> how the kidney and dialysis machine works

> how and why the urine concentration varies

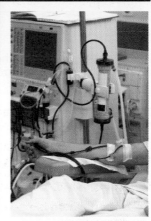

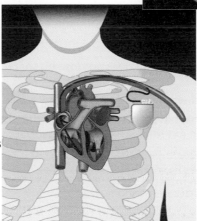

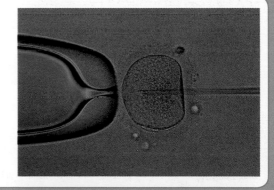

> how hormones control the menstruation cycle

> what causes infertility and how it can be treated

> issues raised by infertility treatments

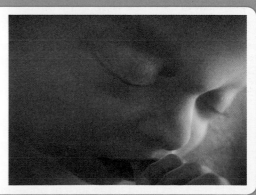

> the importance of checking foetal growth and development

> understanding average growth charts

> having to replace body parts due to disease or trauma

> about the advantages and disadvantages of donor register systems

Skeletons

You will find out:
> about different types of animal skeletons
> what skeletons are made from
> how bones grow

Supporting life

Jack is on holiday by the sea. He goes swimming and sees some colourful jellyfish. (He keeps away from them as he knows they can sting.) Some have been washed on to the beach.

He is disappointed to see they have lost their shape. It is the water that supports them.

FIGURE 1: Living jellyfish are supported by water. What has happened to this stranded jellyfish?

Skeletons

When Jack comes out of the water, he does not lose his shape. He is supported by a hard skeleton.

Animals such as jellyfish and worms do not have a hard skeleton.

Other animals, such as insects, have a hard skeleton on the outside of their bodies. This is called an **external skeleton**. It is made from a material called chitin.

Animals such as fish, frogs, birds and humans have a skeleton inside their bodies. This is called an **internal skeleton**. It is made from **cartilage** or bone.

A human skeleton is made mainly from bone. The outer ear, nose and ends of the long bones are made from cartilage.

A shark skeleton is made just from cartilage. Cartilage is softer and more flexible than bone.

FIGURE 2: A praying mantis has an external skeleton. What is its skeleton made from?

Questions

1 Why does a jellyfish collapse when it is out of water?

2 What type of skeleton do insects have?

3 What is most of our skeleton made from?

4 How is cartilage different from bone?

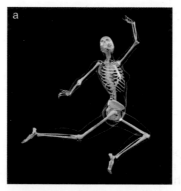

FIGURE 3: a A human skeleton and **b** a shark skeleton. Can you think why it helps a shark to have a skeleton made from cartilage rather than bone?

Comparing an internal and an external skeleton

Jack thinks his internal skeleton is better than an insect's external skeleton. This is because his skeleton:

> forms a framework for his body

> grows as he grows

> allows muscles to be easily attached

> has many joints to give flexibility

> is made from living tissue such as bone cells, cartilage and blood cells.

Q X-rays of fractures   antagonistic muscles

Structure of a long bone

The long bones in Jack's arms and legs are hollow. This makes them stronger and lighter and less likely to break than if they were solid.

A long bone has a main shaft. Inside is **bone marrow**, which contains blood vessels. Each end of a long bone is called the head and is covered with cartilage. Cartilage:

> absorbs shock

> helps bones to slide over each other in joints.

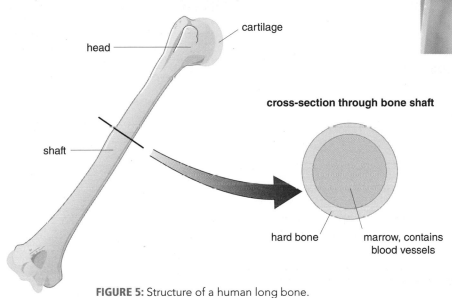

cross-section through bone shaft

FIGURE 5: Structure of a human long bone. Why is it essential that bone is supplied with blood?

FIGURE 4: An insect moulting. Why must the external skeleton be shed before the insect can grow?

Questions

5 What cells are found in bone?

6 Why does an insect not grow continuously?

7 When an insect moults it usually eats its old skeleton. Suggest why.

8 Explain why you do not break your bones when you jump up and down.

Repair and growth of bone

When bones are broken or cartilage is damaged they can become infected with microorganisms. During surgery on bones, like in any other operation, there must be sterile conditions. However, since bone and cartilage are living tissue, they quickly grow and repair themselves.

Jack wonders if he has stopped growing. Before he was born his bones were all made from cartilage. As he ages, calcium salts and phosphates are deposited in his bones, making them hard. This process is called **ossification**.

Jack finds out that an X-ray image of his fingers shows to what extent bone has replaced cartilage. If there is no cartilage between the head of the bone and the main shaft, he has stopped growing.

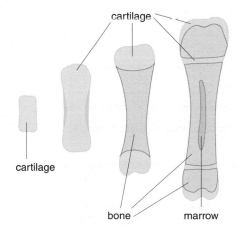

FIGURE 6: The process of ossification. Why does a 'pad' of cartilage remain at the ends of bones in adults?

Questions

9 Suggest why it might be important for Jack to know if he has stopped growing.

10 Why is it important for teenagers to have a diet rich in calcium?

Q external skeletons

Broken bones

Jack has an accident. He thinks he has broken a bone in his arm. A broken bone is called a **fracture**. Jack goes to hospital for an X-ray. X-rays take pictures of the inside of the body. Bones are clearly seen.

There are different kinds of fractures:

> a **greenstick fracture** is when a bone is not completely broken

> a **simple fracture** is when a bone is cleanly broken

> a **compound fracture** is when the broken bone breaks through the skin.

Joints

A joint is when two or more bones meet.

There are:

> fixed joints, found in the skull, where the bones cannot move

> hinge joints, found in the elbow and knee

> ball and socket joints, found in the shoulder and hip.

Instead of breaking a bone, Jack could have damaged:

> a **ligament** which connects bones to each other

> a **tendon** which connects a muscle to a bone (muscles move bones).

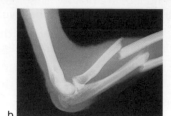

FIGURE 7: a The three kinds of fractures in bones. **b** What type of fracture does this X-ray picture show?

b

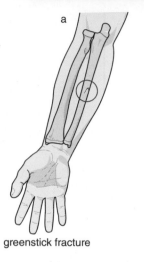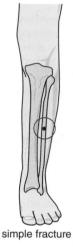

a

greenstick fracture   simple fracture   compound fracture

Questions

11 What is a fracture?

12 Which type of fracture mends quickly?

13 What type of joint is found in the skull?

14 Jack hopes he has not broken the largest bone in his arm. What is the name of this bone?

FIGURE 8: The main structures in the arm. What do ligaments join and what do tendons join?

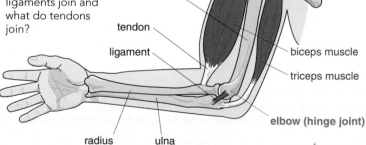

shoulder (ball and socket joint)

humerus

tendon

ligament

biceps muscle

triceps muscle

elbow (hinge joint)

radius   ulna

More on bones

Despite being very strong, bones can easily be broken by a sharp knock, which concentrates the force on a small area.

Jack's elderly grandparents take special care. Their bones are much more brittle due to **osteoporosis**. This is a condition in which bone degenerates. Any fall could easily break their bones. They make sure that their diets are rich in calcium and vitamin D. They visit the hospital to have a bone-density test. Jack's grandparents may eventually need to have their knee or hip joints replaced.

The hip is a ball and socket joint and the knee is a hinge joint. They are called **synovial joints** because they contain synovial fluid.

Remember!
Tendons join muscles to bone; ligaments join bones together.

Q ossification of bone   artificial hip joints

Ball and socket joint

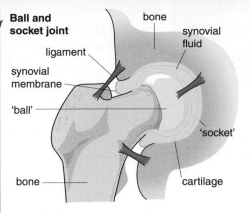

bone
synovial fluid
ligament
synovial membrane
'ball'
bone
'socket'
cartilage

Hinge joint

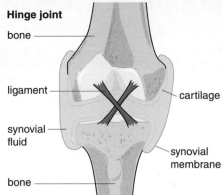

bone
ligament
cartilage
synovial fluid
synovial membrane
bone

FIGURE 9: Two types of joints. The hinge joint can move only up or down. The ball and socket joint can rotate in almost all directions. Apart from the hip, where in the body is there a ball and socket joint?

How the arm moves

Jack has been doing exercises. He is proud of his biceps and triceps muscles. They work as a pair of muscles called **antagonistic muscles** that bend or straighten his arm. As one muscle contracts the other relaxes. This produces precise movements to do complex activities.

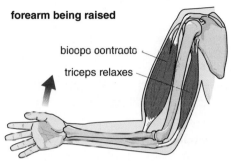

forearm being raised

biceps contracts
triceps relaxes

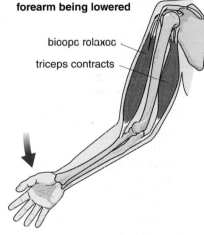

forearm being lowered

biceps relaxes
triceps contracts

FIGURE 10: Raising and lowering the forearm. How do the biceps and triceps muscles work to bring about movement?

Questions

15 What do you understand by a bone-density test?

16 When the arm is being bent, does the biceps muscle get longer and thinner or shorter and fatter? Explain your choice.

17 Suggest why knee joints and hip joints sometimes need replacing.

Caring for a fracture patient

When Jack had his accident, he knew it was dangerous to move a suspected broken arm. Blood vessels and nerves could be damaged. With suspected spinal injuries, specialised first aid is needed. It is very important not to move a person who has suspected damage to their spine. This is because the spinal cord could be damaged, resulting in paralysis or death.

Synovial joints

Synovial fluid in joints is important because it absorbs shock and acts as a lubricant. The synovial membrane keeps the synovial fluid within the joint, preventing any leakage. Cartilage on the ends of bones also helps to absorb shock. Ligaments connect bones, preventing dislocations.

Artificial knee and hip replacements are now common operations. After surgery, patients enjoy an active, pain-free life. However, as with all transplants, there is a danger of rejection and infection.

Levers

The arm movement is an example of a **lever**. The elbow acts as a fulcrum (pivot). As the biceps muscle contracts, it exerts an upward force on the arm bones. Although the muscle contracts for only a short distance (about 9 cm) the hand moves much further (about 60 cm).

Questions

18 Describe what you should do if you find someone with a suspected spinal injury.

19 Your grandmother has been advised that she needs a hip replacement, but she is not convinced. Describe the advantages of having a hip replacement.

Circulatory systems and the cardiac cycle

You will find out:
> about different types of circulatory system
> about open and closed circulatory systems

Misunderstanding about the heart

In the Middle Ages, the heart was believed to control emotions. Blood was thought to be one of four 'body humours'. The others were 'phlegm', 'bile' and 'black bile'. An imbalance of these humours was thought to cause illnesses. Being a doctor was not very scientific!

FIGURE 1: The four humours, as shown in an ancient manuscript. Suggest reasons why doctors were ignorant about the heart and blood.

Ways of circulating oxygen in the body

Some animals, such as amoeba, are very small.

They are so small that all parts of their body can get oxygen directly from the water they live in.

They do not need to have a blood circulatory system.

Some animals, such as insects, do not have lungs for taking in air. The oxygen is taken straight to cells in the body in special tubes.

Insect blood does not carry oxygen. The blood does not travel in blood vessels. It flows slowly round the body cavity. This is called an **open circulatory system**.

Shabeena, like all humans, has a **closed circulatory system**.

Blood flows in blood vessels called arteries, veins and capillaries.

The blood carries oxygen and nutrients all around the body.

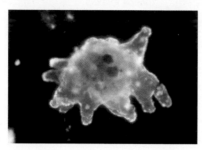

FIGURE 2: Why does an amoeba not need a circulatory system?

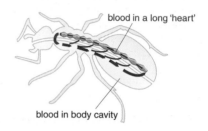

blood in a long 'heart'

blood in body cavity

FIGURE 3: An insect has an open circulatory system. Does an insect have lungs?

FIGURE 4: Arteries in Shabeena's hand. What do arteries carry?

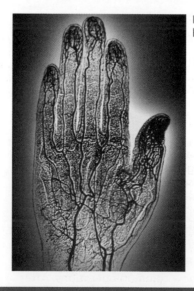

Questions

1 Insect blood does not carry oxygen. Suggest what it does carry.

2 What is a closed circulatory system?

Closed circulatory systems

Many animals have specialised organs in which oxygen enters and carbon dioxide leaves. However, all living body cells need a supply of oxygen and need to dispose of waste carbon dioxide. This means that many animals need a blood circulatory system for transport.

Fish have a closed circulatory system. Their blood picks up oxygen from the gills. The blood goes around a single circuit from the heart. The fish heart therefore has two chambers, one

Q blood circulatory systems   Galen's discoveries about blood to receive blood and one to push it round the body. This is a **single circulatory system**.

Like all humans, Shabeena has a **double circulatory system**. The blood goes round two circuits:

> from the heart to the lungs

> from the heart to the rest of the body.

The human heart therefore has four chambers, two to receive blood and two to push it to the lungs/body.

Changes in blood pressure

As blood flows from arteries through arterioles (small arteries), capillaries and veins, its pressure decreases. The high blood pressure in arteries is caused by the heart muscles contracting to force the blood around the body.

Heart rate

The rate at which Shabeena's heart beats (heart rate) is also affected by hormones, such as **adrenaline**. Adrenaline prepares Shabeena's body for 'flight or fight'. By increasing her heart rate, more energy is supplied to her muscles.

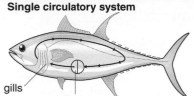

Single circulatory system

FIGURE 5: Fish have a single circulatory system.

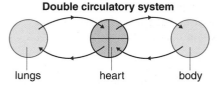

Double circulatory system

FIGURE 6: Mammals have a double circulatory system.

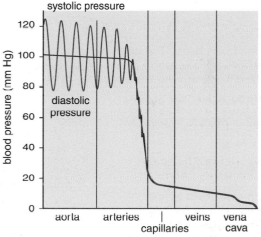

FIGURE 7: Pressure changes in blood vessels. Why does the blood pressure drop in veins?

Questions

3 Why are there large fluctuations in pressure in arteries?

4 Why is the blood pressure high in arteries?

Heart structure and different circulatory systems

> The heart in a single circulatory system has two chambers, one to receive blood and one to pump blood out. This means that the blood pressure is not high, because it has to flow through a meshwork of capillaries in the gills before it reaches the muscles of the body.

> The heart in a double circulatory system has four chambers, and the blood travels through two circuits. This means that the blood pressure in the circuit around the body is maintained at a high level, resulting in a much quicker transport of materials. The right side of the heart contains blood without oxygen (deoxygenated blood) and the left side contains blood with oxygen (oxygenated blood). The valves prevent backward flow of blood.

Question

5 List the sequence of events from blood entering the right atrium to it leaving the left ventricle. Include the names of all the structures involved.

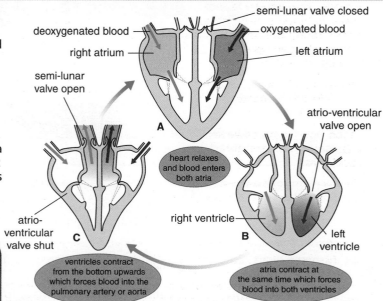

FIGURE 8: The cardiac cycle. Contraction of the heart is called systole (diagrams B and C); relaxation between beats is called diastole. The cardiac cycle takes 0.8 seconds.

Heart muscle

Shabeena's heart is made from powerful muscle.

It is a very special type of muscle that contracts and relaxes without getting tired. It uses a lot of energy.

Oxygen and food (including glucose) are carried in Shabeena's blood. Both glucose and oxygen are needed for respiration.

The **coronary artery** provides a constant supply of food and oxygen to Shabeena's heart. It is seen on the outside of the heart.

Pulse rate

Shabeena measures her pulse rate. The pulse is caused by blood spurting through arteries when the heart beats. The heart muscle contractions put the blood under pressure and cause it to move around the body.

Shabeena can feel her pulse:

> on the underside of her wrist

> in her neck

> by the side of her temple

> by a probe on her ear.

She counts the number of beats per minute.

FIGURE 9: Can you see the coronary artery? What does it do?

FIGURE 10: Measuring pulse in different parts of the body. Can you find your pulse?

underside of wrist

neck

temple

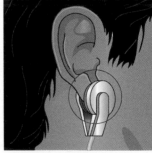

ear lobe

Questions

6 How is Shabeena's heart muscle different from muscles in her arm?

7 What causes a pulse?

8 What happens to the blood when the heart muscles contract?

9 Name four places on the body where the pulse can be measured.

Heart rate

When Shabeena exercises, her heart rate increases. When she rests, the rate drops.

An increased heart rate supplies more blood containing oxygen and food to her muscles.

Her heart rate is controlled automatically. She does not have to worry about forgetting to alter it! Small groups of cells called a **pacemaker** produce a small electric current. This stimulates heart muscle, which then contracts.

Did you know?

Fitting an artificial pacemaker requires only a local anaesthetic and takes about 1 hour. People with a pacemaker carry an identity card to avoid going through a metal detector at airports.

Shabeena's father has been referred to a hospital with a heart problem. Doctors have used:

> an ECG (electrocardiogram) to record the nerve impulses in his heart

> an echocardiogram, which uses ultrasound to produce pictures of his working heart.

Shabeena's father needs to have an artificial pacemaker fitted. It will send electrical impulses to his heart to regulate his heartbeat.

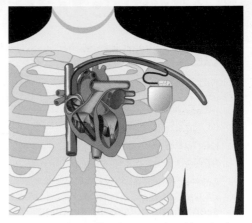

FIGURE 11: An artificial pacemaker. How is the pacemaker connected to the heart?

Questions

10 Why does Shabeena's heart rate increase when she runs?

11 Why does Shabeena not have to worry about changing her pulse rate?

Understanding blood circulation

Our understanding of how the heart works is based on two theories. In the 2nd century, Galen was the first doctor to realise the importance of the pulse in medicine. However, he believed that the liver made blood, which was then pumped around the body by the heart in a backwards and forwards movement.

William Harvey's theory on blood circulation was published in the 17th century. He explained that the heart had four chambers and that the blood travelled through arteries and veins. He believed that they were joined by tiny blood vessels, but the microscopes at that time were not good enough for him to see capillaries.

The sino-atrial node (SAN) group of cells is very unusual because it generates electrical impulses. These spread across the atria, causing both atria to contract. When the impulses reach the atrio-ventricular node (AVN), more impulses spread across the ventricles, causing them to contract. Impulses from the vagus and sympathetic nerve can modify the action of the pacemaker cells.

Electrocardiogram (ECG)

An ECG shows the changes in electrical impulses in the heart muscles. Electrodes attached to the patient's chest sense these changes. Problems with parts of the heart can be identified.

Echocardiogram

The video of an echocardiogram shows if any parts of the heart, such as valves, are not working properly. Use a web search (using the term 'echocardiogram video') to see a normal and damaged heart in action.

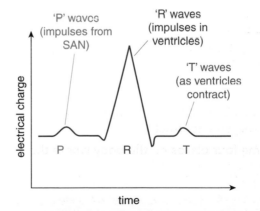

FIGURE 13: An ECG trace of a normal heart. What does the 'P' wave represent?

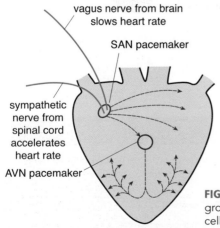

FIGURE 12: How groups of pacemaker cells control heart contraction.

Questions

12 Explain why electric shocks can stop or start the heart beat.

13 An ECG trace shows an abnormal 'R' wave. What could this show?

14 Suggest why some of Galen's ideas were incorrect.

Running repairs

You will find out:
> about what blood is made from
> about the functions of the blood components
> how heart problems can be treated

Donating blood

Rhys is a blood donor. He goes to his local blood donation centre. He donates 300 cm³ of blood at each visit. His body takes 24 hours to replace the lost blood. The blood is stored at various centres throughout the country.

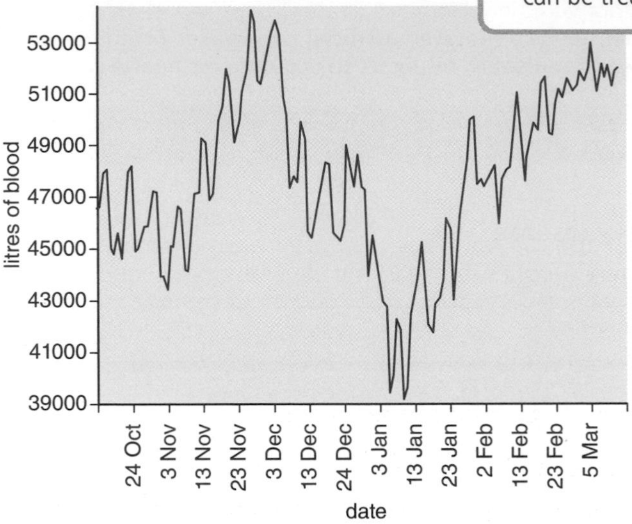

FIGURE 1: Blood stocks in England 2010/2011. Suggest why blood stocks vary throughout the year.

(Graph: litres of blood (y-axis, 39000 to 53000) vs date (x-axis, 24 Oct to 5 Mar))

Parts of the blood

Rhys identifies the parts of his blood.

> The liquid part is called the **plasma**.

> In the plasma there are **red blood cells** and **white blood cells**. There are many more red blood cells than white cells.

> **Platelets** are small pieces (fragments) of cells.

Heart problems

Rhys finds out that there are many types of heart problems.

> An irregular heart beat. This is when the heart does not beat properly.

> A hole in the heart. This is caused by a hole in the heart wall between its left and right sides.

> Damaged or weak heart valves between the **atria** (top chambers in the heart) and ventricles.

> Coronary heart disease. This is caused by a blockage in the coronary artery and can lead to a **heart attack**.

FIGURE 2: Parts of the blood. Look how many more red blood cells there are than white blood cells in this colourised scanning electron micrograph. How many white blood cells can you find?

white blood cell

red blood cell

platelet

Did you know?

One adult in the UK dies from heart disease every 4 minutes.

Did you know?

An adult has about 6 dm³ of blood containing 25 billion red blood cells. End to end, the blood cells would stretch four and a half times round the world!

Questions

1 Which part of the blood carries the red and white blood cells?

2 What are platelets?

3 Where are the main valves in the heart?

4 Which blocked artery can cause a heart attack?

heart attack   hole in the heart

Heart problems

'Hole in the heart'

As the heart develops before birth, the division between atria and ventricles is not complete. Sometimes a hole can be left at birth. This 'hole in the heart' allows oxygenated and deoxygenated blood from the left and right sides of the heart to mix (see Figure 4). Therefore blood leaving the heart in the aorta carries a lower amount of oxygen. Babies born like this are called 'blue babies'. Open-heart surgery can repair the heart wall.

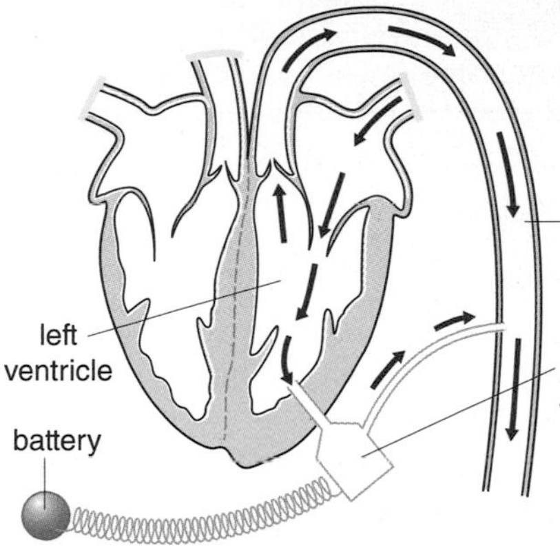

Coronary artery disease

The coronary artery supplies the heart muscle with oxygen and food. If the artery becomes blocked, the heart muscles are not supplied with energy. This can result in a heart attack. During surgery the artery can be by-passed using veins transplanted from other parts of the body. The new veins take blood from the aorta, avoiding the blocked coronary artery.

'Heart-assist devices'

Doctors use 'heart-assist' devices to reduce the work done by heart muscles. They help to pump the blood. This allows the heart muscles to recover, and then the device can be removed.

Problems with heart valves

Valves in the heart prevent back flow of blood. Weak or damaged heart valves affect the blood circulation, slowing down the supply of food and oxygen and affecting heat distribution. Heart valves can be replaced by artificial valves in open-heart surgery.

key:
blood flow

aorta

left ventricle

heart 'back-up' – a heart-assist device

battery

FIGURE 3: Describe how this heart-assist device helps the heart.

Questions

5 Suggest why 'hole in the heart' babies are sometimes called 'blue babies'.

6 In coronary artery by-pass surgery, which major artery do the veins connect to?

Why is there a hole in the heart?

An unborn baby's blood circulation is very complicated, because the blood is oxygenated by the mother's blood in the placenta, not from the lungs. This is because the baby's lungs are not developed enough to work and do not have direct access to air. Therefore there is a hole between the two atria allowing most blood to bypass the lungs. This hole normally closes soon after birth. If this hole remains after birth, the blood (which is now oxygenated by the lungs) mixes with deoxygenated blood from the body. Blood leaving the heart to go around the body now has a lower amount of oxygen available to muscles.

A pacemaker or a heart transplant?

When someone has a serious heart problem, they may have the choice of having:

> repairs, such as replacement artificial valves and a pacemaker

> a heart transplant.

Using replacement valves or a pacemaker means a less traumatic and difficult operation and less risk of the body rejecting them. However, they have a greater risk of needing to be replaced. People receiving a heart transplant need to take immuno-suppressive drugs for the rest of their lives to avoid the transplant being recognised as 'foreign' and rejected.

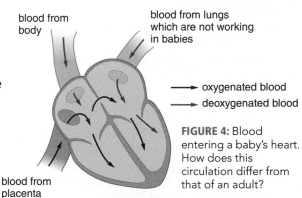

blood from body

blood from lungs which are not working in babies

→ oxygenated blood
→ deoxygenated blood

blood from placenta

FIGURE 4: Blood entering a baby's heart. How does this circulation differ from that of an adult?

Questions

7 Explain why a 'hole in the heart' does not affect an unborn baby.

8 The hole may be between the two atria or the two ventricles. Would the position of the hole have any different consequences?

Q heart pacemakers

You will find out:

> how blood is grouped
> about blood donation
> how blood clots
> about blood transfusions

Blood clotting

Rhys cuts his finger. It bleeds for a few minutes and he loses some blood. His blood quickly clots and seals the cut. Sometimes blood clots inside blood vessels and blocks them. This causes problems inside the body.

Donating blood

Rhys is a **blood donor**. This means he donates blood.

Anti-coagulant drugs are added to the collected blood. This stops it from clotting while it is being stored.

Rhys's blood is tested to find out which blood type it is. Blood can be group A, B, AB or O. Each group can also be **Rhesus positive** or **Rhesus negative**. Rhys has group O, Rhesus-positive blood.

Rhys donates blood to help other people. The blood is given to people during operations or after accidents.

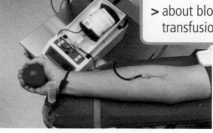

FIGURE 5: Donating blood. What is this blood used for?

Did you know?

The first recorded human blood transfusion was by a doctor called Dr Denys in 1667. He transfused blood from a sheep into a boy. The boy died and Dr Denys was accused of murder.

Questions

9 What happens when Rhys cuts his finger?

10 What are the names of the main blood groups?

11 What type of blood does Rhys have?

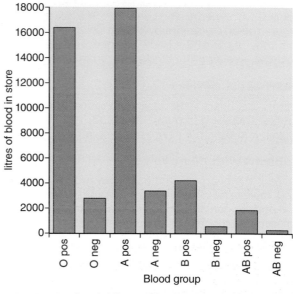

FIGURE 6: Stocks of different blood groups (March 2011). Suggest why the amounts are different.

Blood donation and transfusion

To donate blood, a person has to be aged over 18 years. Before giving blood the donor's haemoglobin level is checked. If their haemoglobin level is low they may become **anaemic** after giving blood. The donor's blood pressure is also checked and their blood group is recorded.

Once the blood is collected, an anti-coagulant is added to prevent it from clotting; the blood is labelled and then stored in a temperature-controlled environment. The process of giving blood and packing it for storage takes less than an hour.

Blood transfusion

Before a blood transfusion is carried out, the new blood is checked to ensure that it does not react with the patient's blood. The blood is then slowly introduced into the patient via a vein. It is important that no air bubbles get into the blood, as this can cause a blockage.

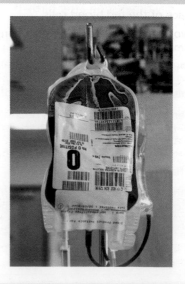

FIGURE 7: Blood ready for transfusion. Which blood group is it?

Q blood donation   blood transfusions

Blood clotting

Blood clots in order to seal wounds to prevent the entry of pathogens and reduce blood loss. However, alcohol slows down the clotting of blood.

Substances in the diet such as vitamin K (from green vegetables and cranberries) are important for the blood to clot.

Doctors use drugs such as warfarin, heparin and aspirin to prevent clotting.

Without these drugs the clots could:

> block the coronary artery and cause a heart attack

> block blood vessels in the brain and cause a **stroke**.

Queen Victoria had an inherited condition called **haemophilia**. This is when the blood does not clot. People with haemophilia are at risk of internal bleeding from the slightest knock.

The process of blood clotting

The process of blood clotting is called a cascade process because it involves many steps. When platelets are exposed to the air at a wound site, this triggers a complex sequence of chemical reactions, eventually forming a meshwork of fibrin fibres (clot).

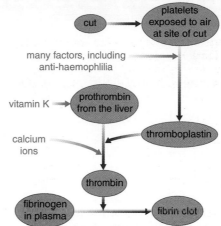

FIGURE 8: Cascade process in blood clotting. In this process, why are vitamin K and calcium important?

Questions

12 Why is it important for blood to clot?

13 Why are some people prescribed warfarin or heparin?

14 Explain why people with haemophilia have to take special precautions.

Blood groups

Early attempts to carry out blood transfusions were not very successful.

Mixing blood from two people often caused blood clumping (**agglutination**). The problem was solved by Karl Landsteiner in 1901. He discovered that there are four main blood groups, which he called A, B, AB and O. The groups depend on the presence or absence of agglutinins, which consist of:

> two proteins, antigen A or antigen B, on the surface of the red blood cells

> two antibodies, anti-A or anti-B, in blood plasma.

These agglutinins determine how blood groups react and whether a blood transfusion will be successful.

Blood group A produces antibody B, but it doesn't produce antibody A. If blood group B is donated to blood group A, the B antibodies in the recipient will react with the B antigens causing agglutination.

Blood group O does not contain any antigens so it can be safely given to all other groups. However, the other groups contain antigens, so they cannot be donated to people with blood group O.

Blood can be further subdivided into Rhesus positive or Rhesus negative.

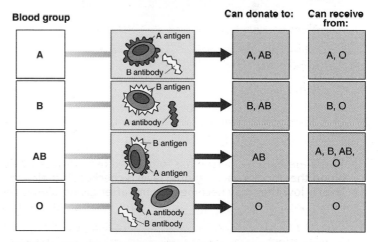

FIGURE 9: Blood group agglutinins, showing donor and recipient possibilities for blood transfusions.

Questions

15 Explain why people with Group O blood are sometimes called 'universal donors'.

16 Explain why people with Group O blood can safely receive only Group O blood.

Respiratory systems

You will find out:
> how different animals get oxygen
> about the structure of the human respiratory system
> how we breathe in and out

Free-diving

Tanya Streeter holds one of the world's free-diving records for women. She went underwater to a depth of 160 m in the sea. She was underwater for more than six minutes.

Tanya did not use any air tanks to help her.

FIGURE 1: Free-diving. Why is free-diving a dangerous sport?

Breathing

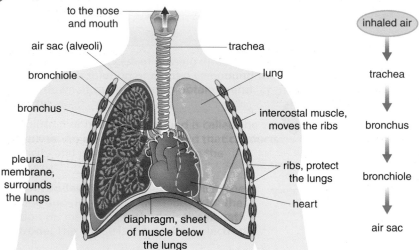

- to the nose and mouth
- air sac (alveoli)
- bronchiole
- bronchus
- pleural membrane, surrounds the lungs
- diaphragm, sheet of muscle below the lungs
- trachea
- lung
- intercostal muscle, moves the ribs
- ribs, protect the lungs
- heart

inhaled air → trachea → bronchus → bronchiole → air sac

FIGURE 2: Our respiratory system. Try to remember the structures that air passes through as it travels into the lungs.

Questions

1 Why do earthworms go deeper in the soil in dry weather?

2 What is respiration?

3 What is the scientific name for 'windpipe'?

4 What two structures protect the lungs?

All living things need oxygen from the air. This is called **gaseous exchange**. Air is breathed in during inspiration (inhalation) and breathed out during expiration (exhalation). Living things use the oxygen to release energy from food. This is called **respiration**.

Small, simple organisms such as amoeba and earthworms exchange gases through their permeable skin. Their skin must be kept moist for oxygen to dissolve in or they will die.

Structure of human lungs

Many animals are much larger and more complex and need special organs to breathe. Our respiratory system has special organs called lungs. Fish have special organs called gills.

Remember!
Breathing is the exchange of gases.
Respiration is the release of energy from food.

Lung capacity

Hiten has asthma. He wants to find out about his lung capacity.

A spirometer is used to measure how much air is breathed in and out. Hiten's total lung capacity is about 5 dm³. When he is resting and breathing normally he is exchanging tidal air.

The overall amount of air that Hiten can use is called the vital capacity. There is a small amount of air that he cannot breathe out. This is the residual capacity.

Q asthma UK   breathing in fish

Breathing in and out

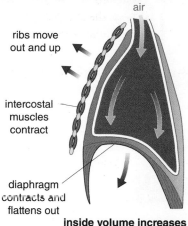

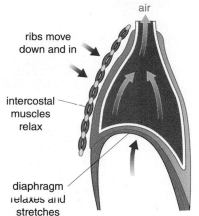

FIGURE 3: Inhaling and exhaling, showing a side view of the chest.

ribs move out and up

intercostal muscles contract

diaphragm contracts and flattens out

inside volume increases and pressure decreases so air is breathed in

ribs move down and in

intercostal muscles relax

diaphragm relaxes and stretches

inside volume decreases and pressure increases so air is breathed out

air

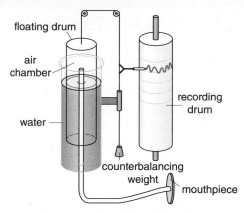

FIGURE 4: A spirometer. What is it used for?

floating drum

air chamber

water

recording drum

counterbalancing weight

mouthpiece

The importance of diffusion

As the air entering the lungs has a higher amount of oxygen than that in the deoxygenated blood flowing through the lungs, efficient diffusion takes place. Also, because the deoxygenated blood contains a higher amount of carbon dioxide, the gas will diffuse out of the blood and into the air so it can be breathed out. Gaseous exchange takes place within alveoli.

Frogs and fish also have special structures to get oxygen into their blood. Frogs' skin and fish gills need to be immersed in water to allow oxygen to diffuse. This means they are restricted to certain habitats: amphibians need moist habitats, and fish gills work only in water.

Questions

5 Polio can paralyse chest muscles. Why is this dangerous?

6 What happens to the ribs when the intercostal muscles contract?

7 Why does the diaphragm flatten when it contracts?

8 Describe the changes in lung volume and pressure when breathing in and out.

Structure related to function

Air sacs in the lungs have bulges called alveoli.

They are adapted to be efficient gas exchangers by having:

> a large surface area

> a moist surface

> a thin permeable lining (only one cell thick)

> a good blood supply.

These adaptations aid diffusion of gases into and out of the many blood capillaries.

FIGURE 5: An alveolus.

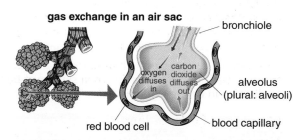

gas exchange in an air sac

bronchiole

oxygen diffuses in

carbon dioxide diffuses out

alveolus (plural: alveoli)

red blood cell

blood capillary

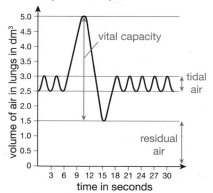

FIGURE 6: Hiten's spirometer trace.

Questions

9 Why does oxygen diffuse into, and carbon dioxide diffuse out of, the blood in human lungs?

10 Look at Hiten's spirometer trace. Work out Hiten's tidal air volume and residual air volume.

11 Look at Hiten's spirometer trace. How many times does he breathe in 30 s?

🔍 lung capacities

Exchange of gases

About one-fifth of the air we breathe in is oxygen. In the lungs the oxygen enters the blood. When the blood travels around the body, the oxygen is released into body tissues.

Body tissues use up the oxygen and produce carbon dioxide. The carbon dioxide enters the blood and is taken back to the lungs. It then leaves the blood so it can be breathed out. The air we breathe out contains more carbon dioxide and less oxygen.

The surface area where gaseous exchange takes place needs to be as large as possible. This makes the exchange of gases more effective.

Asthma

Hiten has **asthma**. He has to take medicine from an inhaler to help him breathe.

Hiten writes in his diary:

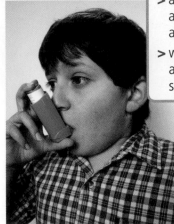

> **You will find out:**
> > how the respiratory system is protected
> > about the symptoms and treatment of asthma
> > what other diseases affect the respiratory system

FIGURE 7: An inhaler used to treat asthma.

Monday

Today is Sports Day at school.
I felt breathless when I woke up so I used my inhaler.
I feel much better now.
I should be OK.
I must remember to take my inhaler to school.
Perhaps I will win my race!

> **Did you know?**
>
> Over 5.2 million people suffer from asthma in the UK.

> **Questions**
>
> **12** What lung condition does Hiten have?
>
> **13** How does Hiten's medicine reach his lungs?
>
> **14** Why must Hiten remember to take his inhaler to school?
>
> **15** Write down what you think Hiten might have written in his diary for Tuesday.

There are other diseases of the respiratory system. Some of them are:

> bronchitis   > cystic fibrosis

> pneumonia (pronounced 'newmonia')   > lung cancer.

More on asthma

Protection against disease

Special cells in the lungs make sticky mucus that traps dust particles and some bacteria.

The linings of the trachea, bronchi and bronchioles are covered by millions of tiny hair-like structures called cilia. The cilia produce a wave-like motion that carries mucus and trapped dust upwards out of the lungs and into the throat.

Wednesday

I had an asthma attack today. I was short of breath, felt a tightness in my chest and started to wheeze when I breathed in and out. My attack could have been triggered by the cold weather or an allergic reaction. It can sometimes be caused by exercise. I used my inhaler. It releases a drug called Ventolin to widen my bronchioles. They constrict in an asthma attack which makes me feel that I can't breathe.

 cilia in trachea   asbestosis causes

Diseases of the respiratory system

Industrial diseases such as asbestosis can be a risk for many working people around the world.

Asbestosis is caused by breathing in fine asbestos fibres. The fibres are trapped in the air sacs, which causes inflammation and scarring and limits the exchange of gases. Asbestos was used as insulation in buildings before scientists realised how dangerous it is to health.

Cystic fibrosis is an inherited condition. Too much mucus is produced in the bronchioles, which causes less exchange of gases. The patient struggles to get enough oxygen.

A poor lifestyle such as smoking is linked to lung cancer. In cancer, cells divide out of control, which reduces the surface area of the alveoli and eventually destroys lung tissue.

FIGURE 8: A physiotherapist massaging a cystic fibrosis patient to help to dislodge the mucus in the patient's lungs.

Questions

16 How does Hiten's inhaler help him?

17 How do mucus and cilia 'clean' the trachea?

More on respiratory systems

Why is the respiratory system prone to disease?

The air sacs in the lungs are a 'dead end'. Any debris that is not removed by mucus and cilia remains, covering and irritating the cells lining the alveoli. These cells are thin, moist and delicate to ensure efficient diffusion of oxygen. They are easily damaged.

This explains why there are so many diseases of the respiratory system.

Causes of asthma

The exact cause of asthma is not clear. It is believed to be a result of a combination of inherited, environmental, infectious and chemical factors. The responses are also rather complex. However, they all lead to less oxygen being available for gas exchange in the lungs.

Gaseous exchange in amphibians and fish

Amphibians are restricted to damp habitats because they rely on a moist skin for respiration. Their permeable skin results in excessive water loss in dry conditions. Fish gills also dry out easily, restricting them to life in water.

Fish gills are also very efficient in exchange of gases. Water first enters the fish's mouth, and then the mouth closes. Water containing dissolved oxygen is then forced over many fine **gill filaments**, which have a large surface area. Gill filaments have a rich blood supply, so they look red. A bony gill bar supports the gill filaments. The herring also has gill rakers for filter feeding. These stop food particles blocking the gills.

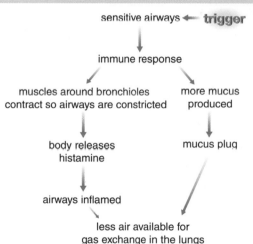

FIGURE 9: Possible pathway that results in an asthma attack.

sensitive airways ← **trigger**
↓
immune response
↓ ↓
muscles around bronchioles contract so airways are constricted / more mucus produced
↓ ↓
body releases histamine / mucus plug
↓
airways inflamed
↓
less air available for gas exchange in the lungs

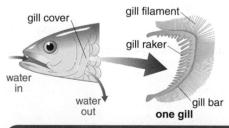

FIGURE 10: Fish gills.

gill cover, gill filament, gill raker, water in, water out, gill bar, **one gill**

Questions

18 If you were designing a lung, what would be your priorities?

19 Write down one advantage of having gills instead of having lungs.

20 Explain why humans would not survive if they had gills instead of lungs.

Q pneumonia

Preparing for assessment: Applying your knowledge

To achieve a good grade in science, you not only have to know and understand scientific ideas, you also need to be able to apply them to other situations and investigations. These tasks will support you in developing these skills.

✻ Breathing out

Zhahid is 15 years old and thinking about what he's going to do when he leaves school. He's thinking about doing A Levels or maybe BTEC. He goes along to the local tertiary college with some of his friends to an open evening. One of the areas they go into is called a Human Performance Laboratory. It doesn't look much like a laboratory; it's more like the fitness suite at the local sports centre, with lots of exercise equipment and staff in tracksuits carrying clipboards.

'Right then', said one of the staff, 'My name's Liz. Let's show you some of the tests we can do with people here.' She held up an empty plastic milk bottle. 'Who can tell me what the volume of this bottle is? How many litres?' No one said anything. Then Zhahid said, 'About one?' 'Right, well done – you've got yourself a job.' Zhahid groaned, and grinned, but he didn't move. All his friends took one step back, and laughed.

'How many litres do you reckon you breathe out in a normal breath? You know, when you're just standing there?' she said. Zhahid shrugged. One of the others said, 'About one?' 'About one? Let's see.'

Liz put a nose clip on Zhahid's nose, and then offered him a pipe linked to a machine. 'This is a spirometer,' she explained. 'It measures the flow of air. Don't breathe in through it, only out. Just breathe out normally. Go on, try it.' Zhahid breathed into the tube. 'There we go,' said Liz, '0.5 litres. That's about right for someone like you. How much air do you think you could breathe out, if you really forced all the air out?'

Zhahid thought hard and looked at the bottle. Then he thought about the 0.5 litres. 'I reckon about two. Maybe two and a half.' Liz handed the tube back to him and he took as deep a breath as he could and breathed out into the tube. 'Well done,' she said, 'very good, 4.7 litres.'

'This kind of study is really important,' Liz said, 'because it shows us how well people are using their lungs. For vigorous exercise you need really efficient gaseous exchange. If you had a restrictive disorder, such as bronchitis or asthma, you wouldn't be able to empty your lungs like that, at least not as quickly as our helper did.'

'That was alright,' thought Zhahid. 'I wouldn't mind doing something on that next year.'

Task 1

What was the spirometer for? Why did Zhahid have to wear a nose clip when he used it?

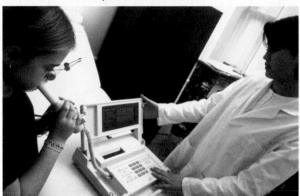

An asthma sufferer using a spirometer.

Task 4

The graph shows air flow in and out of Zhahid's lungs.

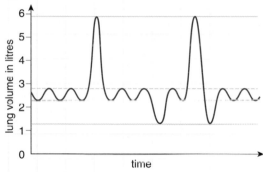

Copy the graph and label the points at which he was:

a breathing in and out whilst standing still

b inhaling deeply

c trying to empty his lungs as Liz had asked him to.

What do you think was happening at the first tall peak on the graph?

Task 5

Where on the graph in Task 4 is the residual air volume shown? How large is it? How might the graph look different if Zhahid had asthma?

Task 2

What is the scientific name for the amount of air you breathe out:

a when you are at rest?

b when you inhale deeply and then breathe out as hard as you can?

When Zhahid had finished doing the second test, he felt he'd completely forced every bit of air out. In fact he hadn't. What is the scientific name given to the amount of air he had left in his lungs?

Task 3

Why are bronchitis and asthma referred to as restrictive disorders? Use a simple diagram to explain your answer. Why do these conditions affect the rate at which the lungs can be emptied?

Maximise your grade

Answer includes showing that you can...	
F	Refer to a condition which affects the respiratory system.
	Refer to several conditions which affect the respiratory system.
	Use with understanding terms such as breathing and respiration.
C	Use one of these terms with understanding: tidal air, vital capacity and residual air.
	Start to interpret data from air-flow graphs such as the one on this page.
	Interpret air-flow graphs and link the patterns to changes in the body.
A	Interpret air-flow graphs and link the patterns to changes in the body with attention to detail and all the features.
	As above, but with particular clarity and detail.

Digestion

You will find out:
> how food is broken down physically
> how food is chemically broken down by enzymes
> about the importance of pH in enzyme action

What can you see, doctor?

In 1700, Alexis St Martin, a Canadian hunter, was shot. The wound did not heal. His doctor was able to see into his stomach!

Pieces of meat were tied together by a thread and then placed into Alexis' stomach. After a few hours the meat had broken down into smaller pieces.

FIGURE 1: What did this experiment prove?

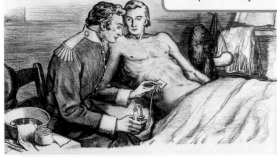

What happens to food?

Charlotte takes care of her teeth. They give her an attractive smile. Her teeth are used to cut and grind food into smaller pieces. This is an example of **physical digestion**.

The food can then easily pass down her **digestive system**. Muscles in the wall of the stomach help to physically digest food by squeezing it. Muscles in other parts of the digestive system also squeeze the food. This keeps it moving.

Charlotte thinks this is like squeezing toothpaste out of its tube. Now she knows how astronauts can eat when floating upside down in space.

Charlotte knows that her food gets into her blood so that it can be carried to all the cells in her body.

She is puzzled because when she cuts her finger, she doesn't see lumps of potato and carrots in her blood.

Parts of the digestive system produce digestive enzymes. These enzymes break down food into smaller and soluble molecules so they can be absorbed by the blood. This is **chemical digestion**.

Did you know?

Sharks have teeth all over their skin as well as in their jaws.
So don't stroke a shark!

Questions

1 What process breaks food down into smaller pieces?

2 Which body system digests food?

3 How long is our digestive system?

4 What do enzymes do to food?

Physical digestion

Physical digestion is important because the food is broken down into smaller pieces, increasing its surface area. This allows it to pass through the digestive system more easily, and to be chemically digested more quickly.

Chemical digestion

Different enzymes break down carbohydrates, proteins and fats into smaller and soluble molecules. These small molecules can then diffuse through the walls of the small intestine. Being soluble, they can dissolve in the blood plasma (carbohydrates and proteins) or lymph (fats) and are carried to the cells. Enzymes are produced in the mouth, stomach and small intestine.

Charlotte does not see lumps of food in her blood because the food has been digested and has dissolved in the blood.

Enzymes are specific in their action.

> **Carbohydrases** break carbohydrates down into simple sugars, such as glucose.

> **Proteases** break proteins down into simpler amino acids.

> **Lipases** break fats down into simpler fatty acids and glycerol.

Charlotte finds out that parts of the digestive system produce:

> enzymes

> acid or alkali conditions so the enzymes can work; for example, the stomach makes hydrochloric acid so that the protease enzyme pepsin can work.

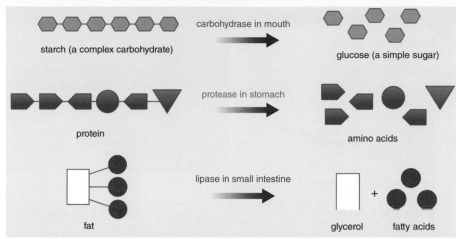

FIGURE 2: What is meant by enzymes being 'specific'?

Questions

5 What types of foods are carried in the blood plasma?

6 How does digested food get through the walls of the small intestine?

Fat digestion

Fats are difficult to digest and absorb because they are not soluble in water. To help with fat digestion, the gall bladder produces **bile**. This emulsifies fats, which increases their surface area for enzymes to act on. Glucose and amino acids diffuse into the blood capillaries. Fatty acids pass into the lacteals, which are part of the lymphatic system (see Figure 7).

Gall stones can form in the gall bladder and may block the bile duct. They are made from bile pigments, cholesterol and calcium salts. Laser surgery or ultrasound can be used to break them down.

Enzyme action

Enzymes have an optimum pH (a pH at which they work best). The protease enzyme in the stomach is called pepsin. It has an optimum of pH 1.6, which is very acidic. Hydrochloric acid, which also kills many harmful microorganisms, provides this acidic condition.

Other enzymes in the digestive system do not need such acidic conditions: salivary amylase (a carbohydrase enzyme) in the mouth has an optimum of about pH 6.7 – 7; lipase from the pancreas has an optimum of pH 8. Therefore there are different pH values in the mouth and small intestine from that in the stomach.

The chemical breakdown of large molecules such as starch requires more than one stage. A starch molecule has hundreds of linked glucose units. Carbohydrase enzymes first break the starch, called a polysaccharide, down to maltose (of two units), called a disaccharide, and then into glucose (of one unit) called a monosaccharide.

FIGURE 3: What are gall stones made of?

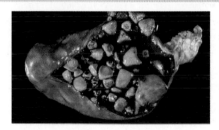

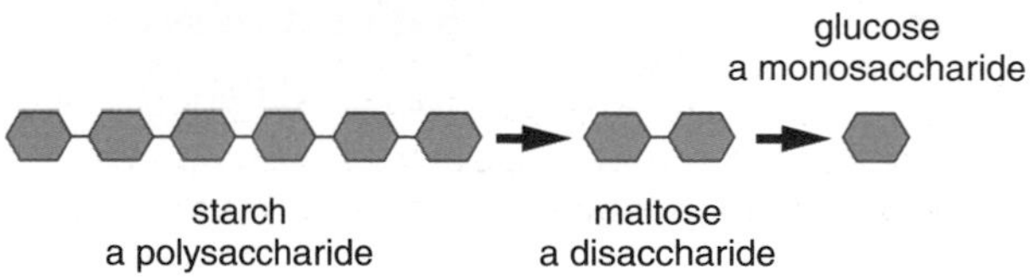

FIGURE 4: Breaking down starch.

Questions

7 Explain the role of the gall bladder in digestion.

8 Suggest a reason why pepsin has a lower optimum pH than other digestive enzymes.

The human digestive system

Food passes down a long tube, from mouth to anus.

Parts are adapted for different functions.

Food is digested by the time it reaches the small intestine. The digested food goes into the blood here, to be taken round the body. The food will leave the blood when it reaches body tissues such as muscles, providing them with energy.

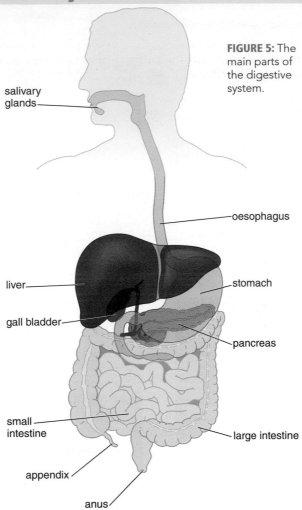

FIGURE 5: The main parts of the digestive system.

- salivary glands
- oesophagus
- liver
- stomach
- gall bladder
- pancreas
- small intestine
- large intestine
- appendix
- anus

Did you know?

Your digestive system is about 9 m long, nearly the length of a bus. This gives time for food to be digested.

Questions

9 Which part of the digestive system joins the mouth and stomach?

10 The small intestine is longer than the large intestine. Suggest why it is called 'small'.

11 Explain why food will reach our stomach even if we are upside down.

12 Name the two main functions of the small intestine.

Specialised parts of the digestive system

Parts of the digestive system have different functions.

> Salivary glands produce saliva which moistens food and contains an enzyme.

> Muscles in the walls of the oesophagus contract and relax, squeezing the food down to the stomach. The waves of muscle action are called **peristalsis**.

> The stomach has strong muscles and produces hydrochloric acid and protease enzymes for physical and chemical digestion.

> The liver produces bile and stores carbohydrates (glycogen).

> The gall bladder stores bile.

> The pancreas produces enzymes and a hormone (insulin).

> The first part of the small intestine produces many enzymes for digestion of carbohydrates, fats and proteins.

> The second part of the small intestine absorbs the digested food.

> The large intestine absorbs water, concentrating the waste.

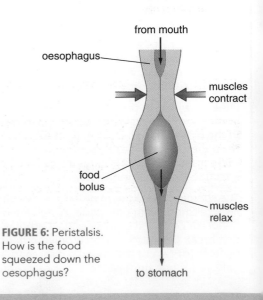

- from mouth
- oesophagus
- muscles contract
- food bolus
- muscles relax
- to stomach

FIGURE 6: Peristalsis. How is the food squeezed down the oesophagus?

Absorption of food

Chemical digestion of food is completed in the small intestine.

Amino acids (from proteins) and simple sugars such as glucose (from starch) are small, soluble molecules. They are able to diffuse through the walls of the small intestine and into the surrounding blood capillaries. Fatty acids from fats diffuse into the lymphatic system.

Question

13 Suggest why fatty acids are not passed into the blood.

More about the small intestine

The small intestine is specially adapted for efficient absorption of digested food:

> it is about 7 m long, allowing time for complete absorption

> it has **villi** and **microvilli,** which give a large surface area

> its lining is thin and permeable, so food molecules can easily pass through

> it is surrounded by a network of blood capillaries to carry the digested food away from the small intestine.

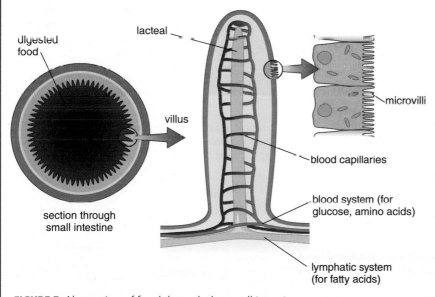

FIGURE 7: Absorption of food through the small intestine.

Questions

14 Explain how and why the small intestine has a large surface area.
15 Explain why a villus contains parts of the blood system and the lymphatic system.

Waste disposal

You will find out:
> which body organs excrete waste
> how the kidneys work
> how a dialysis machine works

Kidney stones

Sometimes the kidney can be blocked by large kidney stones.

The largest known kidney stone weighed 1.36 kg!

Most kidney stones are very small and harmless.

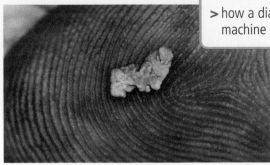

FIGURE 1: A kidney stone placed on a finger. Estimate its size.

 ## Getting rid of waste

Liquid waste

Lauren has kidney problems. She thinks she has kidney stones. She finds out that the kidneys get rid of waste produced by the body.

This is called **excretion**.

These body organs also excrete waste:

> lungs   > liver   > skin.

The wastes from these organs are usually dissolved in water. Lauren knows that kidneys excrete urine, which contains water, salt and **urea**.

Solid waste

The body also gets rid of undigested solid waste (the parts of food which cannot be digested) as faeces through the anus.

This is not excretion, because the waste is not made by the body. Instead, it is called **egestion**.

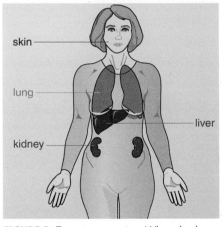

FIGURE 2: Excretory organs. What do the kidneys excrete?

skin

lung

liver

kidney

Questions

1 Name four body organs that excrete waste.

2 What is getting rid of undigested solid waste from the body called?

3 How is waste from kidneys different from waste from the anus?

4 What substances are excreted by the kidneys?

Remember!
Don't muddle up 'excretion' and 'egestion'.

 ## How the kidneys work

Lauren finds out that she has a kidney stone in her ureter (the tube that connects her kidney to her bladder).

> The renal artery brings blood containing waste substances to Lauren's kidneys. One waste substance is urea. It is made in the liver from unwanted amino acids when proteins are broken down. The renal vein takes filtered blood away from her kidneys.

> The kidneys filter the blood, removing the waste. As the blood in the renal vein has a high pressure, the filtering is also done under high pressure. Useful substances such as glucose, some water and some salt are reabsorbed back into the blood before it leaves the kidneys.

The liquid waste from kidneys is called urine.

Lauren is advised to drink a lot of water to increase the amount of urine she produces.

If Lauren does more exercise or is in hot conditions, she loses more water as sweat and less is lost as urine.

It is important to maintain the correct concentration of water in blood plasma to avoid problems with osmosis and damaging red blood cells. You can find out more about osmosis in the Additional Science chapter – Diffusion and osmosis.

FIGURE 3: The structure of the kidneys. What structures does a ureter connect?

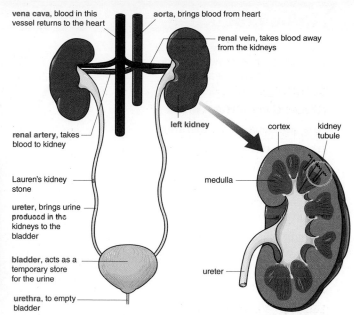

vena cava, blood in this vessel returns to the heart

aorta, brings blood from heart

renal vein, takes blood away from the kidneys

renal artery, takes blood to kidney

left kidney

cortex

kidney tubule

Lauren's kidney stone

medulla

ureter, brings urine produced in the kidneys to the bladder

bladder, acts as a temporary store for the urine

ureter

urethra, to empty bladder

Questions

5 Where in the body is urea produced?

6 Suggest why it is important that the blood in the kidneys is filtered under high pressure.

7 Why is glucose reabsorbed by the blood?

8 Suggest why Lauren is advised to drink a lot of water.

More on the kidney

A kidney contains about half a million kidney tubules, which filter the blood. Each tubule is called a **nephron** and is arranged in a U shape.

Each nephron contains:

> a 'filter unit' made up of a **glomerulus** (collection of blood capillaries) and a surrounding capsule, which remove useful and waste materials from the blood

> a region in which the blood selectively reabsorbs useful substances, such as glucose and some water

> a region that regulates the body's levels of water and salt.

As some of the substances, such as sodium ions, are reabsorbed in the kidney against a concentration gradient, this requires energy and uses an active transport system (not diffusion). A hormone (aldosterone) also increases reabsorption.

A dialysis machine

If Lauren's kidneys stop working (kidney failure) she may have to have dialysis treatment. In an average day, normal kidneys excrete 1.5 dm³ of water, 30 g of urea, 15 g of salt and small amounts of **uric acid**.

Urea and uric acid contain nitrogen and are poisonous at high concentrations. The machine acts as an artificial kidney and removes urea from the blood. As urea molecules are small, they diffuse through the membrane. A dialysis machine also uses different sizes of tubes so slightly increasing pressure during diffusion (ultra-diffusion).

The dialysis fluid contains sodium salts, so it is the same or slightly lower than the desired blood concentration. This maintains or adjusts the sodium levels in the blood. Dialysis tubing is permeable to glucose molecules, so the level of blood glucose can be controlled. Insulin, which normally regulates blood glucose levels, is not removed in dialysis.

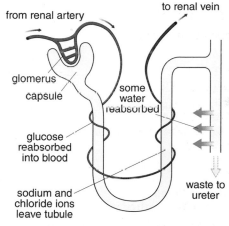

to renal vein

from renal artery

FIGURE 4: A kidney tubule. Which artery brings blood to the kidneys?

glomerus

capsule

some water reabsorbed

glucose reabsorbed into blood

sodium and chloride ions leave tubule

waste to ureter

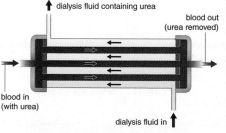

dialysis fluid containing urea

blood out (urea removed)

blood in (with urea)

dialysis fluid in

FIGURE 5: How dialysis works. Which substance diffuses through the membrane in the machine?

Questions

9 Explain why and how some substances are selectively reabsorbed by the blood.

10 Explain why the tubes in a kidney dialysis machine are narrow and made of a semi-permeable membrane.

Urine production

Your kidneys produce urine. The amount and concentration of urine produced is affected by:

> the amount of water drunk

> the surrounding temperature

> the type of exercise you do.

You will produce a small amount of concentrated urine:

> when you drink only a small amount of water

> during vigorous exercise

> when it is hot and you sweat a lot.

The skin

The skin excretes **sweat**. Sweat contains water and salt.

The skin cells are arranged into two layers:

> the outer **epidermis**

> the **dermis** underneath.

Hairs grow in **follicles**. Hair is made from dead material so it does not hurt when it is cut!

Sweat glands produce sweat. The sweat reaches the skin surface at sweat pores.

The lungs are also excretory organs. The carbon dioxide produced in respiration must be excreted from the body. This carbon dioxide is breathed out from the lungs.

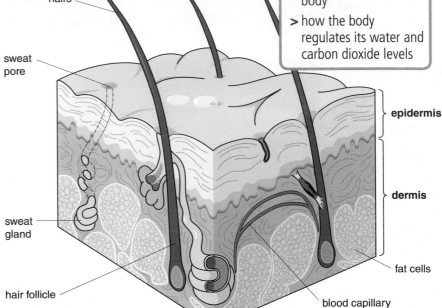

hairs

sweat pore

epidermis

dermis

sweat gland

fat cells

hair follicle

blood capillary

FIGURE 6: What your skin looks like. How does sweat leave the skin?

Questions

11 What does sweat contain?

12 Where is sweat made?

13 What do lungs excrete?

14 Why does it not hurt when your hair is cut?

Why do we sweat?

Although sweating is a form of excretion, it has another important function. It helps the body regulate its temperature. This is an example of homeostasis.

Exercising produces more energy from the oxidation of food, some of which is heat. Therefore more sweat is produced to keep the body temperature constant.

More carbon dioxide is also produced when exercising. This gas must be removed. Carbon dioxide reacts with water in the blood plasma, forming carbonic acid. If it was not removed, the acid would upset the pH level of the blood. Therefore a higher rate of breathing not only takes in more oxygen, it also gets rid of more carbon dioxide waste.

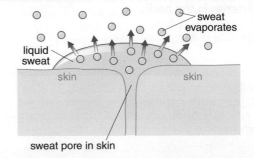

sweat evaporates

liquid sweat

skin

skin

sweat pore in skin

FIGURE 7: The water molecules in sweat use heat energy from the skin to change state from a liquid to a gas by evaporation. Thus the skin is cooled down. About 10 cm³ of evaporating sweat can use 6 J of energy from the skin.

sweating   anti-diuretic hormone

Sweating and urine production

During vigorous exercise or in hot conditions, the body sweats and loses a lot of water. This means that the kidneys don't need to excrete so much water. Then the urine has a higher concentration of salt and urea.

15 Why do you feel cold when you step out of the bath?

16 Explain why Lauren drinks a lot of water after exercising.

17 How does sweating help to regulate body temperature?

18 What conditions will produce a lot of very dilute urine?

Water balance

Lauren notices that when she drinks lots of water her urine is very dilute (pale in colour).

Urine concentration depends on how much water is reabsorbed by the kidney tubules. Reabsorption is controlled by **anti-diuretic hormone (ADH)**, which is made in the pituitary gland in the brain. ADH increases the permeability of the nephrons so that more water is reabsorbed.

A negative feedback mechanism ensures the correct water balance in the blood. The correcting mechanism (ADH production) is switched off when conditions return to normal.

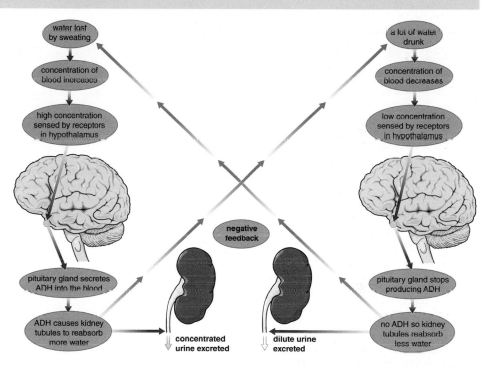

FIGURE 8: Control of water reabsorption by ADH. What type of feedback mechanism is this?

Carbon dioxide balance

The body is more sensitive to the level of carbon dioxide than oxygen. Chemical receptors, in the carotid arteries in the neck, detect increasing levels of carbon dioxide in the blood.

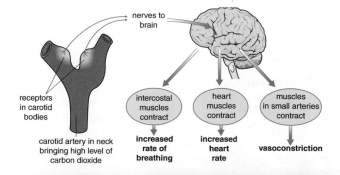

FIGURE 9: Control of carbon dioxide level in the blood by receptors in the carotid artery. What three changes take place to increase the excretion of carbon dioxide?

Did you know?

Birds excrete uric acid instead of urea. The cliffs of Guano Island are covered by bird droppings up to 30 m deep. Guano is used as a fertiliser.

19 Explain what would happen to her urine if Lauren ate a lot of salty crisps.

20 Explain why high levels of carbon dioxide in the blood cause an increase in heart and breathing rates.

Life goes on

You will find out:
> what the menstrual cycle does
> how hormones control the menstrual cycle

Even dinosaurs produced eggs

Female dinosaurs laid eggs. The largest known fossilised dinosaur eggs are from Hypselosaurus. They were laid about 100 million years ago. They are 30 cm long and weigh almost 7 kg. One would make a very big breakfast!

FIGURE 1: Dinosaur egg fossils. Suggest a reason why these dinosaur eggs were big.

Human reproduction

Female humans also produce eggs. The female reproductive system has:

> ovaries that make the eggs

> fallopian tubes to carry the eggs to the uterus, where the baby develops

> a vagina, through which the baby is born.

The male reproductive system has:

> testes to make sperm

> a scrotum to hold the testes outside the body – this temperature will be better for good sperm development

> sperm ducts to carry sperm to the penis.

The ovary and testes are called **endocrine glands**, as they produce hormones. Ovaries produce **oestrogen** and **progesterone**. Testes produce **testosterone**.

The menstrual cycle

Nisha is a teenager. She has reached **puberty** and her ovaries are producing eggs. She has a cycle of stages every 28 days. This is called the **menstrual cycle**. Some of her friends have a slightly longer or shorter cycle.

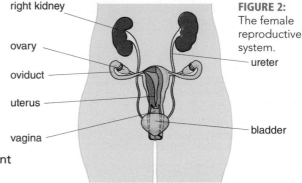

FIGURE 2: The female reproductive system.

right kidney
ovary
oviduct
uterus
vagina
ureter
bladder

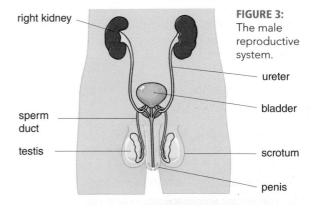

FIGURE 3: The male reproductive system.

right kidney
sperm duct
testis
ureter
bladder
scrotum
penis

Between days 13 and 15 of Nisha's cycle, her ovary releases an egg. This is called **ovulation**. Her uterus lining becomes thicker, with more blood vessels. This will help a fertilised egg to embed in the lining.

The egg may not be fertilised or may not embed in the uterus lining. The uterus lining then breaks down, releasing the broken down cells. This is called menstruation.

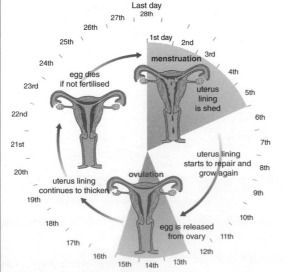

FIGURE 4: The menstrual cycle. For about how long does one cycle last?

Questions

1 Where are: **a** eggs made? **b** sperm made?

2 During which days of the menstrual cycle does menstruation take place?

Q menstrual cycle animation

Hormonal control of the menstrual cycle

The menstrual cycle is controlled by hormones. The cycle is triggered by receptors in the hypothalamus in the brain that cause the pituitary gland to produce two hormones:

> **follicle stimulating hormone (FSH)**, which stimulates an egg to mature inside a follicle in an ovary

> **luteinising hormone (LH)**, which controls the release of an egg (ovulation).

As the follicle in the ovary develops, it releases varying amounts of two hormones: oestrogen and progesterone. These hormones control the growth of uterus cells and therefore the thickness of the uterus lining. Oestrogen causes the repair of the uterus wall, and progesterone maintains the uterus wall after ovulation. Body temperature also changes during the menstrual cycle.

If an egg is fertilised:

> the levels of progesterone remain high

> no FSH is produced

> no more eggs develop or are released

> the uterus lining does not break down.

It is possible to control fertility in humans using sex hormones, the contraceptive pill or fertility drugs.

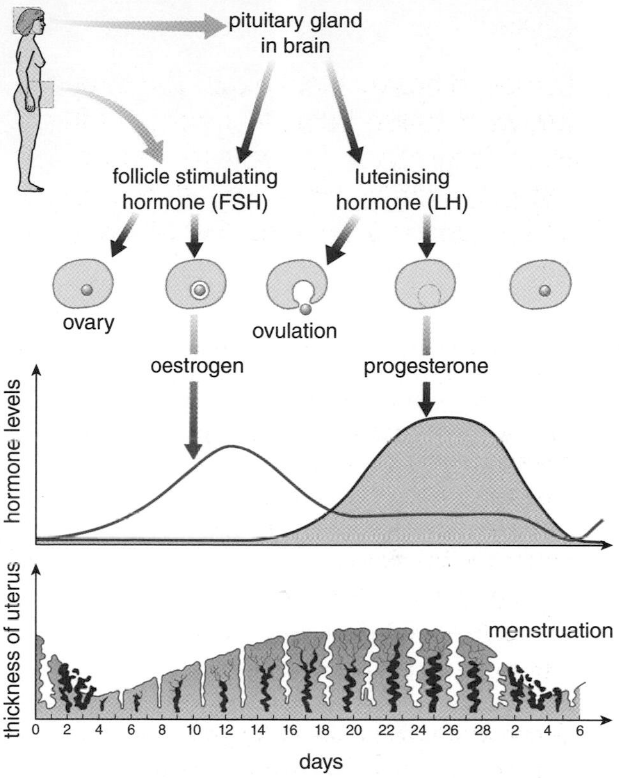

Question

3 Some contraceptive pills contain hormones that stop the release of LH. Explain how this stops an egg from being fertilised.

FIGURE 5: How hormones control the menstrual cycle.

Hormonal feedback

The production of hormones by the pituitary (FSH and LH) is regulated by other hormones (oestrogen and progesterone) produced by the ovaries. For example,

> FSH is produced by the pituitary gland

> FSH stimulates egg maturation and the release of oestrogen from the ovary

> oestrogen acts as a negative feedback mechanism to control FSH production

> oestrogen also stimulates LH production from the pituitary gland

> LH controls egg release and production of progesterone

> progesterone also acts as a negative feedback mechanism on FSH production.

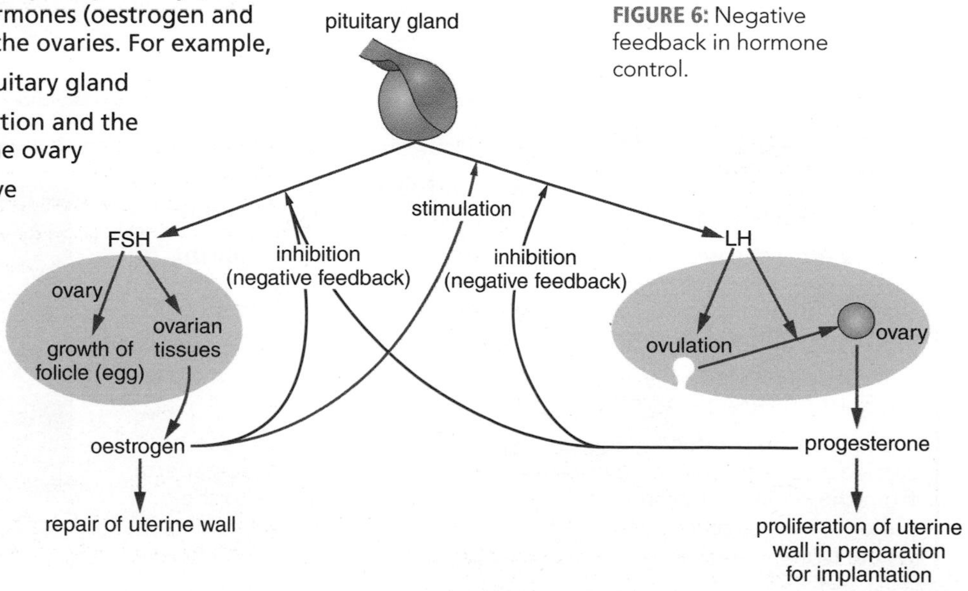

FIGURE 6: Negative feedback in hormone control.

Fertilisation and pregnancy

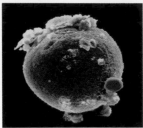

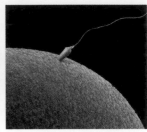

FIGURE 7: Fertilisation. A human egg

Human sperm

One sperm fertilising an egg

A sperm can meet and fuse with an egg. This is called fertilisation. The genetic information of both egg and sperm combine to produce a new individual. The development of the foetus (young baby) takes place during pregnancy.

Despite 133 million babies being born in the world each year, some couples are unable to achieve fertilisation of an egg and pregnancy.

Infertility can be the result of:

> blockage of sperm ducts so no sperm are released

> blockage of fallopian tubes so the eggs don't reach the uterus

> eggs not developing or being released from the ovary

> insufficient fertile sperm being produced by the testes.

However, with special fertility treatment, some of these couples can be successful.

Questions

4 What does 'infertile' mean?

5 Suggest ways in which sperm can be infertile.

Checking foetal development

It is important to check the foetus regularly during pregnancy. The foetus can be checked:

> by a scan to show whether growth and development are in the normal range

> by sampling the fluid surrounding the foetus to check for genetic abnormalities.

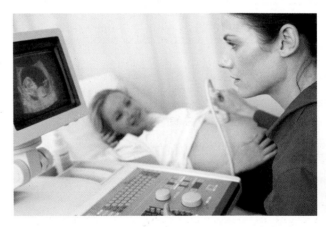

FIGURE 8: A pregnant woman having an ultrasound scan. What sort of image is produced?

Infertility treatments

> Sometimes not enough FSH is produced. The ovary follicles therefore do not develop eggs. Injections of FSH can be given to correct this.

> Sometimes the woman is unable to carry a foetus to term (birth). To correct this, the husband provides the sperm. An egg is harvested from the patient and is fertilised artificially. The fertilised egg is then implanted into a surrogate mother.

> Sometimes the mother is unable to fall pregnant naturally. In-vitro ('in glass') fertilisation (IVF) treatment can be carried out.

> Ovaries stop producing eggs at a young age. Egg donation can be carried out. A relative or friend donates eggs to be fertilised by sperm from the patient's husband. The fertilised egg is then put into the patient. A more recent development involves the transplant of a whole ovary.

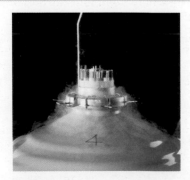

FIGURE 9: Sperm can be frozen and stored for later use.

Q infertility   IVF procedure

> When there is a blockage in the fallopian tubes or the husband has a low sperm count, artificial insemination can be carried out. The husband's sperm (or donated sperm) is put directly into the uterus and oviducts.

Infertility treatments raise ethical issues. Some people think that it interferes with nature and simply further increases problems of over-population. They are concerned that the treatments are expensive with a low rate of success. Other people believe it is right to use advances in science to help every woman who wants a baby to have a successful pregnancy.

Fertility in humans can be controlled using sex hormones. Examples of this are fertility drugs and the contraceptive pill.

Checking foetal development

The development of a foetus can be checked during pregnancy using amniocentesis.

This is often done on pregnant women over 35 years old while they are in their 15–18th week of pregnancy. A thin needle is guided by ultrasound through the abdomen into the amniotic fluid that surrounds the developing baby. A sample of fluid containing loose cells is taken. The cell chromosomes are then examined. These tests can check for up to 150 abnormalities, such as Down's syndrome.

Amniocentesis carries a risk of causing miscarriage in about 1 in 200 cases, and some pregnant mothers may refuse the test. If the test is positive, they and their partners face the difficult decision of whether to proceed with their pregnancy.

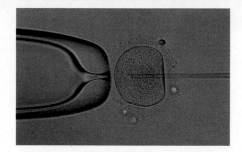

FIGURE 10: An egg being fertilised in vitro.

Did you know?

About one in six couples have difficulty in conceiving naturally. In Britain about 37 000 people are treated at IVF clinics each year. The live birth rate for IVF treatment in Britain is 30%.

Questions

6 Which of the infertility treatments above could be used if the husband is producing only a small number of sperm?

7 Which of these treatments could be used by a woman with cancer?

8 Explain how the treatment 'IVF' got its name.

9 Explain what a surrogate mother is.

Fertility and infertility

The chances of fertilisation can be reduced using female sex hormones. This form of contraception prevents ovulation by inhibiting FSH production, which fools the body into thinking it is pregnant.

If infertility is the result of lack of eggs, FSH can be used to stimulate egg production.

Ethics of infertility treatments

Any infertility treatment carries some risk, however small.

> Injections of FSH are straightforward, and they have been used for many years without major side effects.

> Surrogate mothers may become emotionally attached to the developing baby during pregnancy and then find it difficult to hand over the baby.

> IVF does not have a very high success rate, and the technique is limited to a small number of patients. Multiple births are still a problem, as more than one fertilised egg is used to try to ensure success.

> In egg donation, the egg and resulting baby will not carry the mother's genetic code, but it does carry her husband's.

> Ovary transplants depend on a new technique with possible rejection problems.

> Artificial insemination techniques have been used for a long time without side effects. However, a fertilised egg is not guaranteed to develop when in the uterus.

Foetal investigations

> A blood test is highly recommended for pregnant women. There may not be any known family history of genetic disorders but this does not mean the alleles responsible for disease are not being carried.

Questions

10 Kath is pregnant. Her first blood test shows a possibility of a genetic disorder. Discuss what she should do.

11 In case of a positive Down's syndrome test, make a list of reasons to continue or stop the pregnancy.

Growth and repair

You will find out:
> about the stages of human development
> how growth is measured
> how to use growth charts
> what influences growth

New limbs

Some animals such as starfish can grow new limbs, and many lizards can grow a new tail.

Many scientists are researching how these animals do this.

FIGURE 1: A starfish growing a new arm. Why are scientists trying to find out how some animals can grow new body parts?

Growth

Ben and Charlotte measure their growth.

They find out how much their height and mass increase.

Ben wants to be a soldier in the Royal Marines. He must grow to be at least 163 cm tall.

Charlotte wants to be a model. She hopes to be tall.

How tall and heavy Ben and Charlotte will be depends on many things:

> the information in their genes

> their diet

> the amount of exercise they do

> the amount of **growth hormones** their bodies make

> how healthy they are.

Stages of growth

Humans have different stages in their growth and development, which are shown in Figure 2.

Questions

1 How can growth be measured?

2 In which stage of development are you?

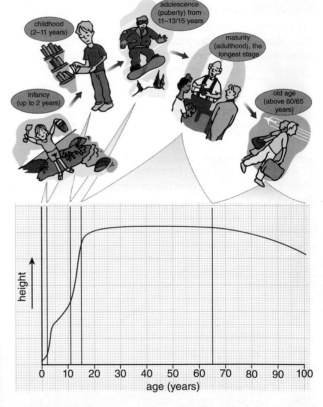

FIGURE 2: Stages in growth and development.

What affects growth?

> Ben and Charlotte have a balanced diet.

> Their diet contains calcium and vitamin D (needed for bones) and proteins (needed for muscle growth).

> They also take regular exercise, as this makes the body release more growth hormones.

Some people are very small (dwarfs) and some very tall (giants). The main causes are either hereditary or the wrong amount of growth hormone.

Growth checks

A health visitor regularly checks a baby's growth. The length, mass and head size of the baby are measured and recorded.

Different parts of the foetus and baby grow at different rates. At birth the head is large and looks out of proportion to the rest of the body.

These measurements are recorded on growth charts to check that the baby's growth is within the normal range.

Q puberty   human growth hormone

Any abnormal measurements indicate growth problems and further medical advice is needed.

Growth charts, such as the one shown in Figure 3 are only a guide to use for comparisons. In the future, with better understanding of diet and environmental factors, the average height and mass could increase and the charts will need to be modified.

mass (kg)
Pre-term
normal range
weeks
32 34 36 38 Birth 2 4 6 8 10 12 14 16 18

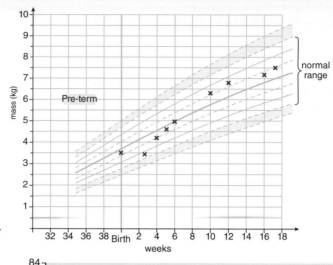

FIGURE 3: Growth chart for a baby boy's mass up to 18 weeks after birth.

Life expectancy

Ben and Charlotte expect to live longer than their parents and grandparents.

They hope to live a lot longer than their ancestors in the 13th century, when life expectancy was about 31 years.

They realise that their long life expectancy is due to:

> modern treatments and cures for many diseases

> fewer deaths from industrial diseases

> healthier diet and lifestyle

> better housing conditions.

life expectancy (years)
year of birth
life expectancy males
life expectancy females

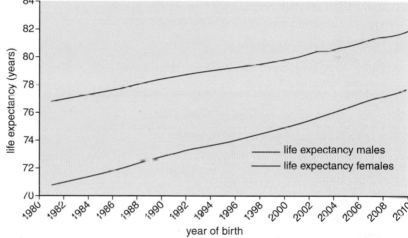

FIGURE 4: Graph to show life expectancy at birth in the UK. What general trends are shown?

Question

3 Suggest why the normal range in growth charts will change in the future.

Growth and old age

The human growth hormone is produced by the pituitary gland in the brain. It has:

> a direct effect of preventing fat being stored, and releasing it for energy and growth

> an indirect effect of stimulating the liver to produce another hormone (IGF-1). This hormone triggers an increased production of cartilage, causing an increase in length of long bones in the arms and legs.

human growth hormone produced by pituitary gland
liver
IGF-1
fat
direct effect
indirect effect

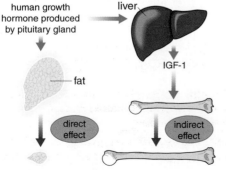

FIGURE 5: Hormone control of growth. Where is IGF-1 produced?

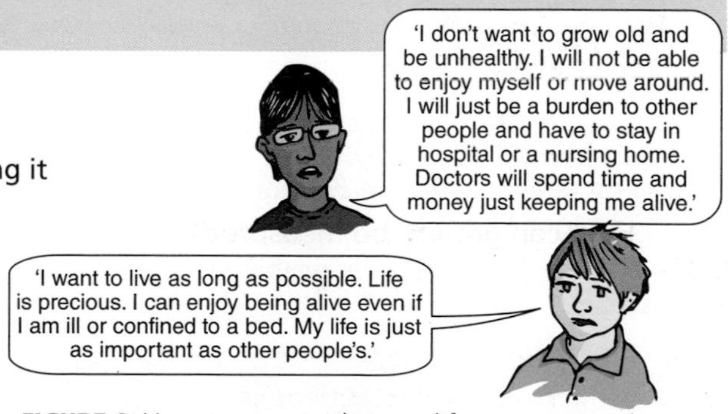

'I don't want to grow old and be unhealthy. I will not be able to enjoy myself or move around. I will just be a burden to other people and have to stay in hospital or a nursing home. Doctors will spend time and money just keeping me alive.'
'I want to live as long as possible. Life is precious. I can enjoy being alive even if I am ill or confined to a bed. My life is just as important as other people's.'

FIGURE 6: Here are some ideas on life expectancy that you may share. Can you think of any others?

Questions

4 Some athletes are thinking about using human growth hormones. Suggest why.

5 Make a list of reasons why you would like to live longer.

Replacement body parts

You will find out:

> which body parts have mechanical or biological replacements
> about the problems of replacements
> about transplant success rates

Certain body parts can be replaced if they are damaged by disease or trauma (accident).

Mechanical replacements

Some body parts can have mechanical replacements. They are not living cells, tissues or organs. They can be transplanted into the body. Some of the parts of the body that can be replaced are:

> the lens in the eye replaced with a plastic one

> the heart replaced with a metal and plastic pump

> worn hip joints replaced with a metal ball and socket

> worn knee joints replaced with a metal hinge joint.

Other mechanical replacements are used outside the body. Examples are mechanical ventilators, kidney dialysis machines and heart and lung machines.

Biological replacements

Some living body parts can be donated to other people. These are biological replacements. These include:

> the cornea of the eye can be transplanted

> the whole heart can be transplanted

> one kidney can be transplanted

> both lungs, sometimes with the heart, can be transplanted

> blood is often transfused during operations

> bone marrow from inside bones can be transplanted.

Body parts, such as blood and bone marrow, can be donated by people when they are still alive.

One kidney can also be donated by a living person, as humans have two kidneys.

Other body parts, such as the cornea, heart and lungs, can be donated by dead donors.

 Questions

6 Name two body parts that have mechanical replacements and two that have biological replacements.

7 Why are some mechanical replacements not used inside the body?

8 Suggest what materials are not used in mechanical replacements fitted inside the body.

Mechanical replacements

Some mechanical replacements are too big to fit inside the body. The first 'iron lung' machine was the size of a small car!

Scientists designing mechanical replacements have to consider many things. This is what their design brief might look like:

> small to fit inside body or to be carried

> lightweight, strong materials

> battery-powered so patient can move around

> inert materials so the body does not react to them.

Because of these problems, mechanical replacements such as the heart and lung machine, kidney dialysis machine, and mechanical ventilators are used outside the body.

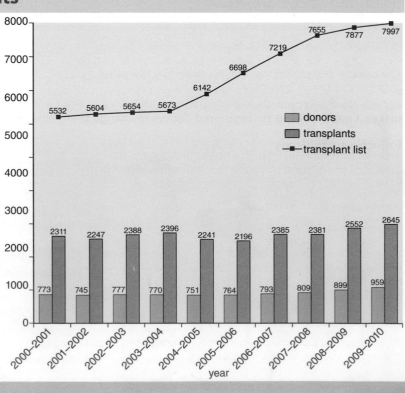

FIGURE 6: Donor and transplant information.

Q donor card

Donors

Rhys carries a donor card. This tells doctors and emergency staff that, if he has a fatal accident, parts of his body can be used to help others. However, about eight thousand people are on the waiting list, and sadly many people will die awaiting a transplant.

The long waiting list is caused by:

> a shortage of donors

> tissues or blood not matching

> size of organs and age of donors/recipients.

Rhys knows that he can donate some body parts, such as one kidney, blood, and bone marrow, while he is still alive. It is important that his tissues and blood match the type of the recipient.

When Rhys dies his organs can be used for transplantation.

> If he is 'brain dead' (suffers irreversible loss of consciousness and inability to breathe unaided owing to damage to the brain stem) his organs and tissues can be donated.

> If he dies from cardio-respiratory failure, only his tissues can be used.

Ethics of body replacements

Consenting to a mechanical or biological replacement may not be straightforward. The patient has to accept that they have a replacement part, either artificial or living.

> Some families also find it difficult to accept that surgical procedures will be carried out on the dead body of the donor.

> Some patients may find it difficult to accept that someone's death has been necessary to provide the transplant. However, many people are now happy to accept that when they die their organs and tissues will help others.

Questions

9 Which do you think is best – a mechanical or biological replacement? Give reasons for your choice.

10 Suggest why a donated organ has to be kept cool and taken quickly to the recipient for transplantation.

A donor register

People over the age of 18 years can register their approval for their organs to be used after their death.

Following a campaign in November 2009 to increase the number of people on the donor register, more than 110 000 people signed up.

The current system uses an 'opt-in' system. People have to declare that they wish to donate body parts. This results in fewer donated organs.

Other countries, such as Belgium and Portugal, use an 'opt-out' system. People have to declare if they do NOT want to donate body parts. This results in more donated organs and causes fewer problems at the death of a donor. However, some people do not agree with this system, saying it is against human rights. It also makes it more difficult for registered donors to change their mind.

Rejection problems

The body's immune system reacts to 'foreign' tissue and rejects it. Even if a donated organ or tissue is carefully matched, the recipient patient has to take immuno-suppressive drugs for the rest of their life. This makes them more prone to other infections.

Tissue type depends on three human leucocyte antigens (HLAs). A leucocyte is a type of white blood cell. The presence of HLAs depends on certain genes on chromosome 6. Fortunately, as families share the same genes they also tend to have the same HLAs. It is therefore much easier to obtain matching tissues and organs from within a family.

Success rates

With advances and experience in new medical techniques, organ transplant success rates are improving.

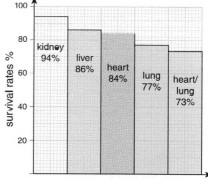

FIGURE 7: Graph to show UK rates of survival 1 year after transplantation in 2008. The survival rate after 5 years drops by about 10 per cent.

Questions

11 Suggest why survival rates after a transplant reduce between 1 and 5 years.

12 Suggest why heart and lung transplants have a lower rate of success than kidney transplants.

13 Look at Figure 6. Suggest why the number of transplants remain fairly static.

Preparing for assessment: Research and collecting secondary data

To achieve a good grade in science, you not only have to know and understand scientific ideas, but you need to be able to apply them to other situations and investigations. These tasks will support you in developing these skills.

✳ Task

> What is the pH range for human salivary amylase?

✳ Context

Roberto visits his dentist. The dentist tells Roberto that his teeth are in excellent condition. The dentist also tells Roberto that his saliva is acidic (pH 5).

Roberto does a web search and finds out that saliva acts as a lubricant to help swallow his food. His search also produces information about the range of pH found in human saliva.

He looks in a science text book and finds out that saliva contains an enzyme called salivary amylase. Amylase digests starch into a sugar called maltose.

Another text book gives information about the effects of different pH values on enzyme action.

Roberto is worried that since his saliva is acidic, he may not be digesting starch in his mouth.

✳ Planning

Plan how you are going to collect this information. You need to:

1 Write down how you found the information.

2 Write down a list of all the sources of your information.

3 Clearly present the information so it could be used to plan an actual investigation.

✳ General rules

1 You may work with other students but your written work should be done on your own.

2 You cannot get detailed help from your teacher.

3 You are not allowed to redraft your work.

4 Your work can be hand written or word processed.

5 It is expected that you complete this task in two hours.

6 You are allowed to do this research outside the laboratory.

7 You must be aware of and mention any health and safety issues.

✹ Research and collecting secondary data

What do you need to research?

To research:

- the known range of pH in human saliva
- the action of salivary amylase on starch
- the importance of digestion of starch.

Where would you find this information?

Information sources:

- Science text books in the school library. Find out about enzymes in the human digestive system and why there are so many different enzymes.
- The internet. Using the search terms human salivary glands and optimum pH for enzymes.
- A dentist.

What do you find out about?

Research found:

- The pH range in saliva
- Different optimum pH ranges
- Pairs of salivary glands
- Enzymes are specific in their action, with amylase breaking down starch into smaller molecules ready for further digestion.

Range of pH of saliva from five different web addresses...
5-8
6.3-6.6
6.0-7.4
7.35-7.45
6.4-7.5

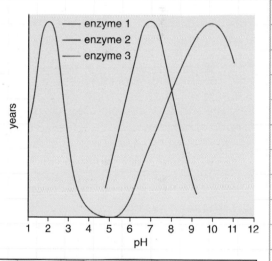

How do you record all this information?

Put together or print off a final account of your research and hand it to your teacher.

Record:

- Use a notebook to write down the information and where it came from.
- Use a mobile phone/voice recorder to record a conversation with a dentist.
- Use a computer to record images, conversations and diagrams.

B5 Checklist

To achieve your forecast grade in the exam you'll need to revise

Use this checklist to see what you can do now. It gives you many of the important points you will need to know. Refer back to the relevant pages in this book if you're not sure and to see if there is anything else you need to know. Look across the three columns to see how you can progress.

Remember you'll need to be able to use these ideas in various ways, such as:

> interpreting pictures, diagrams and graphs
> applying ideas to new situations
> explaining ethical implications
> suggesting some benefits and risks to society
> drawing conclusions from evidence you've been given.

Look at pages 272–294 for more information about exams and how you'll be assessed.

To aim for a grade E	To aim for a grade C	To aim for a grade A
know that some animals have an external skeleton, others have an internal skeleton of cartilage and bone **describe** different types of fracture **know** that the skull is a fixed joint, the elbow and knee are hinge joints, the shoulder and hip are ball and socket joints **identify** the main bones and muscles in a human arm	**understand** that cartilage and bone are living tissues **describe** the structure of a long bone **know** that bones can easily be broken and this occurs more often with age **describe** the structure, types and range of movement of synovial joints **describe** how the arm is moved by antagonistic muscles (biceps and triceps)	**describe** how, during growth, cartilage is replaced by bone in the process of ossification; this process can be used to indicate whether a person is still growing **explain** why it is dangerous to move someone with a broken bone **explain** the functions of parts of a synovial joint **explain** how the arm movement operates as a lever
know that some animals have no blood circulatory system, some have an open system, some have a double circulatory system where blood travels in arteries, veins and capillaries **describe** what the pulse is and where it can be measured	**describe** single and double circulatory blood systems **compare** the circulatory system in fish (single system using gills) with that in mammals (double system involving lungs) **understand** how heart muscle contraction is controlled by pacemaker cells and that artificial pacemakers can be implanted in the human body	**describe** how the discoveries of Galen and Harvey helped understanding of blood circulation **explain** why fish have a two-chambered heart and mammals have a four-chambered heart **describe** the sequence of events in the cardiac cycle **describe** the action of pacemaker cells (called SAN and AVN)
recognise that the heart may have conditions such as an irregular heartbeat, a hole in the heart, damaged or weak valves and coronary heart disease **know** the four different blood groups, A, B, AB and O which can be further divided into Rhesus positive and negative groups **know** that blood clots can seal wounds or block blood vessels	**explain** the consequences in terms of blood flow of these heart conditions **recognise** that heart transplants and 'heart assist' devices can be used **describe** the processes of blood donation and blood transfusion **describe** the process of blood clotting as a series of reactions	**understand** why unborn babies have a hole in their heart structure and why it usually closes at birth **explain** how agglutinins control the success or failure of transfusions

To aim for a grade E

describe the functions of the main parts of the human respiratory system

explain the terms breathing, respiration, inspiration and expiration

know that there are main diseases of the human respiratory system such as asthma, bronchitis, pneumonia and lung cancer

describe the position and function of parts of the human digestive system

describe the process of physical and chemical digestion

know that digested food enters the blood around the small intestine and leaves the blood in body tissues

describe the position and function of the main organs of excretion (lungs, skin and kidneys)

know that carbon dioxide produced by respiration is excreted by the lungs

describe the main stages of the menstrual cycle as menstruation, thickening of the uterus lining and ovulation

understand the possible causes of infertility and recognise that some can be corrected by infertility treatments

describe the main stages of human growth (infancy, childhood, adolescence, maturity and old age)

know that some body parts can be replaced by living or mechanical structures

To aim for a grade C

explain that the process of breathing in and out depends on changes in volume and pressure

describe how the respiratory system protects itself from disease

describe asbestosis, cystic fibrosis and lung cancer

explain the action of carbohydrase, protease and lipase enzymes

understand how digested food is absorbed into the blood plasma (glucose and amino acids) or lymph (fatty acids)

describe the main structures of a kidney and its blood vessels

explain how the kidneys work

explain that the toxicity of carbon dioxide makes its removal necessary

describe the action of hormones (oestrogen, progesterone, FSH and LH)

explain treatments for infertility (artificial insemination, use of FSH, IVF, egg donation, surrogacy and ovary transplants)

explain possible causes of the increase in life expectancy

explain the problems of using biological replacements (such as lack of donors, tissue matching) and mechanical replacements (such as size, power supply)

To aim for a grade A

explain how gaseous exchange surfaces are adapted for efficiency

interpret data from experiments using a spirometer

explain why the respiratory system is prone to many diseases

describe what happens in an asthma attack

explain why the pH of the stomach is acidic, while inside the mouth and small intestine is alkaline

explain how the small intestine is adapted for efficient food absorption

explain the principle of a dialysis machine

explain the action of ADH

explain how increased levels of carbon dioxide are detected by the brain and the reaction produced

explain how negative feedback is used in control of the menstrual cycle

evaluate infertility treatments

describe possible consequences of most people living longer

describe the advantages and disadvantages of a donor register

Foundation Tier

AO1 **1 (a)** Write down the numbers and letters to match up types of skeleton with a correct example.

skeleton	example
1. internal	A. humans
2. external	B. worms
3. no skeleton	C. insects [2]

AO1 **(b)** Write down **one** way in which an external skeleton is not as good as an internal skeleton. [1]

[Total: 3]

2 Look at the diagram of an adult human leg bone.

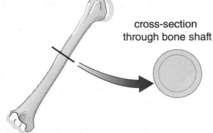

cross-section through bone shaft

AO1 **(a)** Where would you find cartilage? [1]

AO1 **(b)** What are the advantages of having hollow bones? [2]

AO2 **(c)** Bird bones are different from human bones. Suggest how and why they are different. [2]

[Total: 5]

3 Look at graph showing blood pressure changes.

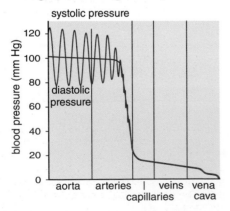

AO1 **(a)** Why does the blood pressure go up and down in the arteries? [2]

AO2 **(b)** What general pattern is shown in the graph? [1]

AO3 **(c)** Heart muscle can be damaged or diseased. Suggest what changes you would expect to see in the graph. [2]

[Total: 5]

4 Rick needs dialysis treatment since his damaged kidneys are not working properly. He finds this information on the web.

Method	Cost in first year in £s	Cost in each following year in £s
Successful transplant	20 000	6500
Dialysis at home	26 000	7000
Dialysis in hospital	29 000	29 000

AO2 **(a)**
AO3 Which method do you think is best? Use information from the table to explain your answer. [2]

AO1 **(b)** Write down two reasons, apart from cost, why some people needing a kidney transplant do not receive one. [2]

[Total: 4]

5 Look at the graph showing life expectancy in the UK.

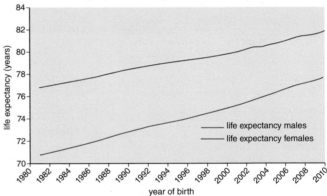

year of birth

AO2 **(a)** What general trend is shown by both males and females? [1]

AO1 **(b)** One reason for this trend is the better treatment of disease. Write down three other possible reasons for this trend. [1]

[Total: 2]

AO1 **6** Complete the sentences about the menstruation cycle. Use words from this list.

fallopian tubes, FSH, LT, oestrogen, ovary, progesterone, sperm ducts, testes, uterus.

Human eggs are produced by the .... and travel down the ....

Egg production is stimulated by ....

If the egg is not fertilised, the wall of the .... breaks down.

Repair of this wall is controlled by .... [5]

AO1 **7** Describe possible treatments for infertility. Include arguments for and against fertility treatments and specific treatments. [6]

AO1 recall the science   AO2 apply your knowledge   AO3 evaluate and analyse the evidence

✳ Worked Example – Foundation Tier

(a) The diagram shows different types of fractures.

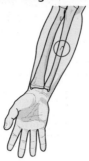 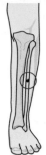 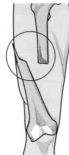

greenstick fracture   simple fracture   compound fracture
    A   B   C

Which diagram, A, B or C, shows the most serious fracture?

Explain your answer. [3]

*Diagram C is the most serious because it is in the
largest bone of the leg.*

(b) (i) Complete the diagram showing the main bones and
muscles in a human arm. [2]

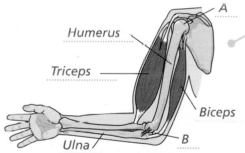

- A
- Humerus
- Triceps
- Biceps
- Ulna
- B

(ii) The labels A and B point to two joints, called synovial
joints. Why are they called synovial joints? [1]

*Labels A and B show a ball and socket joint and a
hinge joint.*

(c) Bones contain living tissue. Which statements support this
fact? Put a tick (✓) in the boxes next to the correct
statements. [2]

Bones can repair themselves. ☐

Bones are part of an external skeleton. ☑

Long bones are hollow. ☑

Bones can easily be broken by a sharp knock. ☑

Bone contains blood cells. ☐

How to raise your grade!
Take note of these comments –
they will help you to raise your grade.

The student correctly identified
diagram C. The student has scored
another mark by realising it is in the
largest bone. However, with a total of
three available marks, a better
explanation is expected. This
compound fracture shows that the
broken bone has pierced the skin, so
infection is a high risk.   2/3

The student has incorrectly labelled
the biceps and triceps muscles. To
help you remember, at the shoulder
the triceps muscle has three
tendons, the biceps has only two.
The bones are correctly labelled. 1/2

The student has shown a common
examination error...the answer does
not fit the question.

They are called synovial joints
because they contain synovial fluid
between the bones.   0/1

This is a common type of
examination question. Sometimes
there is no information about how
many ticks are required – this makes
the question harder since a mark is
deducted for each incorrect answer.
The first and last boxes should be
ticked.   1/2

This student has scored 4 marks
out of a possible 8. This is below
the standard of Grade C. With
more care the student could have
achieved a Grade C.

B5 Exam-style questions

Higher Tier

AO2 **1** Hundreds of years ago, doctors treated sick people by using leeches to remove blood. Suggest why this would cause problems rather than cure them. [1]

Look at the diagram of blood types.

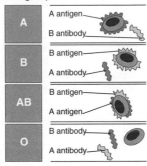

Yousef needs a blood transfusion. He has blood group O.

AO1 **(a)** Which blood group can he receive safely? Explain your choice. [2]

AO1 **(b)** Yousef recovers from his illness and wants to donate some blood. Which blood groups could safely receive his blood? [1]

AO1 **(c)** Explain what will happen **in the blood** if a patient received the wrong blood type. [2]

[Total: 6]

2 The diagram shows relative growth in humans.

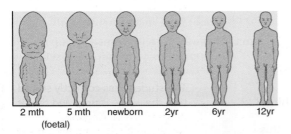

AO1 **(a)** What general patterns are shown? [2]
AO2

Human growth is affected by genes and lifestyle.

AO1 **(b)** Describe what else controls growth in humans. [2]

[Total: 4]

3 Look at the spirometer trace.

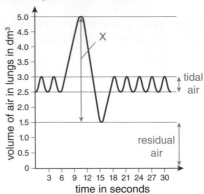

AO2 **(a) (i)** What does X measure? [1]

AO2 **(ii)** How is this part of the trace produced? [1]

AO2 **(b)** Suggest why this measurement is used to test patients with breathing problems. [1]

AO2 **(c)** What is volume of tidal air? Explain when this volume would increase. [2]

[Total: 5]

AO1 **4** Fish have a single circulatory system; mammals have a double circulatory system. Explain the advantages and disadvantages of these systems. [6]

AO1 **5** Look at the diagram of the small intestine.

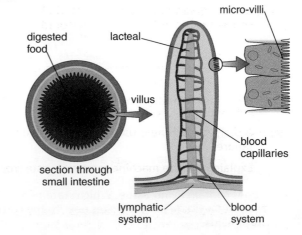

Describe how the small intestine is adapted for efficient absorption of different foods. [4]

AO1 recall the science   AO2 apply your knowledge   AO3 evaluate and analyse the evidence

✳ Worked Example – Higher Tier

The diagram shows what happens to urine concentration when too much water is lost from the body by sweating.

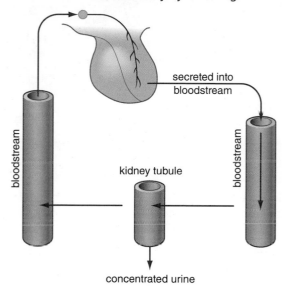

secreted into bloodstream

kidney tubule

bloodstream

bloodstream

concentrated urine

How to raise your grade!
Take note of these comments – they will help you to raise your grade.

(a) (i) Which hormone is involved in this process and where is it produced? [2]

Anti-diuretic hormone (ADH) is involved. It is produced in the brain.

The student has correctly named the hormone but it is produced by the pituitary gland. The word 'brain' is too vague an answer. 1/2

(ii) Use the diagram to help you explain how the water concentration in the blood is restored. [4]

Changes in the blood concentration are detected by the brain. ADH is secreted and is carried by the blood to the kidney. ADH increases the permeability of the kidney. Therefore more water is reabsorbed back into the blood.

The student's answer is too vague. The answer should have referred to a high blood concentration, i.e. too little water present.

The student has referred to the kidney rather than the kidney tubules.

This part is correct. 2/4

(b) A dialysis machine is used when a patient has kidney failure.

Explain how the machine removes waste from the blood. [3]

The artificial kidney removes urea from the blood. The urea molecules diffuse out of the blood and into fluid in the machine.

The student has correctly stated that the waste is urea and that it diffuses out of the blood. However, this is not a full answer. The answer should have included information about partially permeable membranes allowing the diffusion of small molecules such as urea. 2/3

This student has scored 5 marks out of a possible 9. This is below the standard of Grade A. With more care the student could have achieved a Grade A.

B6 Beyond the microscope

Ideas you've met before

Microbes and disease

Bacteria and viruses are microorganisms that cause disease.

Viruses cause colds, flu, chickenpox and measles.

Bacteria are larger than viruses. Food poisoning is caused by bacteria growing on food.

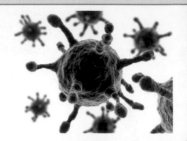

 Describe one way to prevent bacteria causing food poisoning.

Life in soil and water

Plants grow in soil. The soil provides water, minerals and anchorage for the roots. Small animals live in soil where they feed on decaying plant material.

Microscopic plants called phytoplankton live in water where they use the Sun's energy to make food. These plants provide food for animals living in water.

 Why are phytoplankton so important to fish?

Enzymes

Enzymes speed up reactions inside the body.

The digestive system uses enzymes to digest food.

 Suggest why enzymes are called biological catalysts.

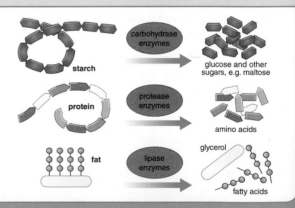

Food webs and pollution

Food chains show what eats what.

When pesticides get into the food chain they get passed from one animal to another. As the poison moves along it gets more concentrated.

The top carnivore has the highest concentration and suffers the most.

 Explain how pesticides sprayed on crops can harm a predator like the peregrine falcon.

In B6 you will find out about...

> how fungi can also cause disease

> how bacteria and fungi can be used to make useful products

> how diseases can be treated by antibiotics

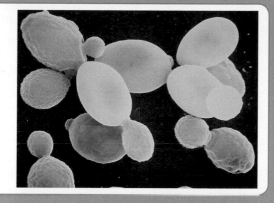

> detritivores, herbivores and carnivores that live on the soil

> the importance of earthworms to soil fertility

> the advantages and disadvantages of living in water

> how enzymes are used to make sweeter sugar

> how enzymes can be immobilised

> how enzymes are used in genetic engineering to cut the DNA

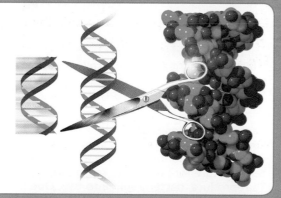

> marine food webs

> the effects of PCBs and DDT on marine life

> how sewage can cause rapid growth of algae, which leads to the death of animals in the water

Understanding microbes

You will find out:
> about bacteria
> about the structure of a bacterial cell
> how bacteria feed
> how bacteria reproduce

Bean sprouts and food poisoning

In September 2010 The Food Standards Agency started to advise people about cooking bean sprouts. This followed an outbreak of food poisoning caused by the bacteria *Salmonella*. *Salmonella* bacteria had invaded bean sprouts. Eating uncooked bean sprouts could result in diarrhoea, stomach pains and fever.

When cooking bean sprouts they should be left to simmer for a few minutes. This should kill the bacteria, leaving them safe to eat.

FIGURE 1: Why is it better to cook bean sprouts before you eat them?

Bacteria

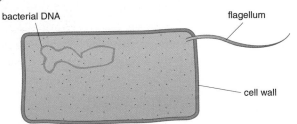

bacterial DNA   flagellum

cell wall

FIGURE 2: A bacterium. What is the flagellum used for?

Bacteria are made from just one cell. The bacterial cell is just a few micrometres long (1 micrometre = 0.001 mm). Plant and animal cells are about 10 times bigger. Bacteria come in all shapes and sizes. Scientists use shape to classify bacteria. *E. coli* bacteria are rod shaped.

Bacteria are often used to make things such as yoghurt. Bacteria are put in large fermenters to grow. The fermenters are kept at the right temperature for bacteria to reproduce rapidly. To reproduce, bacteria split into two. One bacterial cell can turn into two in just half an hour. After 12 hours there will be 16 million of them!

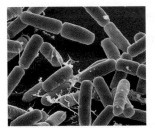

FIGURE 3: *E. coli* bacteria magnified. What shape are they?

Did you know?

The Black Death of 1348 killed nearly half the population of the UK. It was the result of infection by bacteria.

Questions

1 Which is larger, a plant cell or a bacterial cell?

2 What do scientists use to classify bacteria?

3 a How many bacteria could you fit along a line 10 mm in length?

b How many plant cells could you fit on the same line?

Structure of bacteria

Some important parts of a bacterial cell:

Part	Function
flagellum	helps bacterium to move
cell wall	helps bacterium keep its shape and prevents it bursting
bacterial DNA	controls bacterium's activities and its replication

The shape of bacteria

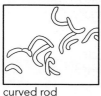 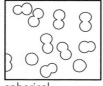 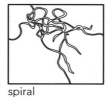 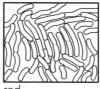

curved rod   spherical   spiral   rod

FIGURE 4: Bacteria can be classified into four different shapes.

Bacterial reproduction

Bacteria can be grown in the lab on agar plates using **aseptic techniques**. The agar is sterilised and poured into a sterile Petri dish. A few bacteria can then be placed onto the agar. The lid is then sealed to stop any microorganisms getting in from the air.

The bacteria grow in number using a type of **asexual reproduction** called **binary fission** (the process of splitting into two). After a few days the agar will be covered in bacterial colonies.

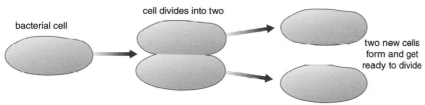

bacterial cell   cell divides into two   two new cells form and get ready to divide

FIGURE 5: Bacteria reproduce by binary fission. What does this mean?

Handling bacteria safely

When using bacteria it is always good practice to use aseptic techniques.

> If possible, wear disposable gloves while handling bacteria. Or wash hands before and after working with the bacteria.

> Disinfect working areas.

> Sterilise all equipment before and after use.

> Never leave the lids off containers.

Questions

4 Explain why bacteria need cell walls.

5 What are the four main shapes of bacteria?

6 Write down one aseptic technique used when growing bacteria.

7 Bacteria can reproduce every half an hour. If you start with two bacteria, how many would you have after 5 hours?

How bacteria feed

Most bacteria feed in the same way as animals. They consume organic nutrients such as carbohydrates and proteins. However, there are bacteria that can make their own food. Some use light energy for photosynthesis. Others use chemicals such as hydrogen sulfide or ammonia to provide the energy to make food. This use of many different energy sources enables bacteria to live in a wide range of habitats.

The problems with bacteria

Bacteria reproduce so fast it is very difficult to stop them. When they grow on food, the food quickly spoils and can become dangerous to eat. If bacteria get inside our bodies they can reproduce. This can cause diseases such as food poisoning, cholera, whooping cough, typhoid and tetanus.

FIGURE 6: Suggest which chemical is used by bacteria living in sulfur springs for food.

FIGURE 7: An agar dish with a bacterial colony on it. Which of the handling precautions has the research scientist taken?

Questions

8 Photosynthetic bacteria make their own food. Suggest which energy source they use.

9 Explain why it is important to wash hands after handling bacteria.

10 Explain why all equipment should be sterilised after use.

11 Suggest one reason why it is dangerous to eat food after its sell-by date.

🔍 binary fission   aseptic technique

Yeast is a very important fungus

You will find out:

> about yeast

> about viruses

Yeast is a single-celled **fungus** that has many uses, such as making bread and alcohol.

Yeast cells reproduce asexually by **budding**.

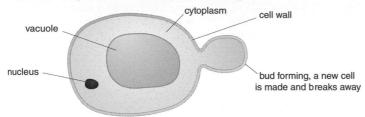

FIGURE 8: A yeast cell. Describe how yeast cells reproduce.

Questions

12 Name the process by which yeast cells reproduce.

13 Write down two structures found in yeast and plant cells but not animal cells.

14 How are viruses different to bacteria?

Viruses

Many illnesses are caused by a group of microorganisms called viruses. Viruses are much smaller than bacteria and fungi. Scientists do not class viruses as living cells.

Growing yeast

When dried yeast is placed in sugar solution it starts to grow, very slowly at first. However, it then grows rapidly. To maintain optimum growth the correct conditions need to be provided:

> food

> optimum temperature

> optimum pH

> removal of waste products.

Virus structure

Viruses are made up of strands of genetic material surrounded by a protein coat.

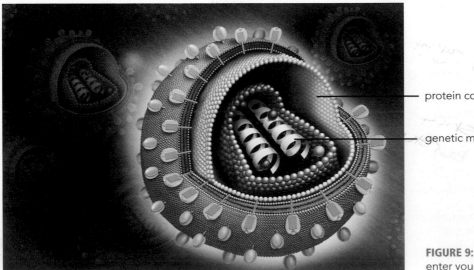

— protein coat

— genetic material

FIGURE 9: A virus cell. Which virus can enter your skin cells causing harm?

🔍 yeast structure   yeast reproduction

They have to enter another living cell before they can reproduce. The living cell could be from a plant, animal or even a bacterial cell.

Different viruses attack and enter different types of cells. For example, the AIDS virus enters white blood cells, whereas the chickenpox virus will enter skin cells.

Questions

15 Suggest one reason why the wrong pH could slow down growth in yeast.

16 What is the coat surrounding a virus made from?

17 A plant virus would not harm a human. Suggest a reason why.

The effect of temperature on yeast

The rate of yeast growth doubles for every 10 °C rise in temperature until the optimum temperature is reached. Above this temperature the yeast cells start to die.

Virus reproduction

Viruses reproduce inside living cells, using the cell to make new proteins.

1 The virus attaches to a specific cell, called the host cell.

2 The genetic material from the virus is injected into the host cell where it enters the nucleus to make mRNA.

3 New viruses are made using material inside the host cell and the genetic material from the virus. This happens in the cytoplasm of the cell.

4 The cell splits open, allowing the viruses to leave but killing the host cell.

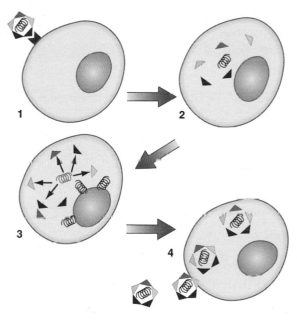

FIGURE 10: Describe the stages in viral reproduction.

Remember!

Always use the word 'microorganism' or 'bacterium' or 'virus' in an exam. Never talk about 'germs'.

Questions

18 Yeast cells start to die at high temperatures. Use ideas about enzymes to explain why.

19 Viruses are not living cells. Use ideas about their reproduction to explain why.

20 Viruses use material inside the host cell to make proteins. Find out which materials they use.

Q virus reproduction

Harmful microorganisms

You will find out:

> how diseases can enter the body

> about the different microorganisms that cause disease

> about the stages of infectious diseases

Swine flu pandemic

Pandemics are diseases that spread across the world. In 2009, governments were worried that a disease called swine flu would spread around the world and kill millions of people.

Swine flu is a new variety of flu virus containing genetic material from other flu viruses. It was first discovered in Mexico and then spread as infected people travelled to other countries. By the end of February 2010 the virus had caused 15 921 deaths worldwide.

The number of new cases started to slow down, and in August 2010 it was no longer classed as a pandemic.

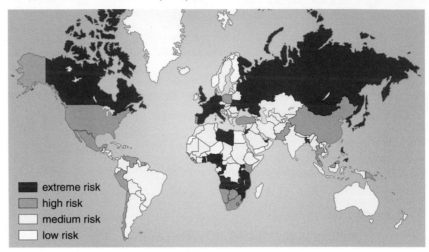

extreme risk
high risk
medium risk
low risk

FIGURE 1: World map showing which countries were at risk from swine flu in 2009. What was the chance of getting swine flu in the UK?

Pathogens

Microorganisms that cause disease are called **pathogens**. The pathogens get inside our bodies, where they reproduce and cause us to become ill. Pathogens can enter our bodies in different ways. These are shown in Figure 2.

Different microorganisms

There are different types of microorganisms that cause diseases.

Type of microorganism	Diseases
bacteria	cholera, food poisoning
virus	influenza, chickenpox
fungus	athlete's foot

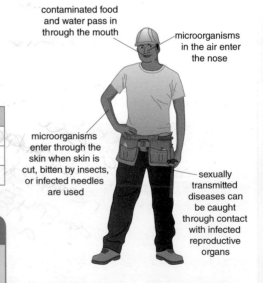

contaminated food and water pass in through the mouth

microorganisms in the air enter the nose

microorganisms enter through the skin when skin is cut, bitten by insects, or infected needles are used

sexually transmitted diseases can be caught through contact with infected reproductive organs

FIGURE 2: How diseases can enter our bodies.

Questions

1 How can diseases enter the body through the skin?

2 Which type of microorganism causes food poisoning?

3 Name one disease caused by a virus.

4 Suggest one way that the influenza virus might enter the body.

Q food poisoning   athlete's foot

Becoming sick

People with chickenpox will not get any spots until 2 weeks after they were first infected. Infectious diseases often take time to go through all their stages.

> First the microorganism enters the body.

> It then reproduces to build up numbers. This is called the **incubation period**.

> The microorganism then starts to make harmful toxins.

> Symptoms start to appear, such as fever.

Preventing the transmission of disease

People become infected with cholera when they drink water containing the bacteria that causes cholera. It is important that drinking water is free from contamination.

FIGURE 3: Food poisoning can be the result of eating food infected with bacteria. How can making sure food is cooked properly prevent infection?

FIGURE 4: Influenza is passed on when a sufferer sneezes. Someone else will then breathe in the airborne droplets that contain the virus. How can you reduce the spread of influenza when you sneeze?

Some diseases are passed on by contact with an infected person. These include sexually transmitted diseases. Using a condom is just one way to reduce the risk of contracting such a disease.

Questions

5 Suggest one precaution you could take to prevent catching cholera while on holiday in an affected area.

6 Describe one way to reduce the spread of influenza.

7 What is happening inside the body between the time of infection and the appearance of the first symptoms?

8 Athlete's foot is passed on by contact with the fungus. Find one other example of a disease that is passed on by contact.

Interpreting data

In the exam you will be asked to interpret data on food poisoning, cholera and influenza. Use question 10 to practise this type of question.

Questions

9 Why are you unlikely to catch cholera in the UK?

10 The graph shows the number of cholera cases between 1993 and 2003 in the world.

a Describe the change in cholera cases over the 10-year period.

b In which year were most cases reported?

c In 1997 there were 150 000 and in 1998 there were 290 000.

i Calculate the percentage increase in cases between 1997 and 1998.

ii Use ideas about the transmission of cholera to suggest a reason for this increase.

d Describe the relationship between the number of cases and the number of countries reporting cases.

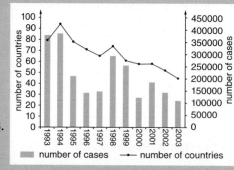

Earthquakes and diseases

When a major earthquake occurs, large areas of land are destroyed. Many people lose their lives when buildings collapse on top of them.

> In the weeks that follow, more people lose their lives as diseases such as cholera and food poisoning spread through the earthquake area. They are caused by damage to water supplies and sewage systems. This damage means water is no longer safe to drink or clean to wash in.

> Sewage contaminates crops after it gets onto the land. The diseases are able to spread because people are forced to live in cramped conditions.

> In an attempt to help them, aid organisations do their best to supply the homeless people with shelter and clean water.

> **Antibiotics** are also needed in an attempt to control the spread of diseases caused by bacteria. The use of antibiotics has to be controlled because the overuse of antibiotics has led to some bacteria becoming resistant. This means they are not destroyed by the antibiotic.

FIGURE 5: These buildings have collapsed because of an earthquake. Which bacteria might be in the water ready to cause disease?

Questions

11 Apart from earthquakes, name another natural disaster that can cause the spread of disease.

12 Write down two diseases that can spread quickly through a disaster area.

13 Why are antibiotics needed in a disaster area?

The rapid spread of disease

Disease spreads very easily after a natural disaster – even in developed countries.

In 2011, a tsunami hit the coastline of Japan. Sewage systems were damaged and water and electrical supplies were lost, making it impossible to keep food and water fresh. Hospitals could not cope with the large number of injured people.

Many people were forced into cramped shelters with poor sanitation.

The only answer was to completely evacuate the area to prevent the spread of disease.

FIGURE 6: Flooding in Japan. Why would a burst sewage pipe be a problem?

Q antibiotic resistance

Finding cures

Diseases such as cholera and dysentery were once common in the UK. The work of scientists has led to them and other diseases almost disappearing.

Pasteur

In the 1860s, Louis Pasteur carried out a series of experiments to prove that microorganisms in the air cause food to go bad. Before then, people believed that the air itself caused food to decay. Pasteur's theory was called the 'germ theory of disease'.

Lister

Joseph Lister noticed that wounds often became infected after operations. In 1865, he started spraying wounds with carbolic acid. This killed the microorganisms, and thus the wounds did not become infected. Lister had discovered the first **antiseptic**. Antiseptics are important tools to help us control disease.

Fleming

In 1928, Alexander Fleming discovered penicillin. He accidentally left some samples of bacteria open to the air. A few days later, he noticed that the samples had gone mouldy. Where the mould grew, the bacteria had stopped growing. This mould was *Penicillium*, and was used to make the world's first antibiotic. Penicillin is still used now to cure internal bacterial infections.

Problems with antibiotics

Antibiotics are useful in controlling the spread of cholera but not influenza. This is because bacteria are affected by antibiotics but viruses are not. Instead you need to take antiviral drugs.

Resistant bacteria

The media sometimes report the discovery of another 'superbug'. These are strains of bacteria that are resistant to antibiotics. The resistance can be caused by a **mutation** in the bacterial DNA or when DNA is passed from one bacterium to another. This provides the bacteria with a defence against the antibiotic. Resistant bacteria spread because they have an advantage over non-resistant bacteria. They are more likely to survive and pass on the resistance in their DNA. This is a modern example of natural selection.

Remember!
To remember who discovered which treatment, try listing the scientists then matching the treatment without looking in the book.

Questions

14 Why does clean water stop the spread of cholera?

15 Which was invented first, antibiotics or antiseptics?

16 Describe one use for each of:

a antiseptics

b antibiotics.

17 Explain why it is important to take antibiotics only if you really need them.

Slowing the spread of resistance

The spread of resistance can be slowed if a few basic procedures are followed:

> doctors should prescribe antibiotics only when they are needed – overuse of antibiotics provides more opportunity for resistance to occur

> if you are prescribed antibiotics, you should complete the course of the drugs to ensure all the bacteria are destroyed

> doctors and nurses should simply wash their hands before and after treating an infected patient.

Question

18 You should not take antibiotics prescribed for someone else. Suggest why.

Useful microorganisms

You will find out:
> about uses of bacteria
> about fermentation

A brief history of beer

It is believed that beer has been brewed for the last 10 000 years. The Egyptians offered what they called barley wine to the gods. During the Middle Ages, monks brewed beer in their monasteries. At this time, cholera bacteria could make local water unsafe to drink. Drinking beer was safer because the bacteria were eliminated during the brewing process. The brewing process moved away from the monasteries in the 18th century and became an industry in its own right. After World War I, many small breweries closed. However, recent years have seen a return to the smaller breweries producing all sorts of beer with very strange names.

FIGURE 1: Which microorganism is used to make the beer in these bottles?

Useful bacteria

Bacteria are used to make lots of important things.

> Yoghurt is made when bacteria turn milk sour.

> Silage is made from fermented grass and is fed to cows.

> Cheese is made from milk using bacteria to help it ripen.

> Vinegar is made when bacteria make acids from alcohol.

> Compost is made by bacteria breaking down dead plant material.

Yeast and fermentation

Yeast is alive. To grow it needs food and water. At school and at home you may have bottles of dried yeast. When the yeast is dry, it stays dormant.

If the yeast is put into a beaker of sugar solution it starts to grow. The yeast breaks down the sugar and turns it into alcohol and carbon dioxide.

Turning sugar into alcohol is called **fermentation**. Fermentation is a special kind of respiration that does not need oxygen.

FIGURE 2: Which chemical turns into an acid when vinegar is made?

Questions

1 Bacteria can turn milk into other foods. Name two of these foods.

2 What food for cows can bacteria turn grass into?

3 What two substances do yeast cells make during fermentation?

4 Why is fermentation a special kind of respiration?

Q fermentation

Making yoghurt

There are five main stages in the production of yoghurt (see Figure 3).

all the equipment is sterilised

milk is pasteurised by heating to 95 °C for 20 minutes, then cooled to 46 °C

Milk

heat

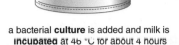

a bacterial **culture** is added and milk is **incubated** at 46 °C for about 4 hours

Milk

samples are taken to find out when the yoghurt is ready

Yoghurt

flavours and colours are added

Yoghurt

yoghurt is cooled and packed ready for sale

FIGURE 3: Making yoghurt. Why is the milk cooled before adding the bacteria?

Fermentation

Fermentation is **anaerobic respiration** in yeast. 'Anaerobic' means it occurs without oxygen.

glucose (sugar) → ethanol (alcohol) + carbon dioxide

Questions

5 Describe how bacteria are used to make yoghurt.

6 How many substrates are required for anaerobic respiration? Name them.

Yoghurt bacteria

The bacteria used to make yoghurt are called *Lactobacillus*. The bacteria are used to break down a sugar in the milk called lactose. *Lactobacillus* converts the lactose into lactic acid. The lactic acid then changes proteins in the milk to give yoghurt its texture.

The fermentation equation

Fermentation produces a type of alcohol called ethanol. The chemical symbol for ethanol is C_2H_5OH.

The balanced chemical equation for fermentation is:

$$C_6H_{12}O_6 \rightarrow 2C_2H_5OH + 2CO_2$$

Aerobic or anaerobic

If yeast is provided with plenty of oxygen, it carries out **aerobic respiration**. During aerobic respiration it converts glucose into carbon dioxide and water. To ensure yeast produces alcohol, it must be grown in the absence of oxygen.

Questions

7 Yeast was allowed to carry out anaerobic respiration at different temperatures. The carbon dioxide produced was collected for 5 minutes. The results are shown in the table.

Temperature in °C	Amount of carbon dioxide collected in 5 minutes in cm³			
	test 1	test 2	test 3	mean
20	0.8	0.9	1.0	
30	23.4	22.8	22.4	
40	39.4	40.8	41.7	
50	39.1	31.9	30.8	
60	3.2	2.8	2.8	

a Identify the anomalous result. Suggest one reason for the anomaly.

b Copy the table and complete it by calculating the mean volume of carbon dioxide for each temperature. Do not include the anomaly.

c Plot a graph to show the mean volume of carbon dioxide collected against temperature.

d Use your graph to describe the effect of temperature on respiration in yeast.

e i Write down the balanced chemical equation for aerobic respiration.

ii Use both respiration equations to explain how the results might change if the yeast were provided with oxygen.

Uses of fermentation

You will find out:

> about uses of fermentation

> about how beer and wine are made

> about how distillation is used to increase alcohol concentration

> about pasteurisation

When yeast is given different sources of sugar, fermentation produces different kinds of drinks, such as wine and beer.

Spirits, such as rum and whisky, are made from the products of fermentation. They have a higher concentration of alcohol, because some of the water has been removed.

Did you know?

In the UK, 17 million working days are lost to hangovers and drink-related illness each year.

FIGURE 4: Fermenting grapes can be turned into wine.

FIGURE 5: Barley grains are allowed to sprout. The sprouting grains are called malted barley. The malted barley is then used to make beer.

Questions

8 Name one drink made by the process of fermentation.

9 What is beer made from?

10 Explain why the alcohol content of whisky is higher than wine.

Brewing beer

Wine is made in a very similar way to beer except it uses grapes in place of the hops and barley.

Spirits

The alcohol content of beer is low compared with spirits, because it contains a lot of water. To increase alcohol content, the alcohol is separated from the water by **distillation**.

> The mixture is heated to evaporate the alcohol.

> The alcohol is then cooled to turn it back into a liquid.

It is legal to brew beer for your own use at home, but spirits have to be distilled on licensed premises.

Did you know?

Alcohol intake per week should be no more than:

> 21 units for men

> 14 units for women.

One unit is equivalent to one small glass of wine, half a pint of beer or one pub measure of spirits.

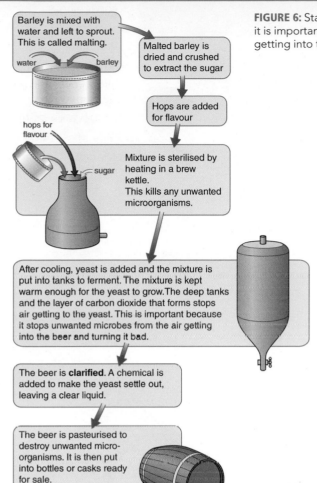

Barley is mixed with water and left to sprout. This is called malting.

water   barley

Malted barley is dried and crushed to extract the sugar

Hops are added for flavour

hops for flavour

sugar

Mixture is sterilised by heating in a brew kettle. This kills any unwanted microorganisms.

After cooling, yeast is added and the mixture is put into tanks to ferment. The mixture is kept warm enough for the yeast to grow. The deep tanks and the layer of carbon dioxide that forms stops air getting to the yeast. This is important because it stops unwanted microbes from the air getting into the beer and turning it bad.

The beer is **clarified**. A chemical is added to make the yeast settle out, leaving a clear liquid.

The beer is pasteurised to destroy unwanted micro-organisms. It is then put into bottles or casks ready for sale.

FIGURE 6: Stages in brewing beer. Explain why it is important to stop air and microorganisms getting into the beer.

Questions

11 Describe how beer is clarified.

12 Explain why beer has to be cooled down before yeast can be added.

13 Explain why wine can be made at home but whisky cannot.

14 What name is given to the process of cooling evaporated alcohol back into a liquid?

Preserving beer

During the 17th century nearly every village had its own brewery. This was because the beer could not be preserved, so it had to be drunk soon after it was made.

At the end of the 19th century, Louis Pasteur discovered that wine could be kept for longer if it was heated. Brewers now heat beer to 72 °C for 15 seconds before it is bottled. This method of preserving beer is called **pasteurisation**.

Yeast strains

There are many different varieties of yeast, known as strains. Different yeast strains give beers or wines different flavours.

Brewers also use different yeast strains because they tolerate different levels of alcohol. The alcohol produced by yeast cells kills them when it reaches a certain level. Those strains that have a greater tolerance can continue fermentation at higher levels of alcohol.

Questions

15 The graph (right) shows the growth of yeast over time.

number of yeast cells

time in hours

a Explain why the graph falls at the end.

b Copy the graph. On your graph draw a line to show a yeast strain with a lower tolerance to alcohol.

16 Explain why it is important to pasteurise beer before bottling it.

Biofuels

You will find out:
> about biogas
> about the bacteria that make biogas
> how biogas can be made on a large scale

Biofuels and chip oil

Countries across the world are trying to find local alternatives to petrol. Brazil grows a lot of sugar cane.

It can turn the sugar cane into alcohol for use in cars. In the UK we eat a lot of chips. The chips are cooked in vegetable oil that is normally thrown away. Scientists have developed cars that can run on used chip oil. This solves two problems: how to replace petrol and what to do with old chip oil.

FIGURE 1: How can eating chips help fuel a car?

What are biofuels?

Green plants use light energy from the Sun to photosynthesise. Photosynthesis produces food. Plants use the energy in the food to grow – this increases their mass.

The mass of the plants is called **biomass**. Biomass can be used to make **biofuels**. **Biogas** and wood are all examples of biofuels.

Alcohol, which is made from yeast, is mixed with petrol to make a biofuel called **gasohol**.

What is biogas?

Bacteria feed on organic material such as dead plants and animals. This causes the organic material to decay. As it rots, it produces a mixture of gases called biogas. The main gas in the mixture is methane. Marshes, septic tanks, and even our own digestive system, produce biogas.

Landfill sites are used to get rid of organic waste. However, this can be dangerous. If methane builds up too much it can explode.

FIGURE 2: What do you think has caused the fire at this landfill site?

Making biogas

Biogas is a biofuel. It can be used as an alternative energy resource to fossil fuels. To do this it has to be made on a large scale in a digester. A digester is a large container full of rotting organic material and bacteria.

Questions

1 What is the main gas in biogas?

2 Why are bacteria important in the production of biogas?

3 Why is methane a problem in landfill sites?

4 What is a digester used for?

FIGURE 3: A digester used to make biogas. What is in the digester?

Q biofuels   biogas

What is in biogas?

Biogas usually contains:

> about 60% methane

> about 40% carbon dioxide

> traces of hydrogen, nitrogen and hydrogen sulfide gas.

Biogas production

The energy in biomass can be released in two ways:

> burning fast-growing trees such as willow – the energy released can be used to heat the water in a power station to make electricity

> using bacteria or yeast to ferment the biomass and produce methane.

Methane is made on a large scale using a digester. Waste is fed into the digester continuously, producing methane and carbon dioxide.

The remaining solids are regularly removed through the outlet tank. This is called a continuous flow method because it carries on without stopping for months, or even years.

What affects biogas production?

Biogas production increases as temperature increases, up to about 45 °C. Above this temperature, production slows down.

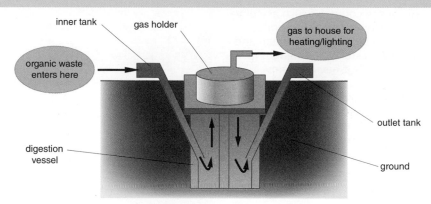

FIGURE 4: How a digester works. What term describes a process that continues for a long time?

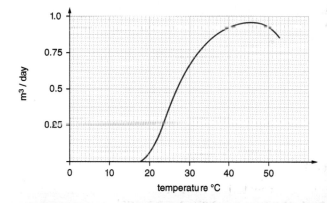

FIGURE 5: Graph showing the effect of temperature on the volume of biogas produced. Can you suggest why above 45 °C production of biogas slows down?

Questions

5 Draw a pie chart to show the gases in biogas.

6 Describe how methane can be produced on a large scale.

7 Why is the continuous flow method a good way to make biogas?

8 What happens to methane production at high temperatures?

The best mixture

Biogas usually contains about 60% methane, although this can vary. To burn easily, the biogas needs to contain more than 50% methane. Care must be taken when using biogas. A mixture of as little as 10% methane in air can be explosive.

Temperature and biogas production

Temperature affects biogas production.

As the temperature increases, the bacteria multiply faster and the enzymes within them work better. Above 45 °C, the enzymes are denatured and the bacteria die.

Questions

9 Explain why homes using biogas need to be well ventilated.

10 Explain what happens to enzymes when they are denatured.

11 Use your knowledge of enzymes to explain why different bacteria are needed to break down fats and fatty acids.

Uses of biofuels

You will find out:

> about different biofuels

> about the uses of different biofuels

> about the advantages of using biofuels

Houses in countries such as Kenya use a 'Home Biogas System' to produce biogas from human and animal waste. Many villages in Kenya do not have a mains electricity supply and use biogas to supply heat and light. This also solves the problem of getting rid of human waste, as there are no sewage systems.

FIGURE 6: A boy and his biodigester in Kenya. Why is biogas so important to people living in remote parts of the world?

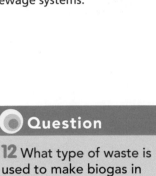

Question

12 What type of waste is used to make biogas in Kenya?

Uses of biogas

When biogas burns it releases energy. The energy can be used in different ways:

> to generate electricity

> to produce hot water and steam for central heating

> as a fuel for vehicles.

Gasohol

Gasohol is a type of biofuel. Yeast is used to turn sugar from sugar cane into a type of alcohol called ethanol. The ethanol is mixed with petrol to make gasohol. Gasohol is then used in cars as a cleaner alternative to petrol.

This is very useful in countries such as Brazil. This is because they have lots of sugar cane but small oil reserves.

FIGURE 7: Harvesting sugar cane in Brazil. What is added to sugar to turn it into alcohol?

Remember!

In the exam you may be asked to evaluate different methods of transferring energy from biomass. Practice exam questions can help you do this.

Why use biofuels?

Biofuels such as gasohol and biogas have advantages:

> they are an alternative to fossil fuels

> burning fossil fuels increases the levels of carbon dioxide in the air; burning biofuels only recycles carbon dioxide used by plants

> no **particulates** (fine particles like soot) are produced when they burn.

FIGURE 8: Why is burning coal not as clean as burning biofuels?

Questions

13 Write down two uses of biogas.

14 Write down the name of the microorganism used to make gasohol.

15 What is gasohol used for?

16 Describe two advantages of using biogas.

17 The diagram shows how a digester can be attached to a house.

a Explain why, after setting up the system, biogas is free.

b Explain how the digester benefits the diet of the people living in the house.

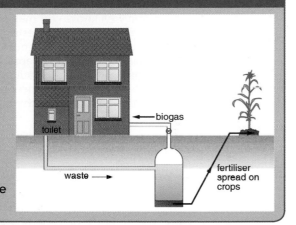

Advantages and disadvantages of biogas

Biogas is a much cleaner fuel than fossil fuels such as diesel and petrol. However, because it is a mixture of gases, it contains less energy than natural gas.

Even though it contains less energy, biogas, along with other biofuels, has advantages over fossil fuels.

> Fossil fuels are running out. Using biofuels means that fossil fuels will last longer.

> Biofuels do not increase the net levels of carbon dioxide in the air if they are burned at the same rate as they are produced and the crops are always grown on the same area.

However, some people still object to the use of biofuels, because large areas of natural habitat are being destroyed to grow crops for biofuels. Destroying natural habitat could lead to the extinction of some species.

Gasohol is more economically viable than fossil fuels in some countries which do not have their own oil reserves but are able to grow crops needed to make gasohol. Using crops as a source of fuel reduces the cost of transporting large quantities of oil from other countries.

FIGURE 9: Using biofuels could help to slow down the melting of icecaps. Explain why.

Questions

18 Explain why biogas – a mixture – contains less energy than natural gas.

19 Explain why biofuels are a sustainable resource.

20 How could the use of biofuels lead to the extinction of some species?

Preparing for assessment: Applying your knowledge

To achieve a good grade in science, you not only have to know and understand scientific ideas, you also need to be able to apply them to other situations and investigations. These tasks will support you in developing these skills.

✳ Bio Willie

Willie Nelson is a name well known for many years, especially in the USA, for country and western music. Having performed to countless fans in venues large and small, and sold millions of records, tapes and CDs, many thought that Willie would gradually wind down and enjoy life. In fact his name is once again being promoted, but this time for something completely different – bio-diesel.

Ordinary diesel fuel is derived from crude oil and although it's widely used the world over it has a number of disadvantages, such as supply and pollution. One of the alternatives is bio-diesel, and Willie Nelson has put his name to it. Willie found out about bio-diesel in Hawaii. He ended up buying a new Mercedes Benz and putting vegetable oil in the tank. "The tailpipe smells like French fries!" he said.

Bio-diesel is derived from crops and is usually blended with conventional diesel. The most common mix is B20, which is 20% bio-diesel and 80% ordinary diesel. It is possible to use pure bio-diesel, but this starts to get slushy at around 0 °C.

Bio-diesel cuts emissions of smog-forming pollutants and global-warming gases. However, it increases the amount of nitrogen oxide released, and huge areas of land would have to be turned over to crops for bio-diesel to impact on our consumption of crude oil.

Nevertheless, Bio Willie is selling well and truckers have been some of the most enthusiastic purchasers. One of the big attractions for Willie and many other Americans is that bio-diesel reduces America's need to buy crude oil from abroad. It may also give a helping hand to farmers as they grow the sunflowers, peanuts or soybeans (some of the crops from which bio-diesel can be made).

Task 1

Why is the supply of diesel running out?

Why is bio-diesel a fuel?

Why is pure bio-diesel not used in cars?

Task 2

Which gas is produced when any hydrocarbon burns?

Why is this gas a problem?

Task 3

Write down the advantages of using bio-fuels such as bio-diesel.

Why is using bio-diesel useful to America's farmers?

Task 4

Bio-diesel needs to be mixed with conventional diesel to make it more efficient. Explain why.

Task 5

Bio-diesel has negative and positive effects on the environment. Evaluate the use of bio-diesel in terms of environmental impact.

Maximise your grade

Answer includes showing that you can...	
 **F**	Understand that bio-diesel is a fuel because it is burnt to release energy.
	Understand that bio-diesel needs to be mixed with diesel so that it can be used at low temperatures.
	Describe the advantages of using bio-diesel in that it is an alternative to fossil fuels, it does not increase the levels of greenhouse gases and produces no particulates.
C	Use the idea that growing crops for bio-diesel will make money for farmers.
	Understand that bio-diesel does not contain as much energy as diesel so it needs to be mixed with conventional diesel.
A	Realise that although the use of bio-diesel reduces pollution, clearing large areas of land for growing bio-fuels could lead to habitat destruction and even extinction of some species.
	As above, but with particular clarity and detail.

71

Life in soil

You will find out:
> what soil contains
> why plants need soil
> about the importance of soil particle size

Soil on Mars

The Mars landings that have taken place since December 2003 have produced lots of data about the planet. Scientists are analysing the soil. The soil on Mars seems to be much the same all over the planet and contains iron-rich clays, magnesium sulfate and iron oxides. However, there are no organic compounds. This, along with the hostile atmosphere on Mars, is probably a reason why Mars does not seem to have ever supported life.

FIGURE 1: View of Mars taken during the 2008 expedition. What is missing from the soil that plants need?

What does Earth's soil contain?

If you look closely at soil with a microscope, you will find it contains:

> different sized mineral particles

> living plants and animals

> decayed remains of dead plants and animals.

The soil also contains:

> air, including oxygen for respiration in the living plants and animals

> water containing dissolved minerals.

Growing in soil

Plants can grow in the soil on Earth because it provides:

> minerals needed for growth

> water to stay alive

> anchorage for the roots.

FIGURE 2: Soil from a garden. Can you see any animal life?

Questions

1 Copy and complete the sentences using the following words.

animals   decayed   mineral   water

Soil contains different sized _____ particles. Mixed in with the soil are _____ remains of dead animals and plants. Soil also contains air, _____, and living plants and _____.

2 Describe three things plants need soil for.

FIGURE 3: A plant growing in the soil. Why do plants have roots?

Q humus   soil types   soil experiments

What makes a good soil?

There are many different types of soil. Soils can be acidic, alkaline or neutral. There are soils that drain well and others that get waterlogged.

> Sandy soils drain well because they have large mineral particles with big air spaces.

> Clay soils have small particles close together, so they easily become waterlogged.

> Loam is a good soil, because it contains clay and sand. Loam usually has large amounts of decomposed material called *humus*.

The types of animals and plants living in the soil depend upon the soil conditions. Plants find it difficult to live in very dry soils because they need water for photosynthesis. Few plants can cope with waterlogged soils. This is because their roots cannot get the oxygen they need for respiration. Humus content is also important. The decomposing material releases nutrients that the plants need. Also soil with high humus content tends to have more air spaces.

Testing soil

Experiments can be carried out to find the water, air and humus content of soil.

To find the air content, use a tin to collect a sample of soil. Push the can into the ground and pull it out full (but not overfull) of soil. Find the mass of the can and soil. Slowly add water and allow the water to soak through the soil, filling the air spaces. Once the water level reaches the top of the can, find the mass of the can, soil and water. The difference in mass is the mass of water added. The volume of water is the same as the mass. If you have 50 g of water, the volume is 50 cm³. This will then equal the volume of air.

To find the water content, weigh a sample of soil. Then slowly dry the soil in an oven until no more mass is lost. The loss in mass is the mass of water in the soil. To find the humus content, take a sample of dry soil. Heat the soil using a Bunsen burner. This will burn off the humus. Reweigh the soil to find the loss in mass.

FIGURE 4: Taking a soil sample. How could the soil be tested for humus content?

Questions

3 What would be the pH of a neutral soil?

4 Some plants find it difficult to grow in waterlogged soils. Explain why.

5 What is humus?

6 Describe how you would find out the humus content of soil.

Improving soil

Adding large sand particles to waterlogged ground separates the smaller soil particles, creating air spaces. This increases the permeability of the soil, allowing water to drain through. **Aerating** and draining soil in this way provides the plants and animals with more oxygen, resulting in a more fertile soil.

Some soils are very acidic. Adding lime to the soil neutralises the acid, providing better conditions for animals and plants.

Questions

7 The diagrams below show particles in two types of soils. The soils were tested for air, water and humus content.

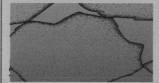 

a Which soil, A or B, would contain the most water? Explain your answer.

b Which soil would contain the most air? Explain your answer.

soil particles   minerals

Soil food web

You will find out:
> about the animals and plants that live in soil
> about soil food webs
> the importance of earthworms

There are many different animals in the soil, all feeding on each other.

> *Herbivores* such as slugs, snails and wire worms feed on plants.

> *Detritivores* such as earthworms, millipedes and springtails feed on dead organic matter.

> *Carnivores* such as centipedes, spiders and ground beetles feed on other animals.

A food web can be used to show the feeding relationships of some of the animals that live in soil and their predators.

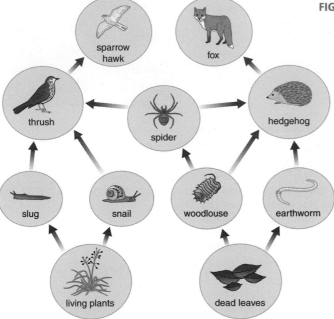

FIGURE 5: A food web. List the herbivores.

Bacteria and fungi

The soil also contains bacteria and fungi. These organisms are **decomposers**. They break down dead plants and animals, releasing elements into the soil.

Earthworms

Animals could not survive in the soil from Mars; there is no water or oxygen to keep them alive. Earth's soil has plenty of animal life. One very common soil animal is the earthworm.

Gardeners like to see plenty of earthworms in their soil. The earthworms help with the structure and fertility of the soil.

If the soil is fertile, the gardener's plants get more minerals and grow better.

Questions

8 Name five things that live in soil.

9 What could you use to identify the different insects living in soil?

10 Write down a food chain that includes woodlice and spiders.

11 What type of food do detritivores eat?

 soil food web   earthworms   detritivores   herbivores   carnivores

The mighty earthworm

The earthworm improves the structure and fertility of soil by:

> burrowing, which aerates the soil and improves drainage

> eating and burrowing through the soil, which mixes the layers in the soil

> taking organic material from the surface and burying it, which allows bacteria and fungi in the soil to decompose it

> adding chalk as it eats the soil, so neutralising acid soils. The chalk comes from glands in the worm's digestive system.

FIGURE 6: Why do gardeners welcome earthworms?

Did you know?

There are lots of different earthworm species. Some species, such as the ones in the Amazon, can grow to nearly one metre in length.

Questions

12 Earthworms take dead leaves into their burrows to eat. Describe what happens to the parts of leaves they do not eat.

13 Explain what is meant by the term 'to aerate'.

14 Farmers plough their fields to mix the soil and make it better to grow crops in. Explain why earthworms have been called 'nature's ploughmen'.

15 Look at Figure 5. Explain what would happen to the number of slugs if all the snails were removed from the soil.

Darwin and the earthworm

Charles Darwin is famous for his theories of evolution. He also spent a lot of time studying earthworms. He calculated that there were 530 000 earthworms in an area the size of a football pitch! The sub-soil earthworms bring to the surface mixes with decaying leaves and manure. In this way the layer of fertile soil becomes deeper and deeper. Darwin realised that without the earthworms, this layer would be very thin and few plants would grow.

Questions

16 Explain the importance of earthworms to agriculture.

17 Earthworms aerate the soil. Explain why this improves the soil.

Microscopic life in water

You will find out:
> about the wide variety of microscopic life in water
> that there are advantages and disadvantages of living in water
> about plankton

Dangerous algal bloom

In 2010 a satellite image showed high levels of blue-green algae growing in the Baltic Sea. The algae covered an area of 377 00 km². When such high numbers of algae appear they are called algal blooms. Algal blooms are dangerous because they can release toxins into the water as they decompose. The toxins can kill fish and cause skin irritation in humans.

FIGURE 1: An algal bloom in the Baltic Sea. Why might high levels of fertilisers in the water increase the size of the algal bloom?

The pond habitat

Katie and Kal went pond dipping in their school pond. They used a net to collect animals living in the water. Mixed in with the pondweed, they found snails, mayfly larvae and freshwater shrimps.

Kal suggested there might be other animals in the water that they could not see. They collected a sample of water to take back to the classroom. In the classroom, Katie and Kal looked at the water using a microscope. They saw lots of microscopic organisms called **plankton**. They used the Internet to find out more about plankton.

> There are two types of plankton – **phytoplankton** and **zooplankton**.

> Phytoplankton are microscopic plants.

> Zooplankton are microscopic animals.

> Plankton find it difficult to move on their own. They move with the water currents.

> Phytoplankton are producers because they can photosynthesise to make food.

> Aquatic food chains and webs start with phytoplankton.

> There are more plankton in the summer when it is warmer and the daylight is brighter.

> In winter the water gets cold and the population of plankton falls.

> The number of plankton depends on the minerals in the water.

> Life in water is very different to that on land. Plankton die if they are left out of water.

FIGURE 2: Plankton seen under a microscope. Why do some plankton need light to survive?

Questions

1 Write down the names of the two types of plankton.

b Which type of plankton can photosynthesise?

c Describe how the population of plankton changes in the winter.

Life in water

There are advantages and disadvantages to living in water.

Advantages

> Organisms do not dry out. There is plenty of water to stop dehydration.

> In a large volume of water, there is less variation in temperature during the year.

> Organisms can grow bigger, as the water supports their weight.

> Organisms can dispose of wastes without losing water.

phytoplankton   zooplankton   contractile vacuole

Disadvantages

> Organisms find it difficult to maintain water balance in their cells.

> Some organisms find it difficult to move against the currents.

Phytoplankton

Phytoplankton are plants – they can photosynthesise. Their growth is limited by changes in temperature, light and the availability of minerals such as nitrates and phosphates. The ability of phytoplankton to photosynthesise is reduced in deeper water, where there is less light.

Zooplankton feed on the phytoplankton. When the population of phytoplankton increases during warm sunny weather, so does the zooplankton population.

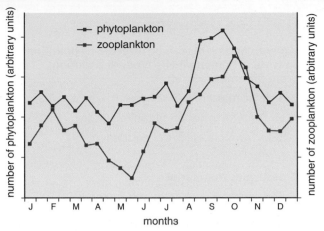

FIGURE 3: Plankton populations. How can they change during the year?

Questions

2 Look at Figure 3.

a In which month is the population of phytoplankton the highest?

b Suggest two reasons why the population increases.

c Describe the relationship shown in the graph between the population of phytoplankton and zooplankton.

d Explain why the population of zooplankton falls in November.

3 Explain why aquatic animals are less likely to become dehydrated than animals living on land.

4 Describe what would happen to the plankton population if nitrate levels in water increased.

Water balance and amoeba

An amoeba is a single-celled organism that lives in fresh water. Its cytoplasm is a concentrated solution of proteins and salts. The water in which an amoeba lives is more dilute. This means water moves into an amoeba by **osmosis**.

The amoeba has no way of stopping the movement, so it has to remove the excess water before it bursts. To do this, the amoeba uses a contractile vacuole. Energy is used to pump water into the **contractile vacuole**. When it is full, it opens to empty the water out of the amoeba.

FIGURE 4: Removal of excess water in amoeba. By what process does water enter an amoeba?

water enters continuously by osmosis

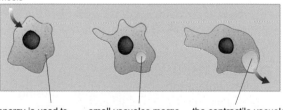

energy is used to collect excess water in small vacuoles

small vacuoles merge to form a large contractile vacuole

the contractile vacuole bursts emptying its contents to the outside

Marine food webs

Food webs that start with green plants or phytoplankton are called 'grazing food webs'. Grazing food webs are common in the ocean, but there are food webs that start with other factors.

> 'Marine snow' – this is organic matter that falls from the surface, including dead and dying matter and faeces. Zooplankton feeds on the marine snow instead of relying on phytoplankton.

> Bacteria – some bacteria found deep in the ocean can use hydrogen sulfide to make organic food. This means food webs can exist deep down, where there is no light.

Questions

5 Explain why mitochondria are found around a contractile vacuole in amoeba.

6 Suggest what would happen if you put an amoeba into salt water.

7 Some bacteria are called chemosynthetic bacteria. Suggest why.

Q marine snow   grazing food webs   chemosynthetic bacteria

Water pollution

You will find out:
> about different sources of water pollution
> what eutrophication is
> about toxic build-up in food chains

Animals and plants can be affected by pollution in water. There are many types of pollution from different sources:

> organic matter from sewage
> oil from factories or oil tankers at sea
> PCBs from old electrical equipment
> fertilisers from farms
> pesticides from farms
> detergents from homes and factories.

Even air pollution can affect aquatic life. Burning fossil fuels can cause acid rain. This can turn lakes acidic and kill the animals in the lake.

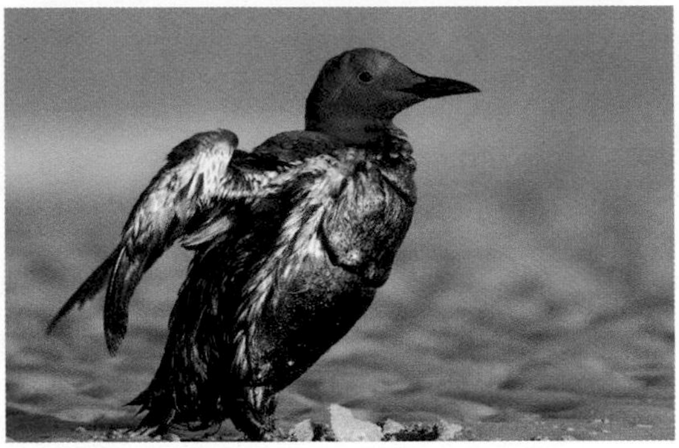

FIGURE 5: This seabird's feathers are covered in oil spilt from an oil tanker out to sea. Suggest ways in which the bird could be helped.

Questions

8 The table shows some different water samples and the pollution they contain.

Water sample	Pollution
A	fertilisers and pesticides
B	organic matter
C	oil and PCBs from electrical equipment
D	detergents
E	oil
F	organic matter and detergents

a Suggest which water samples were polluted by:

i a sewage works

ii a factory

iii a farm.

b Sample E came from a beach. Suggest one way the oil might have reached the beach.

Algae out of control

Phytoplankton such as algae need nutrients to grow. When water is polluted by sewage or fertilisers, it contains a lot of nutrients. The algae grow rapidly. If the algae grow too fast they cause **eutrophication**.

Biological indicators

Pollution levels in water are monitored using **biological indicators**. Caddis fly larvae can survive only in clean water with plenty of oxygen. However, bloodworms can survive in low levels of oxygen. Therefore, polluted water contains bloodworms but not caddis fly larvae.

Pollution can also change the pH of the water. Few organisms can survive in low levels of pH. The absence of salmon and trout in some lakes and rivers could indicate acidic water.

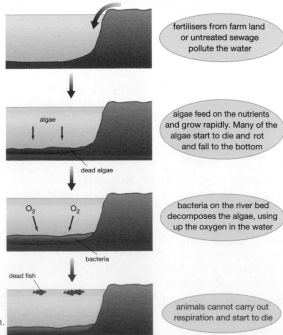

fertilisers from farm land or untreated sewage pollute the water
algae
dead algae
algae feed on the nutrients and grow rapidly. Many of the algae start to die and rot and fall to the bottom
O_2
O_2
bacteria
bacteria on the river bed decomposes the algae, using up the oxygen in the water
dead fish
animals cannot carry out respiration and start to die

FIGURE 6: The stages in eutrophication. Why do the fish die?

eutrophication biological indicators PCBs

Questions

9 Describe the main stages of eutrophication.

10 Name one of the elements found in NPK fertilisers. (*Hint*: the initials 'NPK' are the symbols for the elements.)

11 Explain why caddis fly larvae are not found in water polluted by sewage.

12 Suggest one way in which lakes can become acidic.

Food webs and pollution

Seawater can become contaminated with many different types of pollution.

> DDT is an insecticide that was used to control insect pests. Its use is now banned in most countries.

> PCBs are chemicals that were used to insulate electrical equipment.

Both these chemicals have two major problems. They are toxic and are not broken down in the body so they accumulate over time.

Small amounts of PCBs in plankton are eaten by krill. Penguins feed on the krill and killer whales feed on the penguins. The PCBs build up in their fat store, because they are not broken down. As whales eat more penguins, the concentration of PCBs increases, eventually killing them.

Remember!

In the exam you may be asked to interpret data on marine food webs. Question 14 is an example of the type of questions you might be asked.

Questions

13 DDT has been banned in most countries for a number of years. Suggest one reason why it is sometimes still found in soil samples.

14 Look at the food web below.

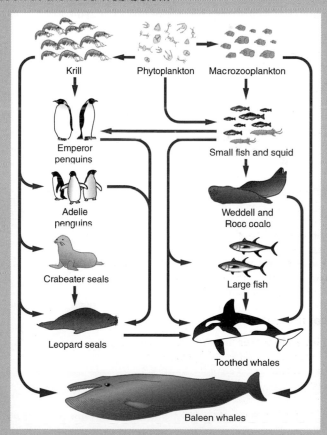

Krill   Phytoplankton   Macrozooplankton

Emperor penguins

Small fish and squid

Adelie penguins

Weddell and Ross seals

Crabeater seals

Large fish

Leopard seals

Toothed whales

Baleen whales

The habitat described by the food web becomes polluted by PCBs.

a Explain why emperor penguins would eventually have higher concentrations of PCBs in their body compared with krill.

b Which animal is more likely to be harmed by the PCBs in the long term? Explain your answer.

Enzymes in action

You will find out:
> about everyday uses of enzymes
> how enzymes work in biological washing powders
> about enzymes and sugar

Lactose problems

If you give your cat too much cow's milk to drink the cat could develop a wind problem. This is because adult cats lose the ability to digest lactose sugar in milk – a problem shared by many humans. Enzymes are used to solve this problem. When an enzyme called lactase is added to milk it digests the sugar for the cat and so prevents wind. This is just one example of many uses of enzymes.

FIGURE 1: Why should you not feed your cat cow's milk?

Using enzymes

Enzymes are biological catalysts. They speed up reactions. Many products you use each day make use of enzymes.

Some examples:

> biological washing powders and stain removers contain enzymes that clean your clothes

> enzymes are needed to make cheese

> juice is removed from fruit using enzymes

> the flavours of some foods are changed using enzymes

> enzymes on special reagent strip sticks are used to test urine for the presence of glucose.

FIGURE 2: Biological washing powder. Why are biological powders good at cleaning clothes?

How enzymes help

When food stains get onto clothes they are difficult to remove. Before biological washing powders were developed, people had to clean their clothes using very hot water. Enzymes in biological washing powders remove the stains at low temperatures and neutral pH, which is gentler on cloth. Using very hot acidic water would stop the enzyme from working.

Questions

1 Write down three uses of enzymes.

2 Copy and complete the following sentence.

Biological w_____ powders work best at l_____ temperatures and n_____ pH.

3 How are enzymes used to test urine for glucose?

How biological washing powders work

There are lots of different types of stains you can get on clothes but only three groups of enzymes to remove them. The enzymes digest the stains so they wash off the clothes.

> Stains from pasta contain a carbohydrate called starch. Washing powders containing **amylases** remove the stain.

> Fatty stains such as butter and grease need **lipases** to break them down.

> Blood stains contain proteins. Proteins can be digested using a group of enzymes called **proteases**.

Enzymes and temperature

Enzymes work best in warm water. If the water is too cold, the molecules move too slowly, so it takes longer to break down the stain. If the temperature is too hot, the enzyme is destroyed and no longer breaks down the stain.

Enzymes and sugar

There are many different types of sugar. Sucrose is the most common one used in food products. Sucrase (invertase) is an enzyme used to break down sucrose. The result is a much 'sweeter' sugar that can be used by the food industry.

sucrose has two molecules joined together

sucrase

sucrase breaks the molecules apart to make 'sweeter' sugars

FIGURE 4: The action of sucrase on sucrose. What would happen to the sucrose at high temperatures?

Remember!
Sucrase has another name; it is sometimes called 'invertase'.

FIGURE 3: Which enzyme would clean this shirt?

Questions

4 Copy the lists of stains and enzymes below and draw a line to link each stain with the enzyme that digests it.

Stain		Enzyme
fat		protease
carbohydrate		lipase
protein		amylase

5 Find out which part of your body makes amylase.

Enzymes and pH

Enzymes are sensitive to pH. If the water is too acidic or too alkaline they do not work. This is because the enzymes **denature**. If the enzymes stop working, the washing powder may not clean the clothes as well.

Making stains soluble

Stains are difficult to get out of clothes because they are insoluble. Enzymes break the stains down into smaller, soluble molecules. The soluble molecules can then easily be washed away by the water.

Insoluble stain	Enyzme	Soluble molecule
carbohydrate	amylase	sugars
fats	lipase	fatty acids and glycerol
protein	protease	amino acids

Sweeter sugar – fewer calories

Sucrase (invertase) breaks down sucrose into two different sugars called glucose and fructose.

$$\text{sucrose} \xrightarrow{\text{sucrase}} \text{glucose} + \text{fructose}$$

Glucose and fructose are much sweeter sugars than sucrose. This means less is needed to make food sweet. Sugar contains energy. If less is put in, there is less energy in the food. This is useful if a person is dieting and counting calories.

Questions

6 Find out a place in your body where proteins are turned into amino acids.

7 Draw a diagram to show the effect of amylase on starch.

8 Explain why someone on a slimming diet should use glucose instead of sucrose in cooking.

🔍 protease   sucrose   sucrase   digestive enzymes

Diabetes and enzymes

People with a condition called diabetes have to control the level of glucose in their blood. They can find out if there is too much glucose in their blood by testing their urine.

There are two ways to do this.

> You can heat the urine with Benedict's reagent – glucose turns the blue reagent brick-red.

> You can use a reagent stick with an enzyme on it – glucose turns the pink strip purple.

The invention of the reagent stick has made it easier for people with diabetes to test their own urine. This helps them to keep the glucose in their blood at a safe level.

Immobilising enzymes

Enzymes are very sensitive chemicals. To protect them and make them easier to use they are **immobilised**. This is done by trapping them inside gel beads.

The enzymes used on a reagent stick are immobilised by sticking them onto the stick.

> **You will find out:**
>
> > how enzymes are immobilised
>
> > about the advantages of immobilising enzymes
>
> > about the uses of immobilised enzymes

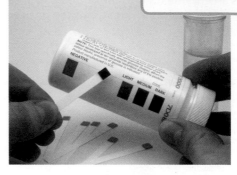

FIGURE 5: Reagent strip sticks. What can they be used for?

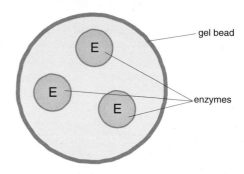

FIGURE 6: These enzymes have been made into gel beads. What is the name given to this process?

Questions

9 The table shows the colour changes caused by glucose. Copy and complete the table.

Test	Starting colour	Finishing colour
Benedict's reagent	blue	
reagent strip		purple

10 Describe how enzymes can be immobilised.

Immobilising enzymes

To make gel beads, the enzyme is first mixed with a chemical called **alginate**. The mixture is then dropped into calcium chloride solution. A reaction takes place, forming a coat of calcium alginate around the enzyme.

Advantages of immobilising enzymes

Enzymes speed up the rate of reactions without becoming part of the product. This means they can be used again and again.

Unfortunately, most enzyme-catalysed reactions take place in solution, and it is difficult to separate the enzyme from the water and products. However, immobilising an enzyme means it is easier to separate it from solution. This prevents the enzyme contaminating a product.

Q diabetes mellitus   lactose intolerance   immobilised enzymes

Alginate beads are useful in industrial processes that involve continuous flow processing. The substrates for a reaction are added to a container as fast as the products are removed. The immobilised enzymes remain inside the container and are reused again and again.

Advantages of the use of immobilised enzymes	Disadvantages of the use of immobilised enzymes
easier to separate enzyme from the products	shape of enzyme may be changed when it is immobilised, so shape of its active site may be changed and the enzyme may not work as well as a result
products need less purification because they do not contain enzyme used in the reaction	enzyme can become detached from its solid support
enzymes can be reused again and again	cost of developing immobilised enzymes can be very high
reaction is continuous, so the process can be automated and therefore cheaper to run	

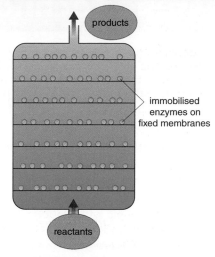

FIGURE 7: Immobilised enzymes in a continuous flow reactor. How are the immobilised enzymes prevented from being removed with the product?

Questions

11 Describe two advantages of using immobilised enzymes.

12 Using immobilised enzymes can be more expensive or cheaper. Explain why.

13 When biological washing powders were first used, many people were allergic to the protease enzyme in them. Suggest one reason why the introduction of immobilised enzymes has resulted in fewer allergic reactions.

Did you know?

An estimated 75% of the world's population is intolerant to lactose.

Lactose intolerance

Lactose is a sugar found in milk. It cannot be absorbed into the blood, so it needs to be digested. Many people lack the enzyme lactase needed to digest it. The lactose remains in their gut, where bacteria ferment it. The result is a build-up of gas that causes diarrhoea and wind.

People with lactose intolerance require lactose-free milk. To produce this, immobilised lactase is added to milk.

The lactase breaks the lactose down into glucose and galactose.

lactase
lactose → glucose + galactose

The glucose and galactose produced are then easier to absorb.

Questions

14 Some people can digest lactose and others cannot. Explain why.

15 Plan a simple experiment to show that lactase digests the lactose in milk to glucose.

Q reagent sticks   alginate beads

Gene technology

You will find out:
> about genes and DNA
> how scientists use genetic engineering
> about the main stages in genetic engineering

The firefly – a weapon against cancer

Fireflies are tiny insects that light up to attract a mate. Scientists have taken the gene that causes the light and inserted it into cancer cells. The cancer cells then light up. The light triggers a reaction that kills the cancer cells. Healthy cells are unharmed because they have no internal light source. It is hoped that this may offer a possible treatment in years to come.

FIGURE 1: Fireflies light up. How can their genes be used to treat cancer?

The genetic code

The nucleus of a body cell contains **chromosomes**. These chromosomes are made from **DNA**. The DNA is a code called the genetic code.

Everyone has their own genetic code. It carries the information that makes your body what it is.

Scientists have found a way to change the genetic code. They call it **genetic engineering**.

Changing the code

To change the genetic code, scientists:

> choose a section of the code called a gene

> take the chosen gene out of the organism and put it into another organism's DNA.

The gene then works in the new organism.

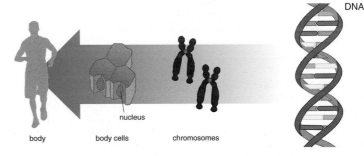

FIGURE 2: DNA carries your genetic code. How can DNA be changed?

Questions

1 Write out the following parts of the body in order of size. Start with the smallest.

cell   chromosome   gene   nucleus

2 Describe how the genetic code of an organism can be changed.

3 Copy and complete the sentence below.

Changing the genetic code is called g_____ e_____ .

Transgenic organisms

Genetic engineering changes the genetic code of an organism. It involves putting a new gene in the organism's existing DNA. The altered organism is called a **transgenic organism**.

The stages in genetic engineering

Scientists can alter the genes of sheep so that they produce milk that contains a special protein called AAT. People who suffer from a lung disease called emphysema need AAT.

To make the transgenic sheep, scientists:

> identify the gene for the protein AAT in human DNA

> remove the gene from human DNA using an enzyme

> cut open the DNA in a bacterium using the same enzyme

> add the human gene to the bacterium's DNA using a different enzyme

> put the bacterium's DNA into a fertilised egg cell of a sheep

> place the fertilised egg into a female sheep so it can develop into a lamb.

When the lamb becomes an adult, it produces milk containing the protein AAT.

Q genetic engineering   bacterial plasmids   plasmid vector

Cloning a transgenic organism

When human DNA has been added to a fertilised egg, clones are made. The egg is allowed to divide into a number of cells. The cells are then separated and placed in separate surrogate mothers. In this way, a higher number of identical transgenic sheep can be produced.

Questions

4 What is meant by the term 'transgenic organism'?

5 Which organism's DNA is used to carry the human gene into the fertilised sheep egg?

6 What is meant by the term 'clone'?

FIGURE 3: Tracy was the first transgenic sheep. What human protein can Tracy make?

Genetic engineering and enzymes

To cut out the DNA, an enzyme is needed. The enzymes used are called **restriction enzymes**. There are many different types. Each one cuts open DNA at a different site. The cut ends are called **sticky ends** because they can be stuck onto other bits of DNA. **Ligase** is another enzyme. It is used to join the 'sticky ends' of the human DNA strands onto the 'sticky ends' of a piece of circular DNA from the bacteria. This circular DNA is called a **plasmid**.

Genes from one organism work in another because the DNA they are made from is the same for all living organisms. All DNA relies on the same four bases (A, T, C and G) to code for amino acids. The code for each amino acid is the same in every organism. For example the base sequence CAA codes for an amino acid called valine in all living organisms. Therefore bacteria containing human DNA will use the code to make the same protein that humans would make.

How do scientists check a gene has stuck in a plasmid's DNA?

Genes are put into bacterial plasmids so that the bacteria will make the proteins. The plasmids with the genes are taken back into the bacterial cell. The plasmids are called **vectors** because they carry the DNA. The scientists need to know that the gene has stuck in the bacteria before they clone them. To find out, they also add genes that make the bacteria resistant to antibiotics. The bacteria are then grown on nutrient agar containing the antibiotic. If the bacteria grow, scientists know the gene has stuck. This technique is called **assaying**.

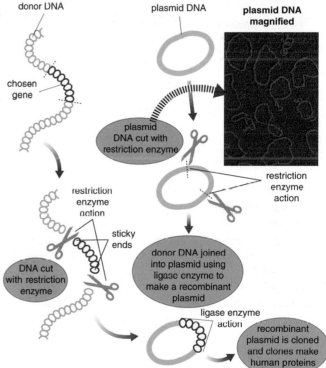

FIGURE 4: Steps in transferring a chosen human gene into a loop of bacterial DNA called a plasmid. What two types of enzymes are used in the process?

Questions

7 Explain the difference between a ligase enzyme and a restriction enzyme.

8 Suggest one reason why scientists need to know the DNA has stuck before they clone the bacteria.

9 Explain what effect an antibiotic has on bacteria without any resistance.

10 What name is given to a loop of bacterial DNA?

restriction enzyme   ligase   transgenic organisms

Genetic engineering and its uses

You will find out:
> about uses of genetic engineering
> about DNA fingerprinting
> how bacteria are used to produce insulin

Bacteria can be genetically engineered to make:

> insulin for people with diabetes

> human growth hormone for people who can't make enough of the hormone themselves.

These bacteria are grown in large fermenters to make large amounts of proteins.

DNA fingerprinting

DNA has other uses apart from **genetic engineering**. Scientists can use DNA to produce a **DNA fingerprint**. Everyone usually has their own unique DNA 'fingerprint', so it can be used to identify someone.

FIGURE 5: Some people with diabetes use insulin made by genetically engineered bacteria.

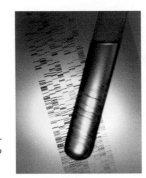

FIGURE 6: Why might the police want to know your DNA 'fingerprint'?

Questions

11 Name the type of organism that is genetically engineered to make human insulin.

12 Usually we all have different DNA 'fingerprints'. Which type of people might share the same DNA 'fingerprint'?

Uses of genetic engineering

Making human insulin

People with diabetes cannot control their blood sugar levels. This is because their bodies do not produce enough of a hormone called insulin. Many diabetics need to inject insulin, so the hormone has to be produced artificially. Large quantities can be produced using bacteria that contain the human gene for making insulin.

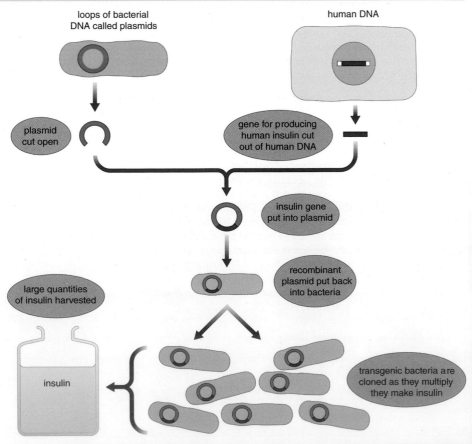

loops of bacterial DNA called plasmids

human DNA

plasmid cut open

gene for producing human insulin cut out of human DNA

insulin gene put into plasmid

recombinant plasmid put back into bacteria

transgenic bacteria are cloned as they multiply they make insulin

large quantities of insulin harvested

insulin

FIGURE 7: Steps in transferring the human gene for making insulin into bacteria. Why do scientists want to make large quantities of insulin?

Q DNA 'fingerprint'   electrophoresis

Using DNA 'fingerprinting'

In 1985, scientist Alex Jeffreys developed a way of using DNA to identify people, called DNA 'fingerprinting'. The process produces a pattern of bands, like a bar code, which is unique.

DNA 'fingerprints' of criminals are taken and stored for future use. Some people think this is a good way to catch people if they do something wrong. Other people don't want their DNA 'fingerprints' stored in case the information is used for the wrong reasons.

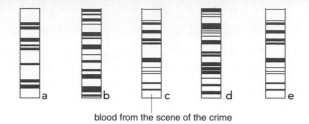

blood from the scene of the crime

FIGURE 8: Look at the bands from people connected to a theft at a jeweller's shop. Who stole the jewels?

Questions

13 Describe how bacteria can be genetically altered to make insulin.

14 Explain how DNA can be used to catch a thief.

15 Suggest one reason why someone may not want their DNA stored.

Making a DNA 'fingerprint'

Figure 9 shows how DNA 'fingerprints' are made. A **radioactive probe** is used to show up the final pattern so it can be matched with other DNA 'fingerprints'.

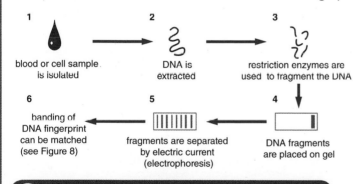

1 blood or cell sample is isolated
2 DNA is extracted
3 restriction enzymes are used to fragment the DNA
4 DNA fragments are placed on gel
5 fragments are separated by electric current (electrophoresis)
6 banding of DNA fingerprint can be matched (see Figure 8)

FIGURE 9: Making a DNA 'fingerprint'. What is used to show up the pattern in the gel?

Questions

16 Describe how a DNA 'fingerprint' is produced.

17 What is meant by the term 'electrophoresis'?

Preparing for assessment:
Analysis and evaluation

To achieve a good grade in science, you not only have to know and understand scientific ideas, but you need to be able to apply them to other situations and investigations. These tasks will support you in developing these skills.

✳ Task

> Find out if temperature can affect the production of ethanol.

✳ Context

Oil supplies are running out so some countries like Brazil have come up with alternative ways of fuelling cars. They use the sugar found in sugar cane.

Yeast can be used to produce biofuels. The yeast cells use the process of fermentation to convert sugar into ethanol. This has to be done at a certain temperature. The ethanol is then mixed with petrol to make gasohol.

✳ How to do this experiment

The growth of yeast can be recorded by measuring the rate of fermentation. The faster the yeast grow, the more carbon dioxide they produce.

✳ Method

1 Place 1 g of yeast into the boiling tube.

2 Add 10 cm³ of glucose solution to the yeast and cover with a layer of oil.

3 Attach the bung and place the test tube into a beaker of water at 40 °C.

4 Fill a measuring cylinder with water and turn it upside down in a container of water.

5 Place the end of the delivery tube into the end of the measuring cylinder.

6 Record how much water is removed over 5 minutes.

7 Repeat the process twice at 20 °C, 30 °C, 40 °C, 50 °C and 60 °C.

✳ Results, analysis and evaluation

A group of students found these results.

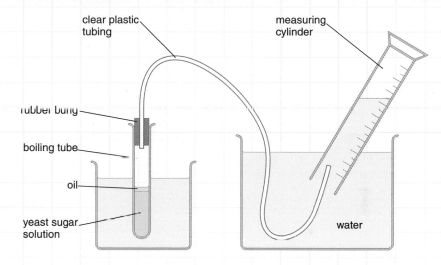

temperature in °C	amount of carbon dioxide collected in 5 minutes (cm³)	
	test 1	test 2
20	0.3	0.4
30	12.6	8.8
40	28.4	28.2
50	17.6	17.7
60	1.2	1.4

1 Calculate the mean for each set of results.

2 Plot a graph of temperature (y-axis) against mean volume (x-axis).

Think about the scale for the axes. Use sensible divisions but make the graph as large as possible.

3 Draw lines of best fit through the points.

Should these be straight lines or smooth curves? Never draw dot to dot.

4 Is there a relationship between temperature and volume of carbon dioxide collected?

Explain your answer.

Hint – look for a trend in the results.

5 How could the students who found the results improve their technique?

6 The students have one possible anomaly at 30 °C. Suggest what they should do to find out which result at this temperature is correct.

Hint – is there a better way of recording the gas produced?

7 Use the data and your scientific knowledge to explain why it is important to control the temperature when producing ethanol.

B6 Checklist

To achieve your forecast grade in the exam you'll need to revise

Use this checklist to see what you can do now. It gives you many of the important points you will need to know. Refer back to the relevant pages in this book if you're not sure and to see if there is anything else you need to know. Look across the three columns to see how you can progress.

Remember you'll need to be able to use these ideas in various ways, such as:

> interpreting pictures, diagrams and graphs
> applying ideas to new situations
> explaining ethical implications

> suggesting some benefits and risks to society
> drawing conclusions from evidence you've been given.

Look at pages 272–294 for more information about exams and how you'll be assessed.

To aim for a grade E

identify and label parts of an *E. coli* bacterium

recognise that bacteria can be grown in large fermenters

describe how yeast reproduces asexually by budding

understand that viruses are not living cells and are much smaller than bacteria and fungi

understand that some microorganisms are pathogens

describe how pathogens can enter the body

recall that diseases such as cholera and food poisoning can be a major problem following natural disasters

understand that bacteria can develop resistance to antibiotics

recall how some bacteria are useful

describe fermentation as the production of alcohol

To aim for a grade C

recall that bacteria reproduce by binary fission

describe aseptic techniques for culturing bacteria

describe how yeast growth rate can be increased

describe the structure of viruses

understand that viruses can only reproduce in other living cells

understand how to prevent transmission of diseases

describe the stages of an infectious disease

explain why natural disasters cause a rapid spread of diseases

describe the work of scientists in the treatment of disease

describe how antiseptics and antibiotics are used in the control of disease

describe the main stages in making yoghurt

recall and use the word equation for fermentation

describe the stages in production of beer, wine and spirits

To aim for a grade A

explain how bacteria can survive in a very wide range of habitats

explain the consequences of very rapid bacterial reproduction

describe how yeast growth rate changes with temperature

explain how a virus reproduces

interpret data on the incidence of influenza, food poisoning and cholera

explain the importance of various procedures in the prevention of antibiotic resistance

describe the role of *Lactobacillus* bacteria in yoghurt making

recall the chemical equation for fermentation

describe what is meant by the term pasteurisation

To aim for a grade E

recall that biogas can be produced using a digester

explain why methane being released from landfill sites is dangerous

recall that alcohol can be used as a biofuel by mixing with petrol

describe the main components of soil

describe the role of bacteria and fungi as decomposers

recognise that earthworms can improve soil structure and fertility

recognise that there are a wide variety of micro-organisms living in water

recognise that plankton are microscopic plants (phytoplankton) and microscopic animals (zooplankton)

analyse data on water pollution to determine pollution source

recall that biological washing powders do not work at high temperature and extremes of pH

describe how people with diabetes test their urine

recall how some enzymes can be immobilised

describe the process of genetic engineering

recall that bacteria can be genetically engineered to produce useful human proteins

recall that a person's DNA can be used to produce a DNA 'fingerprint'

To aim for a grade C

evaluate, given data, different methods of trans-ferring energy from biomass

describe the advantages of using biofuels

describe how biogas production is affected by temperature

describe simple experiments to compare the humus, air and water content of different soils

interpret data on soil food webs

explain the importance of humus, oxygen and water in the soil

explain why earthworms are important to soil structure and fertility

explain the advantages and disadvantages of life in water

interpret data on seasonal fluctuations in plankton

explain what causes eutrophication

describe how organisms are used as biological indicators

explain why biological washing powders work best at moderate temperatures

describe how sucrose can be broken down by invertase

describe how enzymes can be immobilised in gel beads

explain the advantages of immobilising enzymes

describe the main stages in genetic engineering

describe how bacteria can be used in genetic engineering to produce human insulin

interpret data on DNA 'fingerprinting' for identification

To aim for a grade A

explain why the burning of biofuels does not cause a net increase in greenhouse gas levels but can lead to habitat loss and extinction of species

understand why gasohol is more economically viable in some countries than others

explain how particle size affects the air content and permeability of soils

explain why aerating, draining neutralising acid soils and mixing up soil layers is important

recognise the part played by Darwin in highlighting the importance of earthworms

explain the problems of water balance caused by osmosis

understand that some food chains rely on 'marine snow' or bacteria

explain the accumulative, long-term effect of PCBs and DDT on animals

explain why biological washing powders may not work in acidic or alkaline water

explain how foods are sweetened using invertase

explain the condition of lactose intolerance

know the principles behind production of lactose-free milk

explain the role of restriction enzymes and ligase enzyme

explain how plasmids can be used as 'vectors' in genetic engineering

recall that assaying techniques are used to check that the new gene has been transferred

describe the stages in the production of a DNA 'fingerprint'

B6 Exam-style questions

Foundation Tier

1 Carol investigates growth in yeast. She uses a microscope to count the number of yeast cells over time. The graph shows her results.

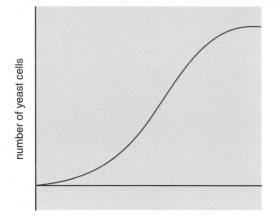

time (hours)

AO2 **(a) (i)** Describe the pattern in the graph. [1]

AO3 **(ii)** Carol thinks that the yeast grow at the same rate during the experiment. Is she right? Explain your answer. [1]

AO2 **(b)** The graph is for yeast cells kept at 20°C. Describe how the graph would change if the cells were kept at 40°C? [1]

AO1 **(c)** Yeast breaks down sugar to produce alcohol. Which gas is made during this process? [1]

[Total: 4]

AO1 **2** Phytoplankton are microscopic organisms that live in water. Describe how phytoplankton are affected by the seasons and the water they live in.

The quality of written communication will be assessed in your answer to this question. [6]

3 Look at the drawing of a bacterial cell.

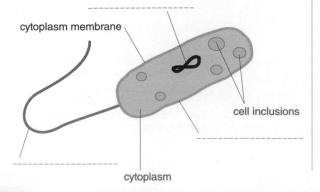

cytoplasm membrane

cell inclusions

cytoplasm

AO1 **(a)** Finish labelling the diagram. [3]

AO1 **(b)** Bacteria and yeast both reproduce asexually. Compare the way that bacteria and yeast reproduce asexually. [3]

[Total: 6]

4 Look at the food web.

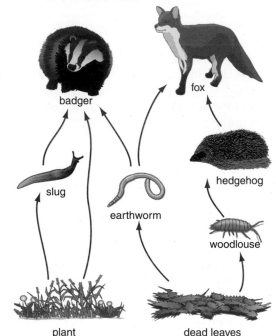

badger

fox

slug

earthworm

hedgehog

woodlouse

plant

dead leaves

AO2 **(a)** Write down the name of two detritivores from this food chain. [2]

AO3 **(b)** One year there were fewer slugs in the area. The number of earthworms went down. Use the food web to suggest why this happened. [1]

AO1 **(c)** Earthworms are important to soil structure and fertility. Explain why. [3]

[Total: 6]

AO1 recall the science   AO2 apply your knowledge   AO3 evaluate and analyse the evidence

✳ Worked Example – Foundation Tier

John has influenza. Look at the graph. It shows John's temperature over 15 days.

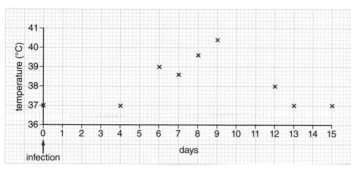

infection / days

(a) (i) How long was the incubation period of this disease? [1]

13 days

(ii) How many days did the fever last? [1]

9 days

(b) The increase in temperature is a response to the infection. Suggest a reason why the body responds to infection by increasing temperature. [1]

The body raises the temperature to try and kill the influenza pathogen.

(c) Someone suffering from influenza is not given antibiotics. Explain why. [2]

It will not work.

(d) John's sister Clare does not want to catch influenza from John.
Explain how she can avoid catching influenza from John. [3]

Don't go near him so you can't breathe in his germs which are in the air.

Remember!
pathogens are bacteria, viruses or fungi, never refer to them as germs

---

How to raise your grade!
Take note of these comments – they will help you to raise your grade.

The incubation period is the time between infection and the signs of symptoms, so the correct answer would be 4 days. 0/1

This is correct. 1/1

This is a good answer; the student applied the knowledge that high temperatures kill organisms. 1/1

The student needs to say why it will not work. This is the idea that influenza is caused by a virus and antibiotics do not affect viruses. 0/2

This should score a mark for the idea that the pathogen is airborne.

However, for a full explanation, the student needs to refer to the way the pathogen travels from John when he sneezes and that Clare could covers her face to prevent entry through her nose. 1/3

This student has scored 3 marks out of a possible 8. This is below the standard of Grade C. With more care the student could have achieved a Grade C.

Higher Tier

1 Look at graph. It shows the oxygen concentration of a river near a sewage outlet.

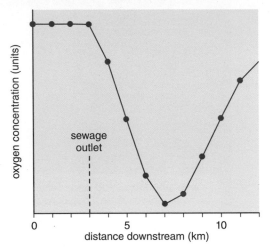

AO2 **(a)** Describe the pattern in the graph. [2]

AO1 **(b)** Explain why oxygen levels change near the sewage outlet. [3]
[Total: 5]

AO2 **2** Salmon are fish that live in both fresh water and sea water. Living in two types of water causes the salmon problems. Compare the problems the salmon might have in the two types of water. [3]

3 Look at the diagram.

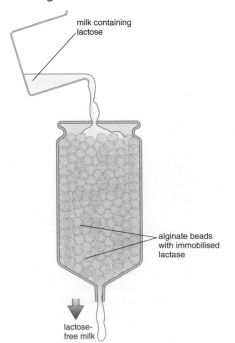

milk containing lactose

alginate beads with immobilised lactase

lactose-free milk

AO1 **(a)** Explain why the milk leaving the container is lactose free. [2]

AO1 **(b)** Immobilised lactase is used to make lactose-free milk instead of using a solution containing lactase. Explain the advantages and disadvantages of using immobilised lactase instead of a solution. The quality of written communication ✐ will be assessed in your answer to this question. [6]
[Total: 8]

4 Look at the table. It shows the DNA fingerprint for a child and three sets of parents.

child	parents A	parents B	parents C

AO3 **(a)** Scientists analysing the data believe that the child belongs to parents C. Are they correct? Explain your answer. [2]

AO1 **(b)** Describe how the DNA fingerprints were produced. [3]
[Total: 5]

AO1 recall the science   AO2 apply your knowledge   AO3 evaluate and analyse the evidence

✳ Worked Example – Higher Tier

Look at the diagram. It shows the action of enzymes in genetic engineering.

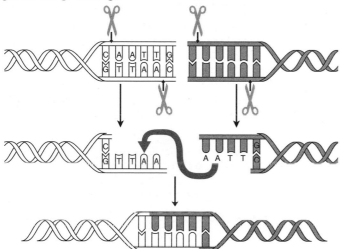

(a) Explain how enzymes are used to insert a human gene into bacterial DNA. [3]

A restriction enzyme is used to cut open the human DNA at a specific point to produce sticky ends. Another enzyme is used to cut open the bacterial DNA. The two sections of DNA are then joined using ligase.

(b) Genetic engineering is used to produce human insulin. Explain the benefits of this to society. [2]

It enables the production of human insulin on a large scale, which is of benefit to diabetics. They no longer need to rely on insulin from animals.

(c) The gene for human insulin is placed into plasmids (loops of bacterial DNA). Bacteria then take up the plasmids and start producing the insulin. Assaying techniques are used to see if the bacteria have taken up the DNA. This is done by placing a gene for antibiotic resistance in the plasmid along with the insulin gene. The bacteria are then grown on agar plates containing the antibiotic. Look at the diagram showing the results of growing bacteria this way.

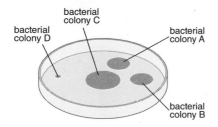

bacterial colony C

bacterial colony D

bacterial colony A

bacterial colony B

What conclusions about the uptake of plasmids can be made from the results? [2]

The bacterial colony C has grown the best.

How to raise your grade!
Take note of these comments – they will help you to raise your grade.

This is almost a correct answer using the required scientific terms. Unfortunately the same enzyme is used to cut open the human and the bacterial DNA. **2/3**

The student has answered appropriately. **2/2**

Although this is true it does not answer the question as there is no link to the uptake of plasmids. **0/2**

This student has scored 4 marks out of a possible 7. This is below the standard of Grade A. With more care the student could have achieved a Grade A.

C5 How much (quantitative analysis)

Ideas you've met before

How much chemical?

A chemical formula gives information on the number and types of atoms.

Relative atomic mass depends on the mass of different isotopes of an element.

Relative formula mass depends on the elements in a compound.

Mass is conserved during reactions.

Working out reacting masses.

 Use the periodic table to find out the relative formula mass of H_2O.

Chemical reactions

Changing concentration affects the reaction rate.

Neutralisation occurs between acids and alkalis.

Food types undergo different types of reactions.

 What happens to the speed of a reaction if an acid is more concentrated?

Analysing data

Drawing line graphs is useful in analysing data.

Graphs can be interpreted when analysing data.

Gradients can be used to find the rate of change.

 Which variable (dependent or independent) is usually plotted on the x-axis?

More reactions

Acids and their acidity are identified using the pH scale.

Precipitation reactions occur when a product is insoluble.

Reversible reactions occur when the reaction can go in both directions.

 What is the pH of pure water?

In C5 you will find out about...

> the mole concept

> calculating the mass of reactants and products

> how to work out empirical formula

> percentage composition

> making up different concentrations of solution

> interpret information on food packaging

> working out guideline daily allowances

> pH curves

> carrying out a titration

> collecting gases accurately

> calculating gas volumes from equations

> reversible reactions and equilibrium

> the contact process for manufacturing sulfuric acid

> strong and weak acids

> the difference between strength and concentration

> ionisation

> writing ionic equations

> the role of spectator ions

Moles and molar mass

You will find out:
> about moles
> about molar mass
> how to work out molar mass from a formula

Counting atoms

Sand is a compound called silicon dioxide. 60 g of sand (about an eggcup full) contains 6.02×10^{23} silicon atoms – that's 6 with 23 zeros after it. It would take you billions of years to count them all. Atoms are tiny!

FIGURE 1: Sand dunes. How many grains of sand are in the picture? How long would it take to count them?

The mole

Single atoms cannot be weighed – they are too small. Instead, they are used in standard 'pre-packed' amounts.

'**The mole**' is a unit for a standard pack of a substance. One mole of any substance contains the same number of particles as one mole of another substance.

The unit for **molar mass** is g/mol.

Calculating molar mass

Sodium chloride has the formula NaCl. Its molar mass is found by adding up all the relative atomic masses in the formula. Ar is the abbreviation used for relative atomic mass.

Example:

> Oxygen gas has the formula O_2.
> The relative atomic mass for oxygen is 16.
> In O_2, there are two oxygen atoms.
> The molar mass for O_2 is 16 + 16 = 32 g/mol.

Example:

> Sodium chloride has the formula NaCl.
> The relative atomic mass for Na is 23 and for Cl it is 35.5.
> In NaCl, the molar mass is
> 23 + 35.5 = 58.5 g/mol.

Iron (III) chloride 270.3g
Copper sulphate 249.7g
Potassium iodide 166.0g
Potassium manganate (VII) 158.0g
Sodium chloride 58.5g
Cobalt nitrate 291.0g

FIGURE 2: Each dish has 1 mole of a different compound. Why are their masses different?

 Question

Use the relative atomic masses in the periodic table on page 164 to help you.

1 Work out the molar mass of:

a nitrogen gas, N_2

b chlorine gas, Cl_2

c zinc oxide, ZnO

d magnesium chloride, $MgCl_2$.

Molar mass calculations

The mass of 1 mole of a substance is called its molar mass.

Example:

> Calcium carbonate has the formula $CaCO_3$.
> Relative atomic masses: Ca = 40, C = 12, O = 16.
> $$\underset{40}{Ca} \quad \underset{12}{C} \quad \underset{(3 \times 16)}{O_3} = 100$$
> 40 + 12 + (3 × 16) = 100
> so the molar mass is 100 g/mol.

Example:

Magnesium nitrate has the formula $Mg(NO_3)_2$.

Ar:

Mg = 24, N = 14, O = 16.

The added complication here is the brackets.

This formula contains:

$1 \times Mg$, $2 \times N$ and $6 \times O$

$= 24 + (2 \times 14) + (6 \times 16) = 148$

So the molar mass is 148 g/mol.

 Question

Use the relative atomic masses (Ar) in the periodic table on page 316 to help you.

2 Work out the molar mass of:

a magnesium carbonate, $MgCO_3$

b aluminium bromide, $AlBr_3$

c ammonium sulfate, $(NH_4)_2SO_4$

d iron(III) sulfate, $Fe_2(SO_4)_3$.

What is the 'relative' in relative atomic mass?

The relative atomic mass of an element is the average mass of an atom of the element compared to the mass of an atom of carbon-12. Values for each element are shown in the periodic table (see page 164).

Originally, chemists compared the masses of all atoms to the mass of the lightest atom, hydrogen. Later, it was found that better results were obtained if everything was compared to one-twelfth of carbon-12.

Number of moles

In chemical reactions, it is useful to work out quantities of the substances.

One way is using mole calculations:

number of moles = $\dfrac{\text{mass of chemical}}{\text{molar mass}}$

Example:

How many moles of hydrogen molecules are there in 6 g of hydrogen?

Answer:

$\dfrac{6\ g}{2\ g}$ = 3 moles of hydrogen molecules

Moles and equations

Moles can also be used to predict masses from equations.

Example:

This is the equation for burning a fuel called heptane, C_7H_{16}:

$C_7H_{16} + 11O_2 \rightarrow 7CO_2 + 8H_2O$

What mass of carbon dioxide, CO_2, is formed when 100 g of C_7H_{16} is burned?

Answer:

Molar mass of C_7H_{16} is $7 \times 12 + 16 \times 1 = 100$ g.

Molar mass of CO_2 is $1 \times 12 + 2 \times 16 = 44$ g

From the equation, 1 mole of heptane produces 7 moles of CO_2.

The mass of 7 moles of CO_2 is $7 \times 44 = 308$ g.

Mass and moles

Example:

What mass of each element is contained in 3 moles of copper nitrate, $Cu(NO_3)_2$?

Answer:

Mass of each element in 1 mole of $Cu(NO_3)_2$ is:

Cu, 64 g; N, 28 g; O, 96 g.

In three moles the mass of each element is:

Cu =192 g; N= 84 g; O = 288 g.

Remember!

You can check your answer by finding the molar mass, and seeing if everything adds up.

Questions

Use the relative atomic masses in the periodic table on page 164 to help you.

3 Calculate the molar mass of each of these compounds:

a $Ca(NO_3)_2$

b $Al_2(SO_4)_3$.

4 How many moles of water are in 36 g of water?

5 How many moles of carbon dioxide are in 11 g of CO_2?

6 What are the masses of sodium and chlorine in 6 moles of NaCl?

Mass changes in reactions

You will find out:
> about mass changes in chemical reactions
> how to determine a reacting amount

In a chemical reaction, all the atoms in the **reactants** are rearranged into **products**. This means that the mass of atoms at the start equals the mass of atoms at the finish. This is called '**conservation of mass**'.

When copper carbonate is heated, it reacts to make copper oxide and carbon dioxide. Where does the carbon dioxide gas go?

Jamie heats 5 g of copper carbonate.
After heating, Jamie measured the mass of the copper oxide.
It had a mass of 3.2 g.
How much gas was made?
The answer is 1.8 g. How did Jamie work this out?

Next Jamie heats 6 g of copper. When it cooled it had a mass of 7 g. Why? The answer is that the copper reacted with 1 g of oxygen to form copper oxide.

FIGURE 3: Thermal decomposition.

Questions

7 zinc carbonate → zinc oxide + carbon dioxide
If 2 g of zinc carbonate makes 1.3 g of zinc oxide, what mass of carbon dioxide is made?

8 Copper sulfate crystals change into copper sulfate powder when heated. Water is given off.
If 2.5 g of copper sulfate crystals make 1.6 g of powder, how much water is made?

9 When magnesium is heated in air it gains mass.
If 2.4 g of magnesium makes 4 g of magnesium oxide, how much oxygen was added from the air?

Reacting masses using ratios

All metal carbonates **thermally decompose** into oxides if they are heated enough:

metal carbonate → metal oxide + carbon dioxide

Lead carbonate, $PbCO_3$, has a M_r of 267 (Pb = 207, C = 12, O = 16).

lead carbonate → lead oxide + carbon dioxide
    267   →   223 +   44

If 267 g of $PbCO_3$ is heated, how much PbO is made?
The answer is 223 g – just use the number in the equation in grams.

If 534 g of $PbCO_3$ is heated, how much PbO is made?

The starting amount is doubled, which means the amount of PbO made is doubled and so 446 g is made.

The mass of CO_2 made will also be doubled.

If 111.5 g of PbO was made, how much $PbCO_3$ was heated?

111.5 g is half the M_r for PbO, so half the M_r of $PbCO_3$ needs to be heated, which gives an answer of 133.5 g.

Question

10 24 g of magnesium + 16 g oxygen → 40 g magnesium oxide.

a How much oxygen is needed to react with 12 g of magnesium?

b How much magnesium is needed to make 10 g of magnesium oxide?

c If 0.6 g of magnesium is heated, what mass of magnesium oxide is made?

Q thermal decomposition   conservation of mass

Reacting masses and the mole concept

The idea of using ratios for calculating reacting masses can be extended by using the mole concept:

number of moles = $\dfrac{\text{mass of chemical}}{\text{molar mass}}$

This equation can be rearranged to find the mass of chemical:

mass of chemical = number of moles × molar mass

Calculate the mass of 0.5 moles of magnesium oxide.

First, find the M_r of MgO (Mg = 24, O = 16) = 40 g/mol.

Then use the equation:

mass of chemical = number of moles × molar mass
$$= \quad 0.5 \quad × \quad 40 \text{ g/mol}$$
$$= \quad 20 \text{ g}$$

Using balanced equations

Sometimes it is easier to use a balanced equation to work out the mass of reactants or products.

Example:

What mass of MgO is made from heating 0.5 moles of $MgCO_3$?
$$MgCO_3 → MgO + CO_2$$
This equation tells us that:
$$1 \text{ mole } MgCO_3 → 1 \text{ mole } MgO + 1 \text{ mole } CO_2$$
so
$$84 \text{ g } MgCO_3 \text{ gives } 40 \text{ g MgO}$$
Thus 0.5 moles of $MgCO_3$ make 20 g (0.5 moles) of MgO.

Remember!
Check equations are balanced by counting atoms on each side.

Example:

What mass of aluminium oxide is needed to make 1 kg of aluminium?
$$2Al_2O_3 → 4Al + 3O_2$$
This equation tells us that:
$$2 \text{ moles } Al_2O_3 → 4 \text{ moles } Al + 3 \text{ moles } O_2$$
so
$$204 \text{ g } Al_2O_3 \text{ gives } 108 \text{ g Al}$$
Thus 204 g of aluminium oxide is needed to make 108 g aluminium.
The mass of aluminium oxide needed to make 1 g aluminium is:
$$\frac{204}{108} = 1.89 \text{ g}$$
Now we just need to change the units. So 1.89 kg of aluminium oxide is needed to make 1 kg of aluminium.

Questions

11 Calculate the mass of:

a 0.4 moles of oxygen gas, O_2

b 3 moles of carbon dioxide, CO_2

c 0.6 moles of magnesium oxide, MgO.

12 $2Mg + O_2 → 2MgO$

a If 1 mole of Mg is heated, what mass of MgO is made?

b How much magnesium is needed to make 1 tonne of magnesium oxide?

Q mole calculations   reacting masses

Percentage composition and empirical formula

You will find out:

> how to work out the mass of an element in a compound

> how to write empirical formula

> how to calculate empirical formula

Mass in reactions

When magnesium burns it combines with oxygen to make a white powder called magnesium oxide.

FIGURE 1: If the magnesium was weighed at the start and at the end, its mass would increase. How can this be true if mass is conserved in a chemical reaction?

Chemical reactions

When chemicals react, the atoms of all the elements are still there at the end of the reaction. The total mass of the reactants equals the total mass of the products.

Ammonia is a compound containing nitrogen and hydrogen.

Example:

> 14 kg of nitrogen and 3 kg of hydrogen react completely to make ammonia.
>
> What mass of ammonia is made?

Answer:

> nitrogen + hydrogen → ammonia
> 14 kg + 3 kg → ?
> So 17 kg are made.

Example:

> 34 g of ammonia contain 28 g of nitrogen.
> What mass of hydrogen is in this amount of ammonia?

Answer:

> 34 g – 28 g = 6 g
> There are 6 g of hydrogen in 34 g of ammonia.

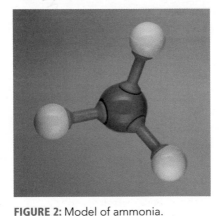

FIGURE 2: Model of ammonia.
How many white hydrogen atoms are in the molecule?
How many blue nitrogen atoms are in the molecule?
If *each* hydrogen is 1 unit, and nitrogen is 14 units, how many units are in ammonia?

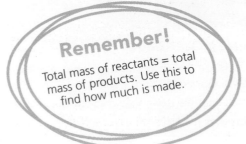

Remember!

Total mass of reactants = total mass of products. Use this to find how much is made.

Questions

1 2 g of hydrogen react with 16 g of oxygen to make water. What mass of water is made?

2 58.5 g of sodium chloride, NaCl, were made from 23 g of sodium. What mass of chlorine was used in the reaction?

Molecular formula

The molecular formula tells you the number and type of each atom in a molecule.

Empirical formula

The **empirical formula** is the simplest way of writing a whole-number ratio of each type of atom inside a molecule.

Some examples are:

> What is the empirical formula of glucose, $C_6H_{12}O_6$? Find the simplest ratio, so CH_2O (divide by 6).

> What is the empirical formula of methanoic acid, HCOOH? Bring each type of element together, so CH_2O_2.

> What is the empirical formula of ethanoic acid, CH_3COOH? Bring each type of element together, so $C_2H_4O_2$. Look for the simplest ratio, so CH_2O.

FIGURE 3: Glucose is a complex molecule.
How many black carbon atoms?
How many red oxygen atoms?
How many white hydrogen atoms?
What is the molecular formula for glucose?

Questions

3 A compound has the molecular formula of C_2H_6. What is the empirical formula?

4 A compound has a molecular formula of $C_{10}H_{16}O_4$. What is the empirical formula?

Empirical formula calculations

It is easy to work out the empirical formula of a compound if you know the mass of each element in the compound.

Working out the empirical formula

Stage 1 Look up the relative atomic mass of each element.

Stage 2 Write down the mass of each element.

Stage 3 Work out how many moles there are of each element.

Stage 4 Choose the element present in the smallest amount.

Stage 5 Divide the moles of each of the other elements in the compound by the moles of the one in the smallest amount.

Calculating empirical formulae from actual mass

Example:

2.45 g of sulfuric acid contains 0.05 g H, 0.8 g S and 1.6 g O. Calculate the empirical formula of sulfuric acid.

Answer:

		H	S	O
Stage 1	mass of each element	0.05 g	0.8 g	1.6 g
Stage 2	relative atomic mass of each element	1	32	16
Stage 3	convert to moles of each element	$\frac{0.05}{1}$	$\frac{0.8}{32}$	$\frac{1.6}{16}$
		0.05 mol	0.025 mol	0.1 mol
Stage 4	identify the element present in the smallest amount – sulfur (0.025 mol)			
Stage 5	divide by the smallest	$\frac{0.05}{0.025}$	$\frac{0.025}{0.025}$	$\frac{0.1}{0.025}$
		2	1	4

The empirical formula of sulfuric acid is H_2SO_4.

Questions

Use the relative atomic masses in the periodic table on page 164 to help you.

5 132 g of a compound contains 36 g C and 96 g O. Calculate its empirical formula.

6 A compound contains 14.4 g C, 2.4 g H and 19.2 g O. Calculate its empirical formula.

7 The empirical formula of a compound is CH_2O, with a molar mass of 30 g/mol. If the compound has a molar mass of 90 g/mol, what is its molecular formula?

Q percentage mass   empirical formula

How much water is in crystals?

What is the difference between the copper sulfate in Figure 4 and that in Figure 5?

Water is needed for crystals to form.
If copper sulfate crystals are heated, water is given off. The blue crystals change into a white powder.

The molar mass of water in the crystals can be found if the masses of the crystals and powder are known.

Mass of crystals before heating, 25 g.
Mass of powder after heating, 16 g.

How much water is present?

25 g – 16 g = 9 g

Recap on molar mass

Molar mass is the relative formula mass in g/mol.

For example, water is H_2O (H = 1, O = 16).
The molar mass of water is $(1 \times 2) + 16 = 18$ g/mol.

FIGURE 4: Copper sulfate crystals.

FIGURE 5: Copper sulfate powder.

Questions

Use these relative atomic masses to help answer the questions:
H = 1, C = 12, N = 14, O = 16

8 Work out the molar mass of: **a** O_2 **b** N_2 **c** CO_2 **d** NO_2 **e** C_2H_6.

9 6.5 g of copper sulfate crystals make 4 g of copper sulfate powder. How much water was released?

Molar mass calculations

Molar mass is the mass of 1 mole of a compound in g/mol. It is used to show equal amounts of particles to be reacted.

Example:

> What is the molar mass of ammonium sulfate, $(NH_4)_2SO_4$?
>
> Answer:
>
> *Stage 1*
>
> Use the periodic table (on page 164) to find the relative atomic mass for each element in the formula:
> N = 14, H = 1, S = 32, O = 16
>
> *Stage 2*
>
> Work out how many atoms of each element are in the formula:
> N × 2, H × 8, S × 1, O × 4
>
> *Stage 3*
>
> Multiply the number of atoms of each element by the relative atomic mass:
> $(2 \times 14) + (8 \times 1) + (1 \times 32) + (4 \times 16)$
>
> *Stage 4*
>
> Add them all together:
> 28 + 8 + 32 + 64 = 132 g/mol

Did you know?

Water (H_2O) has a molar mass of 18 g/mol. Large compounds like proteins have molar masses of thousands of g/mol.

Q water of crystallisation   growing crystals

Percentage mass

The percentage of an element in a compound can be found using the formula:

% mass of element = $\dfrac{\text{total mass of the element in the compound} \times 100}{\text{relative formula mass of compound}}$

Remember!

Use the periodic table to find relative atomic masses. They are always the larger numbers shown for each element.

Example:

What % of carbon and hydrogen are in ethane, C_2H_6 (C = 12, H = 1)?

% carbon = $\dfrac{(12 \times 2) \times 100}{30}$ = 80%

hydrogen = $\dfrac{(1 \times 6) \times 100}{30}$ = 20%

Questions

Use the relative atomic masses in the periodic table on page 164 to help you.

10 Work out the molar mass of these compounds:

a Na_2CO_3 **b** $Al(OH)_3$ **c** $Al_2(SO_4)_3$ **d** $Mg(HCO_3)_2$.

11 Find the percentage mass of:

a copper in copper chloride, $CuCl_2$

b magnesium in magnesium nitrate, $Mg(NO_3)_2$.

Calculating empirical formulae from percentage composition by mass

Percentage composition by mass means that you are starting with 100 g of the compound, so work in exactly the same way as before.

Example:

Sulfuric acid contains 2.04% H, 32.65% S and 65.31% O by mass

Calculate the empirical formula of sulfuric acid.

Answer:

		H	S	O
Stage 1	mass of each element in 100 g of the acid	2.04 g	32.65 g	65.31 g
Stage 2	relative atomic mass of each element	1	32	16
Stage 3	convert to moles of each element	2.04 mol	1.02 mol	4.08 mol
Stage 4	identify the element present in the smallest amount – sulfur (1.02 mol)			
Stage 5	divide by the smallest	2	1	4

The empirical formula of sulfuric acid is H_2SO_4.

Questions

Use the relative atomic masses in the periodic table on page 316 to help you.

12 Calculate the empirical formula of a compound containing: K, 56.5%; C, 8.7%; O, 34.8%.

13 A compound contains 38.7% C and 16.2% H, with the rest being N.

a Find the percentage of N in the compound.

b Calculate the empirical formula of the compound.

c If the compound has a molar mass of 93 g/mol, what is its molecular formula?

14 Calculate the percentage mass of each element in propanol, C_3H_7OH.

Quantitative analysis

You will find out:

> how to measure the concentration of a solution

> about making up solutions at different concentrations

> about using calculations involving concentration

Making solutions

Milk for premature babies in hospital can contain different ingredients, depending on what each baby needs. The concentration also needs to be controlled. Nursing staff need to know about making accurate solutions.

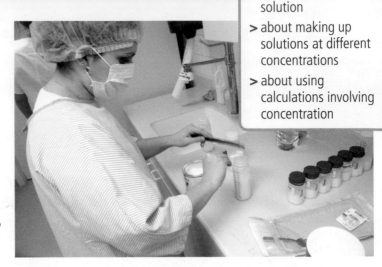

FIGURE 1: A nurse making up a specialised baby milk.

Measuring volume

Volume is measured in cm³ or dm³. The easiest way to measure liquid volume is by using a measuring cylinder.

If the final cylinder in Figure 2 contains 100 cm³ of red liquid, what volume is in all the other cylinders? If the final cylinder contains 1000 cm³ (1 dm³), what volume is in all the other cylinders?

FIGURE 2: What's the volume?

Concentration

Concentration is the amount of **solute** (solid) in grams dissolved in 1 dm³ of solution. The units are g/dm³ (g per dm³).

If 50 g of a solute are dissolved in 1 dm³, the concentration is 50 g/dm³. Another word for dm³ is a litre, so the concentration is 50 g/litre.

Concentration of solutions can also be measured in mol/dm³.

Chemical solutions, such as acids, are usually measured in mol/dm³.

Remember!
1000 cm³ equals 1 dm³ (1 litre).

Diluting solutions

If 100 cm³ of liquid has a concentration of 100 g/dm³, what needs to be done to dilute it to 50 g/dm³? To halve the concentration, double the amount of water, so make up to 200 cm³ by adding water.

If a bottle contains 50 cm³ of a 4 mol/dm³ solution, what needs to be done to make it a 2 mol/dm³ solution? To halve the concentration, double the amount of water so make up to 100 cm³ by adding water.

Questions

1 a What concentration is 20 g dissolved in 1 dm³?

b What concentration is 10 g dissolved in 1 dm³?

c Which one is most dilute and why?

2 How much water needs to be added to dilute 200 cm³ of a 3 mol/dm³ solution to 1.5 mol/dm³?

More on concentration

In Figure 3, what happens to the number of solute particles dissolved as the concentration doubles? What would a diagram representing 3 mol/dm³ look like?

As concentration increases, the solute particles become more crowded.

How to dilute solutions

If the concentration of a solution needs to be reduced, it is done by adding water to dilute it. This increases the volume, so the solute particles become more spread out.

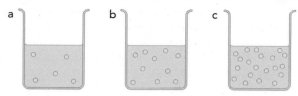

FIGURE 3: Understanding concentration. If (a), with five particles, represents 1 mol/dm³, (b) would be 2 mol/dm³ (10 particles) and (c) would be 4 mol/dm³ (20 particles).

Example:

You have to make a solution that is a tenth as concentrated. You start with 1 cm³. How much water do you add?

Answer:

The starting volume is 1 cm³ and the final volume is 10 cm³. The extra amount of water to be added is 10 − 1 = 9 cm³ of water.

Converting volumes

Sometimes it is useful to covert cm³ into dm³, or vice versa.

To convert cm³ to dm³, divide by 1000.
To convert dm³ to cm³, times by 1000.

Questions

3 Convert 1640 cm³ into dm³.

4 Convert 2.34 dm³ into cm³.

5 What is the solute in a solution of salt in water?

6 20 cm³ of windscreen-wash fluid needs to be diluted from 1.0 mol/dm³ to 0.1 mol/dm³. How much water needs to be added?

Concentration calculations

A relationship exists between the amount in moles, concentration in mol/dm³ and volume in dm³:

> amount in moles = concentration × volume

> concentration = amount in moles ÷ volume

> volume − amount in moles ÷ concentration.

A way to remember this is by using a formula triangle, as in Figure 4.

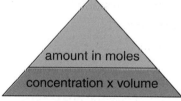

FIGURE 4: Use this by covering up the one needed, and read off the relationship (the horizontal line shows division).

Example:

Find the amount of moles in 500 cm³ of a 0.1 mol/dm³ solution.

Units always need to match so convert cm³ into dm³:
500 cm³ = 0.5 dm³

Then use the formula:

amount in moles = concentration × volume

= 0.1 mol/dm³ × 0.5 dm³

= 0.05 mol

Questions

7 A solution of 0.12 moles has a volume of 0.6 dm³. What is the concentration in mol/dm³?

8 A solution with a concentration of 2 mol/dm³ contains 2 moles of solute. What is the volume, in dm³, of the solution?

9 What is the amount of solute, in moles in 500 cm³ of a solution with a concentration of 3 mol/dm³?

Guideline daily allowance

Many packaged foods show **guideline daily allowances** (GDAs). These depend on if you are a man, woman or child, on your health and weight, and on how active you are. The table below shows some average GDAs.

Who?	Energy (J)	Sugars (g)	Fat (g)	Saturated fat (g)	Salt (g)
Women	8400	90	70	20	6
Men	10 500	120	95	30	6
Children 5–10	7500	85	70	20	4

What is the GDA for salt for a child aged 8? Why do energy needs vary?

Why does concentration matter?

Powdered baby milk must be mixed with the correct amount of water:

> Too much water means the baby does not get enough food.

> Too much powder causes digestive problems for the baby.

Concentrated orange cordial needs to be diluted so it does not taste too strong and to suit personal taste. Some medicines also need to be diluted to avoid an overdose.

Car windscreen wash is diluted before putting it into cars. In summer the solution can be dilute, but in winter the solution must be more concentrated to stop the water freezing.

Questions

10 Use the table to answer these questions:

a Who has the lowest sugar GDA?

b Who should limit their energy intake to 10 500 J?

c Who should limit their fat intake to 70 g?

d How do you work out the amount of unsaturated fats?

Food labels

Many people find food labels difficult to interpret. The government introduced a colour-coded 'traffic light' system to help, but different manufacturers present this in different ways.

In Figures 5 and 6, which label gives most information?

Chilli with beans

Nutrition facts	
Serving size: 1 cup (253 g) Servings per container: 2	
Amount per serving	
Calories 260	Calories from fat 72
	% daily value
Total fat 8 g	13
Saturated fat 3 g	17
Cholesterol 130 mg	44
Sodium 1010 mg	42
Total carbohydrate 22 g	7
Dietary Fibre 9 g	36
Sugars 4 g	
Protein 25 g	

FIGURE 5: A typical food label.

	FAT 7.7 g per serving
LOW	

	SATURATES 2.0 g per serving
LOW	

	SUGAR 42.2 g per serving
HIGH	

	SALT 2.0 g per serving
MED	

Per serving	
FAT	7.7 g
SATURATES	2.0 g
SUGAR	42.2 g
SALT	2.0 g

■ HIGH ■ MED ■ LOW

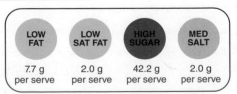

LOW FAT	LOW SAT FAT	HIGH SUGAR	MED SALT
7.7 g per serve	2.0 g per serve	42.2 g per serve	2.0 g per serve

FIGURE 6: The Traffic Light System.

Questions

11 Use the information in Figure 5 to help answer these questions:

a If 44 g of total carbohydrates were eaten, what %GDA does this represent?

b If 8 g of fat is equal to 13% of the GDA, how much fat would equal 100%?

12 What are the advantages and disadvantages of the traffic light system?

 More on guideline daily allowances

Sodium ions are essential in our diets for water balance and nerve responses, but too much can cause high blood pressure and heart disease.

The main source of sodium ions is salt.

Food labels need to be read with care. Some list the amount of 'sodium', others the amount of 'salt'.

Ingredients are listed in order of mass. So the main ingredients come first. If they are high fat ingredients such as butter, cream or oil, then the food would be classed as high fat.

Remember!
Food labels are useful, but they do not tell the whole story.

Example:

> **Converting grams of sodium to grams of salt**
> Sodium chloride is NaCl.
> Relative formula mass of NaCl = 23 + 35.5 = 58.5
> There are 23 g of sodium in 58.5 g of salt.
> There is 1 g of sodium in 58.5 ÷ 23 = 2.5 g of salt.
> So, the mass of salt is 2.5 times the mass of sodium.

Did you know?

Food labelling is not compulsory in every country.

Why is the amount of salt not the best measure of sodium?

Prepared foods usually contain other sodium compounds in small quantities, so their sodium ions must be counted as well.

Food	Sodium content (mg)
Bacon rasher	800
Slice of white bread	200
Packet of plain crisps	270
Tin of cream of tomato soup	1125

 Questions

13 A tin of beans, A, contains 1.2 g sodium. Tin B contains 2.8 g salt. Which tin, A or B, contains more sodium?

14 Convert the sodium content for each food in Figure 5 to the equivalent mass of salt.

15 Some food labels only give the amount of added salt. Explain why this is not an accurate indication of the total amount of sodium.

Titrations

You will find out:
> about pH
> how pH changes during a titration
> how to read a pH titration curve

What are titrations?

A titration is a common method used in chemistry to find the concentration of an unknown reactant. Titrations usually involve adding an acid to an alkali.

FIGURE 1: Carrying out a titration. Why are goggles needed?

Changes in pH number

low pH
acid ⟵ neutral ⟶ alkali
0   7   14
high pH

> Acids have low pH numbers.

> Adding acid to a solution decreases its pH.

> Adding alkali to a solution increases its pH.

Measuring pH

A **pH meter** is used to measure the **pH** of a solution accurately. Universal indicator solution is good for a quick estimate. Different brands of universal indicator are slightly different, so the colour must always be compared with the reference card.

When an acid is added to an alkali, the pH decreases (pH number gets lower). When an alkali is added to an acid, the pH increases (pH number gets higher).

You need to read values from a pH curve (shown in Figure 3) to find:

> pH at a certain volume

> volume at a certain pH.

FIGURE 2: Why do some pH papers not cover the whole pH range?

What are the colours of different indicators?

You need to know the colour changes of these **indicators**.

Indicator	Colour in alkali	Colour in acid
litmus	blue	red
phenolphthalein	pink	colourless

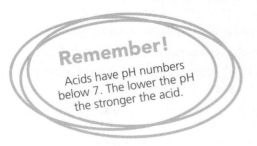

Remember!
Acids have pH numbers below 7. The lower the pH the stronger the acid.

Questions

1 What colour does litmus paper turn in acid?

2 What colour does phenolphthalein turn in alkali?

3 Look at the graph in Figure 3.

a What is the pH of the alkali at the start of the experiment?

b What volume of acid gives a pH of 12?

c When the pH is 9, what volume of acid has been added?

What happens to pH when an acid reacts with an alkali?

When an acid and alkali cancel out each other, it is called **neutralisation**.

This chemical equation represents the reaction:

acid + alkali → salt + water

A titration experiment

> The graph in Figure 3 shows that at the start of the experiment there is just alkali so the pH number is high.

> As acid is added it starts to neutralise the alkali and the pH drops slowly at first.

> The point at which all the alkali has reacted with the acid is called the end point

> At the end point, pH changes very suddenly.

What does a pH of less than 7 indicate?

Deciding which indicator to use

Indicators such as phenolphthalein and litmus have a single, sudden colour change, which is easier to see than the continuous range of colours seen if a universal indicator is used.

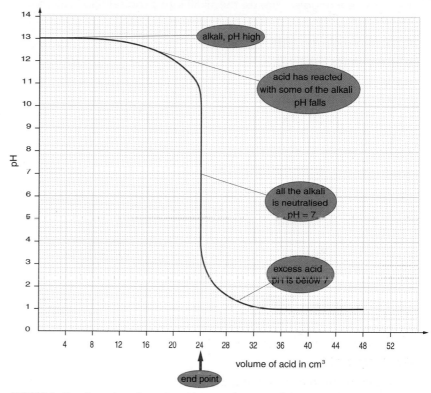

FIGURE 3: Graph to show how the pH curve changes when an acid is added to an alkali.

Questions

4 Use the graph in Figure 3 to estimate the pH after 16 cm³ of acid are added.

5 What volume of acid produces a pH of 4?

6 What volume of acid is needed to neutralise the alkali?

It is often useful to sketch a pH titration curve

For any acid and alkali titration the shape is always the same, but to make the curve the right size it is useful to know the:

> starting pH

> end pH

> end-point volume (the most vertical section of the curve)

> pH range over which the sudden change takes place.

To show the end point a single indicator needs to be used. Mixed indicators, like universal indicators, give continuous colour changes which are harder to see.

Questions

7 These are the results of the titration of a weak ethanoic acid into a strong alkali sodium hydroxide:

start pH, 13.5   final pH, 4.8
end point, 9.0   vertical range, 12 to 5

a Sketch a pH curve using the data above.

b Explain how to estimate the end point from a titration curve.

8 Explain why for titrations a single indicator is better than a mixed indicator.

Carrying out a titration

You will find out:
> how to carry out a titration
> different types of indicators
> how to do titration calculations

1 Wear safety goggles. **Pipette** the alkali into a conical flask using a pipette filler. This is safer than sucking it up by mouth.

2 Put a few drops of indicator into the flask.

3 Fill a **burette** with acid.

4 Add the acid slowly to the alkali, mixing the solution in the conical flask by swirling the flask gently.

5 Stop adding acid when the indicator suddenly changes colour – this is the end point.

6 Repeat steps 1 to 5 until the readings are the same.

Titration results

Volume of acid reacting with 25.0 cm³ of alkali				
	rough	first	second	third
at start (cm³)	1.20	3.20	2.50	3.70
at end (cm³)	28.2	29.6	29.0	30.1
titre (cm³)	27.0			

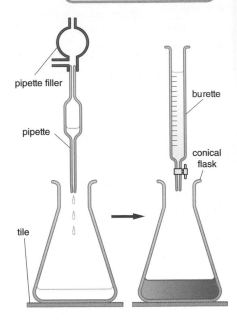

FIGURE 4: Apparatus for carrying out a titration. Suggest why it is important to mix the solution as the acid is added.

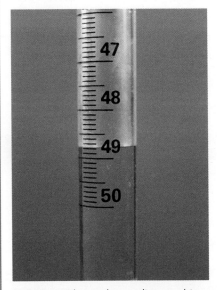

FIGURE 5: What is the reading on this burette?

Remember!
- Always read to the bottom of the meniscus in a burette and a pipette.
- The reading on a burette goes downwards (not upwards).

Questions

9 Look at the table above showing titration results. Work out how much acid is needed for each of the three accurate titrations.

10 Explain why it is important to use a pipette filler when filling a pipette with alkali.

11 What is a burette used for?

🔍 reading a burette   using a pipette

Why repeat a titration?

> **Titration** is an accurate technique, but there is a chance of experimental error. If there are large differences in the readings, then it is a warning that the technique is not accurate.

> If the readings are close, the technique is reliable.

Small differences result from normal experimental error, so an average reading is calculated.

 Question

12 Look at the titration results in the table on page 112. Work out the titration averages. Which titrations should be used to calculate the average?

Using titrations

Titrations are used to find the concentration of an alkali from a known concentration of acid, or vice versa. You need to be able to use and rearrange this equation:

$$\text{concentration} = \frac{\text{number of moles}}{\text{volume in dm}^3}$$

Example:

> 23.8 cm^3 of a solution of 0.11 mol/dm^3 of hydrochloric acid reacts with 25.0 cm^3 of sodium hydroxide solution. What is the concentration of the sodium hydroxide solution?
>
> The equation for this reaction is:
>
> $$HCl + NaOH \rightarrow NaCl + H_2O$$
>
> (Calculations are limited to titrations where 1 mole of acid reacts with 1 mole of alkali.)

Step 1 How many moles of acid were used?

  number of moles acid   = acid concentration × volume in dm^3

  number of moles acid   = 0.11 × 23.8 ÷ 1000

  number of moles acid   = 0.00262

Step 2 How many moles of alkali were used?

  number of moles alkali = alkali concentration × volume in dm^3

  number of moles alkali = alkali concentration × 25.0 ÷ 1000

  number of moles alkali = alkali concentration × 0.0250

Step 3 Link the acid with the alkali.

  number of moles acid   = number of moles alkali

  0.00262   = alkali concentration × 0.0250

  0.00262 ÷ 0.0250   = alkali concentration

The concentration of the sodium hydroxide solution is 0.105 mol/dm^3.

Questions

13 Explain why phenolphthalein is a better indicator than a universal indicator.

14 26.1 cm^3 of hydrochloric acid react with 25.0 cm^3 of 0.095 mol/dm^3 sodium hydroxide solution. Calculate the concentration of the acid.

Preparing for assessment: Applying your knowledge

To achieve a good grade in science, you not only have to know and understand scientific ideas, you also need to be able to apply them to other situations and investigations. These tasks will support you in developing these skills.

✳ Pass the salt!

Rosie is studying the effect that various elements have on the human body and is doing some homework on sodium. Her grandfather is staying with her family for a few days and he's talking to her about when he was a soldier in the Second World War in North Africa.

"I reckon that it was the desert that was the enemy half of the time. I've never been anywhere like that before or since. It was so hot we sweated nearly all the time. Dehydration was the problem. We had to take salt tablets. Great big ones. The officers stood over us and made us take them."

Rosie smiled. Then her dad brought in their dinner so she cleared her books away.

"My favourite," said Granddad, reaching for the salt. He poured a pile of salt on the edge of his dinner plate. He caught her looking and laughed. "Don't mind me," he said, "it's just that I got so used to having to take extra salt all those years ago that food doesn't taste right without it."

After dinner Rosie carried on with her homework; she was gathering information to produce a concept map for her notes. She knew that salt was a compound consisting of two types of ion and that the ion that tended to cause health problems was sodium. The next part of her homework was to read a health advice leaflet on salt in foods and one of the first things she picked up from it was that on average people in the UK consume more than twice the GDA of salt, and sometimes five times as much. It went on to explain that excess sodium causes problems with blood pressure, kidney stones and water retention.

Finally she went on the Internet to add some more details. She found out that in the US the GDA for salt is 2400 mg, as opposed to the 1600 mg in the UK. She also found out that there are two different reasons why food manufacturers add salt: as a flavour enhancer and as a preservative. But what was really fascinating was something on advice to athletes. This site said that if you exercise you sweat but the sweat causes more water to be lost than sodium, so the best drink is plain water.

She finished the concept map and was halfway through putting her papers away and then she stopped. "Granddad," she called, "Can I talk to you about something..."

Task 1

What does the 'mg' mean in 1600 mg? How many grams is that?

What do the letters 'GDA' stand for? What does it mean to have a GDA for sodium?

Sodium is one of the ions in table salt. What is the other ion?

Task 2

Suggest three foods that are rich in salt.

Task 3

Rosie's dad cooked their dinner from fresh ingredients. Why would it be likely to have more salt in it if it was a 'ready meal'?

Task 4

If Rosie manages to persuade Granddad to stop putting salt on his food, why is it likely that he will still be getting some sodium in his diet?

Task 5

When you sweat you lose water and sodium. Why is it that after sweating the sodium in the body is more concentrated than it was before?

Task 6

Why should it be that if the GDA for sodium in the UK is 1600 mg, it is 2400 mg in the US?

Maximise your grade

Answer includes showing that you can...

F

Give examples of foods rich in salt.

Interpret some information about the amount of salt in food.

Suggest why 'ready meals' may have more salt in them.

C

Give several reasons why salt is added to food.

Explain why sweating changes the sodium concentration in the body.

A

Explain that medical opinions may vary with regard to GDAs.

Explain why group 1 metals are sometimes known as the 'alkali earth' metals.

As above, but with particular clarity and detail.

Gas volumes

You will find out:

> about the volume of 1 mole of any gas

> how to measure gas volumes

> about gas volumes at room temperature and pressure

Gas powered

Engines in aeroplanes and cars work by chemical reactions that produce a gas. The more gas that is produced, the more powerful the engine is. Designers need to be able to measure how much gas is made. Reactions that give off gas are easy to monitor.

Gases take up 3000 times as much space as solids.

FIGURE 1: Jet engines produce gas. How does the gas move the aeroplane?

Measuring the gas produced in a reaction

There are three ways to collect gas given off in a reaction. They are shown in Figure 2.

To find the volume of gas, read it off the scale.

Using a balance is another good method of following a reaction. Instead of showing the volume of the gas, the balance shows the drop in mass as a gas is given off.

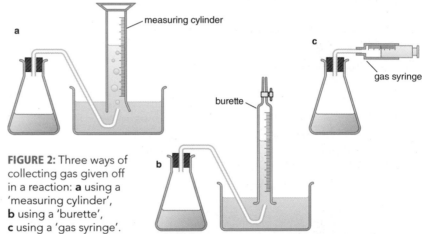

FIGURE 2: Three ways of collecting gas given off in a reaction: **a** using a 'measuring cylinder', **b** using a 'burette', **c** using a 'gas syringe'.

FIGURE 3: Why are some reactions better done in a fume cupboard?

Did you know?

It is often easier to measure the volume of a gas made rather than very small changes in mass.

FIGURE 4: How can a digital balance be used to measure the amount of gas given off?

Questions

1 Name three different pieces of apparatus that are used to measure the volume of a gas given off.

2 Suggest why some reactions that give off gas should only be carried out in a fume cupboard.

3 If you used more of the reactants, what happens to the total amount of gas made?

Experimental details

> If an upturned burette or measuring cylinder is used it must be filled with water before it is turned upside down.

> The volume is read off the scale on the side.

> The scale on a burette goes the opposite way to that on a measuring cylinder.

> If a balance is used, a loose plug of cotton wool placed in the neck of the flask prevents a spray of liquid droplets from escaping, but allows the gas through.

You need to be able to describe an experimental method to measure gas volume when given details about a reaction.

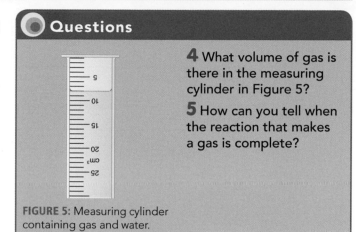

FIGURE 5: Measuring cylinder containing gas and water.

Questions

4 What volume of gas is there in the measuring cylinder in Figure 5?

5 How can you tell when the reaction that makes a gas is complete?

What information does gas volume give?

> Gas volume enables the rate of a reaction to be monitored.

> The number of moles of gas can be found if its volume is known.

> At room temperature and pressure (r.t.p.) 1 mole of particles of a gas has a volume of 24 dm³.

It is the number of particles that matters. It does not matter how large the molecules are. This is because the space between molecules in a gas is enormous.

number of moles = volume of gas in dm³ ÷ 24

FIGURE 6: What volume does 1 mole of each of these particles occupy?

1 mole of helium (He) atoms

1 mole of hydrogen (H_2) molecules

1 mole of methane (CH_4) molecules

Did you know?

Gas molecules are so far apart that, if they were the size shown in Figure 6, this page would only hold three molecules.

Remember!

For calculations, convert all gas volumes to dm³. In the exam, you will be told that 1 mole of any gas takes up 24 dm³.

Questions

6 A reaction gave off 48 cm³ of carbon dioxide. How many moles of carbon dioxide were produced?

7 Mohammed weighed a flask of magnesium and acid as it reacted. He found that it produced 0.00082 g of hydrogen gas (H_2).

a How many moles of hydrogen gas were produced?

b How many cm³ of the gas were produced?

What can you tell from graphs of gas volumes?

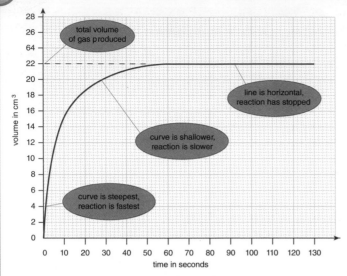

FIGURE 7: Graph to show the volume of gas given off against time. What does the horizontal part of the line tell you?

Reactions start fast and then slow down. Reactions only stop completely when one of the reactants is used up. For a reaction that gives off a gas it is very easy to tell if it has stopped – no more gas is given off. When the gradient (slope) of the graph is steeper, the reaction is taking place faster.

Increasing the amount of reactant increases the total amount of gas given off.

Questions

8 Look at the graph in Figure 7. How much gas was produced after 20 seconds?

9 What is the total volume of gas produced in the reaction?

10 At what time did the reaction stop?

11 How would the graph change if more of the reactants were used?

12 Anita measured the volume of gas given off during a reaction.

time in seconds	0	20	40	60	80	100	120
total volume in cm³	0	16	28	36	40	42	42

a When was the reaction fastest?

b When did the reaction stop?

Changing the amount of a reactant

When magnesium ribbon reacts with acid, the magnesium dissolves, the acid is neutralised and hydrogen gas is given off. If exactly the right amounts of reactants are used, no magnesium ribbon is left at the end of the reaction and all the acid is neutralised.

If the same amount of acid but half the amount of magnesium is used, half the volume of gas is produced. This is because there is not enough magnesium. In this case, the magnesium is called the **limiting reactant**:

> The reactant used up first is the limiting reactant.

FIGURE 8: Reaction 2 has half the limiting reactant as reaction 1. The reaction does not stop sooner, but when it does stop it has produced only half the amount of gas. How does the graph show this?

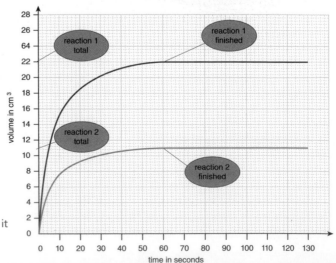

🔍 reading data from graphs   drawing scientific graphs

Remember!

The volume of gas produced is proportional to the limiting reactant, providing the other reactant is in excess.

Questions

13 Isabel reacts marble chips with acid. When the reaction stops there are still some marble chips in the flask. Which reactant is the limiting reactant?

14 Look at the graph in Figure 8.

a How long does it take reaction 1 to produce 16 cm³ of gas?

b How much gas does reaction 2 produce in 20 seconds?

c What is the final volume in each reaction?

d How much gas would be produced if the limiting reactant was double the amount used in reaction 1?

Changing the rate of reaction

It is the gradient of a curve that shows the rate of a reaction. The steeper the gradient, the faster the reaction. The amount of product made depends on the amount of reactants used.

If the reaction produced a gas, doubling the amount of reactant particles would double the amount of the gaseous product made.

➤ The total amount of gas produced is directly proportional to the amount of the limiting reactant.

Questions

15 Use Figure 9 to calculate the average rate of gas production in:

a the first 10 seconds

b the next 10 seconds.

16 The gas in the experiment above was measured using a gas syringe. Sketch the shape of the graph if you had measured the gas by using a balance.

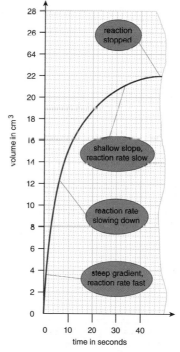

FIGURE 9: When does the reaction stop? What was the total volume collected? What does the gradient of a curve show?

Equilibria

You will find out:
> about reversible reactions
> how to recognise and change equilibrium positions

Forwards and backwards

We always think of reactions as going from left to right.

Some reactions can go in both directions, even at the same time!

Reversible reactions

Nitrogen and hydrogen react to form ammonia. The ammonia made in the forward reaction breaks down to form nitrogen and hydrogen again.

nitrogen + hydrogen $\rightleftharpoons$ ammonia

In a **reversible reaction**, there is a forward and a backward reaction. Sometimes these take place at the same time. The $\rightleftharpoons$ symbol is used to show that a reaction is reversible.

Equilibrium

When a balance is reached between the amount of reactions and products present, the reaction has reached **equilibrium**.

Two factors can alter the equilibrium composition. One is temperature.

Temperature in °C	Percentage of reactants at equilibrium	Percentage of products at equilibrium
20	35	65
30	42	58
40	48	52
50	53	47

What is the composition at 40 °C?

The other factor that can alter the equilibrium composition is pressure.

Pressure in atm	Percentage of reactants at equilibrium	Percentage of products at equilibrium
1	34	46
5	45	55
10	36	64
15	27	73

What is the composition at 10 atmospheres (atm) pressure?

As the pressure increases, what happens to the percentage product?

FIGURE 1: Ammonia fertiliser. Which two chemicals react to form ammonia?

Questions

1 What are reactions that go forwards and backwards called?

2 Which two equations show reversible reactions?
a copper + oxygen $\rightarrow$ copper oxide
b $N_2 + 3H_2 \rightleftharpoons 2NH_3$
c $Zn + 2HCl \rightarrow ZnCl_2 + H_2$
d sulfur dioxide + oxygen $\rightleftharpoons$ sulfur trioxide

3 Name two conditions that can alter the equilibrium.

Q pressure in gases   chemical reactions

At equilibrium

> The rate of the forward reaction equals the rate of the backward reaction.

> The concentrations of reactants and products do not change.

Equilibrium position

When the backward and forward rates balance, the concentrations of reactants and products are not equal. There is usually more of one than the other.

> If the concentration of reactants is greater than the concentration of products, we say that the position of equilibrium is on the left.

> If the concentration of the reactants is less than the concentration of the products, the position of equilibrium is on the right.

 Questions

4 How do you know when a reversible reaction has reached equilibrium?

5 At equilibrium, the reaction:

$$A + B \rightleftharpoons C + D$$

has 5 moles of A, 5 moles of B, 1 mole of C and 1 mole of D.

What can you say about the position of equilibrium?

Reaching equilibrium

This only works if the chemicals cannot escape – if it is a closed system. Initially, the forward reaction rate is fast, but then it slows as the reactants are used up.

At the same time, the rate of the backward reaction increases as more products are available to react. Eventually the backward reaction is as fast as the forward reaction. Equilibrium has been reached.

Manipulating the equilibrium

In industry, the equilibrium composition can be altered to obtain more of the desired product.

One product made by a reversible reaction is dinitrogen tetroxide, N_2O_4, used as a propellant in the Space Shuttle. The reaction to produce dinitrogen tetroxide is:

$$2NO_{2(g)} \rightleftharpoons N_2O_{4(g)}$$

The (g) following the formula is called a state symbol, and stands for gas.

If N_2O_4 is removed as it is formed, the equilibrium moves to the right, giving a higher yield as more is made to replace it.

If more NO_2 is introduced, the equilibrium moves to the right, making more product.

The equation shows 2 moles of NO_2 produce 1 mole of N_2O_4. Increasing the pressure moves the equilibrium to the right to reduce the pressure as there are fewer gas particles on the right.

The reaction $2NO_2 \rightleftharpoons N_2O_4$ produces heat, so is **exothermic**. If the temperature is increased, the equilibrium moves to the left, making more NO.

The reverse reaction $N_2O_4 \rightleftharpoons 2NO_2$ is **endothermic**, so lowering the temperature moves the equilibrium to the right, making more N_2O_4.

These are examples of Le Chatelier's Principle, which states that if a change is made in a closed system, the equilibrium will shift in a way to reduce the effect of the change.

FIGURE 2: Discovery launch.

 Question

6 Explain why it is necessary to react chemicals in a closed system when finding the equilibrium point of a reaction.

🔍 reversible reactions   chemical equilibrium

The Contact Process

Sulfuric acid is used to make fertilisers, paint, plastics and soap. Even some clothes use sulfuric acid in their manufacture.

The Contact Process is used to make sulfuric acid.

Raw materials

Sulfuric acid is made from three raw materials:

> sulfur

> air

> water.

How sulfuric acid is manufactured

> Sulfur is burnt in air to make sulfur dioxide.

> Sulfur dioxide is reacted with oxygen to make sulfur trioxide.

> Sulfur trioxide is reacted with water to make sulfuric acid.

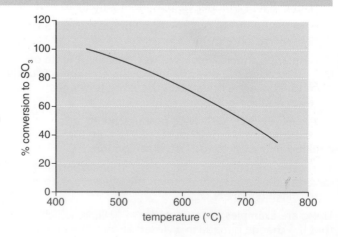

FIGURE 3: Discuss how the pictures could be connected to sulfuric acid.

Questions

7 What is made by the Contact Process?

8 What raw materials are used in the Contact Process?

9 How is sulfuric acid made?

Altering the equilibrium

Changing temperature, pressure or concentration can all alter the position of the equilibrium.

An example is changing the temperature in the Contact Process to convert sulfur dioxide to sulfur trioxide.

The graph shows that lowering the temperature increases the % conversion by altering the position on the equilibrium.

Graphs can also be used to show how pressure affects the position of the equilibrium.

FIGURE 4: Graph showing sulfur trioxide conversion. What is the % conversion at 600 °C? What effect does raising the temperature have?

making sulfuric acid   uses of sulfuric acid

Conditions for the Contact Process

In the **Contact Process**, the reaction between sulfur dioxide and oxygen is reversible:

sulfur dioxide + oxygen ⇌ sulfur trioxide

$$2SO_{2(g)} + O_{2(g)} \rightleftharpoons 2SO_{3(g)}$$

To obtain the most economic yield, the reaction is carried out:

> at around 450 °C

> at atmospheric pressure

> using a **catalyst** of vanadium pentoxide, V_2O_5.

(A catalyst is a small amount of a chemical that speeds up a particular reaction, without being used up.)

Questions

10 What can change the equilibrium position in a reaction?

11 What is a catalyst?

12 Give the three conditions used in the Contact Process.

Contact Process explained

The equations for the three stages of the Contact Process are:

Stage 1
$$S + O_2 \rightarrow SO_2$$

Stage 2
$$2SO_2 + O_2 \rightleftharpoons 2SO_3$$

Stage 3
$$SO_3 + H_2O \rightarrow H_2SO_4$$

The equilibrium in Stage 2 needs to be moved to the right to increase the yield.

Conditions for Stage 2

450 °C

This temperature is a compromise. The forward reaction is exothermic, so high temperatures reduce the yield and drive the equilibrium to the left. However, high temperatures increase the rate of reaction, so the chemical is produced faster.

Atmospheric pressure

There are three gas molecules on the left of the equation and two on the right, so high pressure increases the yield – this forces the equilibrium further to the right. However, the equilibrium lies to the right anyway, so the cost of equipment with thicker walls to withstand a high pressure is not economical.

Catalysts

Catalysts do not affect the position of the equilibrium.

However, they do make the reaction go faster, so more product is produced every second.

FIGURE 5: A chemical industrial plant. What are the raw materials for the Contact Process?

Questions

13 Explain why the temperature and atmospheric pressure conditions for the Contact Process are a compromise.

14 Why do you think that the Contact Process uses a catalyst?

Strong and weak acids

You will find out:
> about strong and weak acids
> about hydrogen ions
> about the pH scale

Useful chemicals

Acids are very useful chemicals. They are used as rust removers, in batteries, as lime-scale removers and in cooking. Your stomach contains hydrochloric acid with a pH of 1 or 2.

FIGURE 1: What pH is shown by the indicator paper? Does this pH represent acid, neutral or alkali?

Strong and weak acids

Many laboratory acids are **strong acids**.

Examples of strong acids are:

> hydrochloric acid

> sulfuric acid

> nitric acid.

Acids used in the home are weak acids.

An example of a **weak acid** is:

> ethanoic acid (in vinegar).

Strong and weak acids have the same reactions

Weak acids do not seem to be as acidic, yet they react just as much as strong acids:

> Strong acids have a lower pH than weak acids.

> Strong and weak acids react with:
  > magnesium to make hydrogen gas
  > calcium carbonate to make carbon dioxide gas.

> Weak acids react more slowly than strong acids, but make the same volume of gas as strong acids if the same volume and concentration are used.

Remember!
Strong and weak acid will produce the same amount of products, but at different rates.

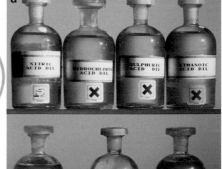

FIGURE 2: a strong acids; **b** weak acids.

Questions

1 Give three examples of strong acids.

2 Which reacts more slowly – ethanoic acid or hydrochloric acid?

3 Which has a lower pH – a weak acid or a strong acid?

Why are acids acidic?

All acids contain hydrogen atoms in their formula. For example, hydrochloric acid is HCl. In water acid molecules ionise. The hydrogen part of an acid forms hydrogen ions, H^+:

acid molecule → hydrogen ions + other ions

Ionisation

Strong acids, such as hydrochloric acid, ionise completely when they are in water:

strong acid → hydrogen ions + other ions

There are lots of hydrogen ions, so the acid seems very acidic and reacts quickly, as lots of collisions occur.

In weak acids, such as ethanoic acid, only a few of the acid molecules ionise in water. A reversible reaction is set up:

weak acid ⇌ hydrogen ions + other ions

The reversible reaction forms an equilibrium mixture. The equilibrium position is on the left. The mixture contains lots of acid molecules, but not many H^+ ions so weak acids do not seem to be so acidic.

Questions

4 Look at the two test tubes in Figure 3. Suggest which tube contains the stronger acid.

FIGURE 3: The tubes contain different acids.

5 Explain, in terms of ionisation, the difference between a strong and a weak acid.

6 What happens to acid molecules in water?

pH and acidity

Acid reactions are caused by hydrogen ions, H^+. The pH scale is related to the concentration of H^+ ions.

low pH number = high concentration of H^+

higher pH number = lower concentration of H^+

Strong acids

Strong acids ionise completely in water:

$HCl → H^+ + Cl^-$

A high concentration of H^+ means that the pH is low.

Weak acids

The formula for ethanoic acid is $CH_3COO\underline{H}$.

The underlined hydrogen is the one that makes it an acid. The equation for the reversible reaction is:

$CH_3COO\underline{H} ⇌ H^+ + CH_3COO^-$

As this equilibrium lies to the left, the concentration of H^+ ions is low and the pH is higher.

Rates of reaction

Rate is a measure of the speed at which a reaction takes place. This can be explained using collision theory covered in Module C3.

When an acid reacts with calcium carbonate, carbon dioxide is produced.

If a strong acid is used, the rate is fast as the high concentration of H^+ ions increase the collision frequency.

With a weak acid, the rate is slower, as few H^+ ions are available to react.

Questions

7 Explain why sulfuric acid is a strong acid.

8 List the characteristics of a weak acid. Include the following terms in your answer:

H^+ ions, pH, equilibrium, reversible reaction

9 Nitric acid, HNO_3, is a strong acid. Citric acid, $C_6H_7O_7H$, is a weak acid. Write equations for each to show how they ionise in water.

Acids, electrolysis and conductivity

All acids produce hydrogen at the negative electrode (cathode) during **electrolysis**.

The same amount and concentration of ethanoic acid and hydrochloric acid produce exactly the same amount of hydrogen.

The only difference is that it is harder for the current to flow through ethanoic acid because its electrical conductivity is lower.

This means ethanoic acid makes hydrogen gas at a slower rate than hydrochloric acid during electrolysis.

Remember!

In electrolysis the negative electrode is called the cathode and the positive electrode the anode.

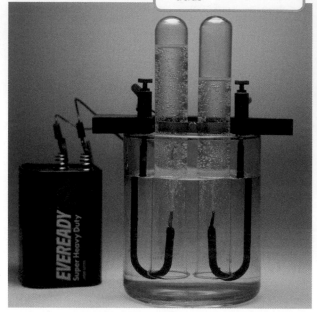

FIGURE 4: Electrolysis of sulfuric acid. The negative electrode is on the right. Which gas forms here?

Questions

10 What gas does ethanoic acid produce at the negative electrode during electrolysis?

11 Does hydrochloric acid produce gas faster or slower than ethanoic acid during electrolysis?

Electrolysis and conductivity

In electrolysis it is the ions that move. Weak acids have fewer hydrogen ions (H^+ ions) to move through the liquid than strong acids do, so weak acids do not conduct as well as strong acids. However, over time the same volume of gas will be made if the same amounts of reactants are used.

All acids form H^+ ions in water. Positive ions always go to the cathode, so during electrolysis all acids produce hydrogen at the cathode.

Descaling agents

The concentration of hydrogen ions in a weak acid is enough to attack lime scale on a kettle element, but it is not enough to attack the kettle element. The higher concentration of hydrogen ions in strong acids means that they would damage the metal of the element itself.

Q electrolysis of sulfuric acid   descaling a kettle

Questions

12 Which of the following affects the volume of gas produced when an acid reacts with magnesium?
a acid strength
b acid concentration
c acid volume.

13 Describe why ethanoic acid is less conductive than hydrochloric acid at the same concentration.

Acid strength versus acid concentration

The two terms tell us totally different things. One tells us how much an acid reacts and the other tells us how fast it reacts.

> The concentration of an acid tells us how many moles of acid there are in a dm³ of solution.

Acid concentration is not the same as H⁺ ion concentration. A concentrated solution of weak acid still has a low concentration of H⁺ ions.

> The **strength** of an acid tells us how much an acid ionises.

Electrical conductivity

Ethanoic acid has a lower concentration of H⁺ ions than the same concentration of hydrochloric acid. It is the H⁺ ions that carry the charge, so ethanoic acid is less conductive.

How useful are weak acids?

A strong acid can be diluted with lots of water to obtain a pH like that of most weak acids. If it is used to descale a kettle, it contains so little acid that it is used up almost immediately.

In a weak acid solution, as H⁺ ions are used up, the non-ionised weak acid produces more hydrogen ions. A weak acid reacts with 100 times more lime scale than the same volume of a diluted strong acid.

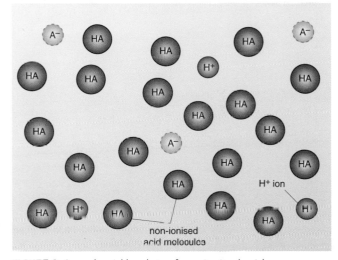

FIGURE 6: A weak acid has lots of non-ionised acid molecules and only a few H⁺ ions. How would the diagram be different for a strong acid?

Questions

14 Explain why the pH of a strong acid is lower than the pH of a weak acid of the same concentration.

15 What experiment shows us that hydrogen ions really do exist in solutions of acids?

16 Explain, using collision theory, why a weak acid takes longer to react with magnesium than the same concentration of a strong acid, but still produces the same amount of product.

Ionic equations and precipitation

You will find out:
> about precipitation reactions
> about state symbols
> how to test for halide and sulfate ions

Crystals

The size of a crystal depends on many factors. Some crystals form as water evaporates, some form as precipitates.

FIGURE 4: Gypsum crystals in a Mexico cave. How big is a crystal?

Precipitation reactions

When two solutions react to make an insoluble substance, the insoluble substance appears suddenly as a solid. This solid is called a **precipitate**. This works best if the compounds are ionic. Ions from one compound react with ions from the other compound to make new substances.

Identifying ions using lead nitrate

Lead ions from lead nitrate combine with iodide ions from sodium iodide to make a bright yellow precipitate.

Lead nitrate solution can be used to test for these halide ions.

FIGURE 2: The reaction of lead nitrate with sodium iodide. What colour is the precipitate in this reaction?

Halide ion	Precipitate colour
Cl⁻	white
Br⁻	cream
I⁻	bright yellow

Another useful precipitation reaction is using **barium chloride** solution. This is used to test for sulfate ions. If they are present, a white precipitate forms.

Questions

1 What word describes a solid made from a reaction between two solutions?

2 Name the chemical used to identify halides?

3 What precipitate colour would lead bromide have?

4 Which ion is barium chloride used to test for?

making salts   insoluble salts

Ionic lattices

Compounds containing metal ions and non-metal ions form **ionic lattices**. Ions are fixed in position within a solid lattice.

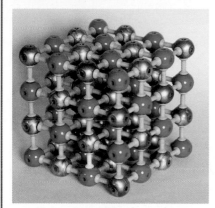

FIGURE 3: A lattice containing sodium ions Na⁺ attracted to chlorine ions Cl⁻. Suggest why the different ions are strongly attracted to each other.

Sodium chloride is soluble. As water molecules collide with it, the lattice breaks apart into separate sodium ions and chloride ions. This is what is meant by dissolving.

In precipitation reactions, the reactant ions must be able to move if they are to collide and react. This is why the reactants need to be soluble.

Questions

5 Look at these results. Use the information in section on precipitation reactions above to help you.

Reaction	Reagent solution	Chemical tested
1	barium chloride	magnesium sulfate
2	barium chloride	magnesium bromide
3	lead nitrate	sodium chloride
4	lead nitrate	potassium bromide
5	lead nitrate	sodium iodide
6	lead nitrate	potassium sulfate

a Which reaction would give a yellow precipitate?

b Which reaction would give a positive result for a bromide?

c Which two results would not give precipitates?

6 How would you show that a solution contains sulfate ions? Give the name of the reagent you would use, and state what you would see.

Precipitation reactions

A single crystal of salt contains trillions of positive and negative ions in an ionic lattice. As the lattice dissolves, the separate positive and negative ions repel each other and spread out. Collisions with the H⁺ and OH⁻ ions in water speed up this process.

In a precipitation reaction that involves mixing two different ionic solutions, the reaction is extremely fast. The trillions of each ion mean there is an extremely large collision frequency between the ions. The result is that both the soluble product and the insoluble precipitate are formed so quickly the reaction appears to be instant.

Questions

7 How is a crystal lattice held together?

8 What happens when a crystal dissolves?

9 Why is the collision frequency high when two ionic solutions are mixed?

precipitation reactions   crystal lattices

State symbols

Sometimes it is useful to show the state of the reactants and products.

solid	(s)
liquid	(l)
gas	(g)
aqueous (dissolved in water)	(aq)

Preparing an insoluble salt by precipitation

Example:

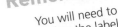

lead nitrate$_{(aq)}$ + sodium iodide$_{(aq)}$ → lead iodide$_{(s)}$ + sodium nitrate$_{(aq)}$

Remember!

You will need to remember the labels in Figure 4, as in the exam you may be given a similar diagram and asked to label it.

Stage 1 Mix the solutions of the reactants

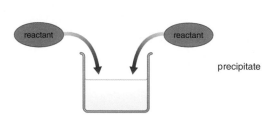

reactant   reactant

Stage 2 Filter the precipitate

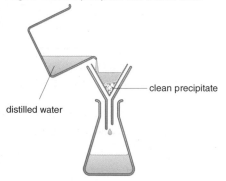

solution and precipitate

precipitate

filter funnel

filter paper

solution

Stage 3 Wash the precipitate with distilled water

clean precipitate

distilled water

Stage 4 Dry the precipitate

evaporating dish

FIGURE 5: Preparing an insoluble salt. What happens at each stage?

Questions

Look at the equation:

$Pb^{2+} + 2Cl^- \rightarrow PbCl_{2(aq)}$

10 a Which are the symbols for the reactants?

b Which is the symbol for the product?

c What state is the product in?

11 a In Figure 5, suggest why the precipitate is washed in distilled water.

b Why is the washed precipitate left in an evaporating dish?

Constructing word equations

Precipitation reactions can be summarised using word equations.

Example:

What is made when barium chloride is added to sodium sulfate?

We know the reactants, so we can work out the products:

barium chloride + sodium sulfate → barium sulfate + sodium chloride

You can think of this as 'swapping partners'.

testing for halide ions   preparing insoluble salts

Preparing a clean and dry sample of an insoluble salt

Stage 1 Mix: this makes the precipitate of lead chloride and a solution of sodium nitrate.

Stage 2 Filter: the precipitate is in the filter paper, together with traces of the sodium nitrate solution.

Stage 3 Wash: distilled water is poured over the precipitate to remove traces of sodium nitrate solution.

Stage 4 Dry: the easiest way of drying most precipitates is to leave them in a warm place for the water to evaporate.

Questions

12 Write a word equation for the precipitation reaction between silver nitrate and potassium chloride.

13 Describe how to prepare a clean and dry sample of barium sulfate from barium nitrate and sodium sulfate. Include a word equation in your answer.

Spectator ions

The reaction between lead nitrate and sodium iodide is really the reaction of lead ions with iodide ions as these are precipitated out of solution.

Ionic equations

An equation can be written out fully in words or symbols:

lead nitrate$_{(aq)}$ + sodium iodide$_{(aq)}$ → lead iodide$_{(s)}$ + sodium nitrate$_{(aq)}$

$$Pb(NO_3)_{2(aq)} + 2NaI_{(aq)} \rightarrow PbI_{2(s)} + 2NaNO_{3(aq)}$$

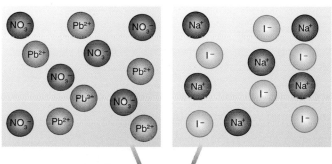

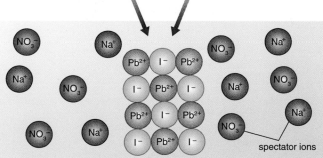

FIGURE 6: How lead iodide is formed.

The sodium ions and nitrate ions do not directly take part in the reaction. They are called **spectator ions**.

If the equation needs to show what is actually reacting, only the reacting ions are shown. The spectator ions are left out.

$$Pb^{2+}_{(aq)} + 2I^-_{(aq)} \rightarrow PbI_{2(s)}$$

This is an ionic equation. Like all equations, it must balance. Remember that charges must balance as well as the numbers of each molecule.

Questions

14 Silver nitrate, $AgNO_3$, is made from Ag^+ ions and NO_3^- ions. Sodium sulfate is made from Na^+ ions and SO_4^{2-} ions. The two solutions react to form a precipitate of silver sulfate, Ag_2SO_4.

a Write a word equation for this reaction.

b Write a full chemical equation, with state symbols.

c Write a balanced ionic equation.

d Write the symbols for the spectator ions.

15 Construct ionic balance symbol equations for the reactions between:

a $Ba^{2+}_{(aq)}$ ions and $SO_4^{2-}_{(aq)}$ ions

b $Pb^{2+}_{(aq)}$ ions and $Cl^-_{(aq)}$ ions.

Preparing for assessment: Planning and collecting primary data

To achieve a good grade in science, you not only have to know and understand scientific ideas, but you need to be able to apply them to other situations and investigations. These tasks will support you in developing these skills.

✸ Tasks

> Plan an investigation to see how the **mass of carbonate** heated is related to the **volume of gas** produced.

> Write a risk assessment for your plan. Check you understand the risk of 'suck back'.

> Once your plan has been approved, perform the investigation, record the results and write a simple conclusion.

✸ Context

The UK Government is committed to reducing greenhouse gases such as carbon dioxide. The Government charge manufacturers a 'carbon tax' for the amount of carbon dioxide released. Is there an easy way to work this out?

Limestone (calcium carbonate) is mined, and then roasted to make lime. This releases carbon dioxide.

Lime is used in agriculture, chemical manufacture, for construction, glass making and removing sulfur from power station waste gases.

✸ Planning your investigation

You can measure the **mass** of carbonate before and after, and measure the **gas volume**.

These are the things you will need to consider when planning your investigation. (You can develop your plan in groups of two or three.)

1 What will you need to measure to obtain the results?

2 How will you change the amount of carbonate used?

3 Does it matter which metal carbonate you use?

4 What do you need to keep the same to make it a fair test?

5 How many different results will you need before you can identify a trend?

6 Will you need to repeat your readings? If so, how many times?

7 You should carry out a risk assessment before you start the investigation. What precautions should you take?

8 Write the plan for the investigation.

Try to write the plan in a logical order and ask yourself if someone can perform the investigation following just your plan.

✳ Performing the investigation

Once your plan has been approved you can perform the investigation.

1 Measure the mass of carbonate used each time.

2 If you repeated any readings, all of these will need to be recorded as well as the average result.

3 Record your results in a table like this. You may need to add extra rows.

mass of carbonate and tube at the start in g	mass of carbonate and tube at the end in g	volume of gas produced in cm³	

> How many sets of readings will you need to take to identify a trend?

4 If you were to complete this as a GCSE controlled assessment, you would go on to plot a graph and evaluate the investigation.

> What graph would you draw? What would the labels be on the axes? How would you use the graph to decide on the answer to the task?

5 Is there any way in which you could have improved on how you performed the investigation?

6 What have you found out about how the mass of carbonate affects the volume of gas produced?

> Think about accuracy and precision. How are they different?

C5 Checklist

To achieve your forecast grade in the exam you'll need to revise

Use this checklist to see what you can do now. It gives you many of the important points you will need to know. Refer back to the relevant pages in this book if you're not sure and to see if there is anything else you need to know. Look across the three columns to see how you can progress.

Remember you'll need to be able to use these ideas in various ways, such as:

> interpreting pictures, diagrams and graphs
> applying ideas to new situations
> explaining ethical implications

> suggesting some benefits and risks to society
> drawing conclusions from evidence you've been given.

Look at pages 272–294 for more information about exams and how you'll be assessed.

To aim for a grade E	To aim for a grade C	To aim for a grade A
recall the unit for the amount of substance is the mole **understand** molar mass is the relative formula mass measured in g/mol **understand** mass is conserved in a chemical reaction	**calculate** the molar mass of a substance using formula containing brackets **use**, given reacting masses, simple ratios to scale how much is made	**use** and rearrange the relationship: *number of moles = mass / molar mass* **recall** that relative atomic mass is based on the average of 1/12th of a carbon-12 atom **calculate** mass of products using the mole concept
find the mass of an element in a compound when given the masses of all the other elements **calculate**, given relative atomic masses, the molar mass of a simple substance from its formula	**understand** empirical formula gives the simplest ratio of each type of atom in a compound **calculate** the percentage mass of elements in a compound when given appropriate information **calculate**, when given relative atomic masses, the molar mass of a compound	**calculate** the empirical formula from: • percentage composition by mass • mass of each element in a sample of the compound **calculate** the percentage mass of an element given its formula and relative atomic masses
recall volume is measured in dm³ or cm³ and that 1000 cm³ = 1 dm³ **recall** concentration of solutions is measured in g/dm³ or mol/dm³ **describe** how to dilute a concentrated solution **explain** why concentrated orange juice, medicines and baby milk need to be at the correct concentrations **interpret** information from food packaging about guideline daily amounts (GDA)	**convert** cm³ into dm³ and dm³ into cm³ and calculate how to dilute solutions **understand** that increasing concentration increases the number of particles in a given volume (more crowded solute particles) **interpret** GDA food package information to calculate percentage in a portion	**recall**, rearrange and use the formula *moles = concentration × volume* **interpret** complex food packaging information, for example converting sodium to the amount of salt **explain** why the above conversion might not be accurate due to sodium ions in other sources
interpret simple pH curves **explain** how universal indication can be used to estimate pH of a solution **identify** the apparatus needed to perform a titration, explain how it is carried out safely, and use results to find the titre (result) **describe** the colour changes in acids and alkalis of universal indicator, litmus and phenolphthalein	**interpret** pH curves to determine the volume of acid or alkali at neutralisation and the exact pH at particular volumes **explain** the need for several titre readings **know** that single colour change indicators are used for titration **describe** the colour changes during titration of universal indicator, litmus and phenolphthalein	**sketch** a pH curve for an acid–alkali titration **calculate** the concentration of an acid (or alkali) from titration results **explain** why single colour change indicators are used for titration

To aim for a grade E

identify different apparatus used to collect a gas – gas syringe, upturned measuring cylinder, upturned burette

recall mass can be used to measure gas loss

explain why a reaction stops

interpret tables or graphs about gas production to find how much is made, when the reaction stops, volume at certain times and comparing the rate from gradients

To aim for a grade C

describe in detail how to measure gas volume, or mass loss, including details of the reaction

understand the role of the limiting reactant – determines the final volume and is used up at the end

interpret data about gas volumes during reactions in tables and charts to find:
- total volume
- when the reaction stops
- volume at particular times
- times at particular volumes
- gas with different amounts of limiting reactants

To aim for a grade A

explain in terms of reacting particles why the amount of product is directly proportional to the amount of limiting reactant used

calculate the volume of a gas at rtp (given 1 mole of any gas = 24 dm^3)

calculate the amount of moles of gas at rtp given the molar gas volume

sketch a volume graph during a reaction when given details

---

understand reversible reactions ($\rightleftharpoons$) go forward and backwards

interpret data from tables and graphs about equilibrium position at particular temperatures and / or pressures

recall the raw materials used in the contact process – sulfur, air, water

describe how sulfuric acid is made by the contact process:

sulfur + oxygen → sulfur dioxide

sulfur dioxide + oxygen → sulfur trioxide

recall that at equilibrium:
- the rate forward and backwards are the same
- the concentration of reactants and products does not change

recall adding extra reactants, removing products, changing temperature, pressure or concentration can change the equilibrium position

describe the conditions needed in the contact process, and recall
$$2SO_{2(g)} + O_{2(g)} \rightleftharpoons 2SO_{3(g)}$$

explain why reversible reactions reach equilibrium in a closed system

explain how the equilibrium can be altered changing factors such as temperature, pressure and amounts of reactants or products

explain the conditions used in the contact process (temperature and pressure) in terms of moving the equilibrium position

understand the catalyst increases the rate but not the equilibrium position

---

recall ethanoic acid is a weak acid and hydrochloric, nitric and sulfuric acids are strong acids

understand strong acids have a lower pH than weak acids

recall all acids react with magnesium to give hydrogen, and with carbonates to give carbon dioxide, but weak acids of the same concentration react slower and have lower electrical conductivity

understand strong acids ionise completely in water giving H$^+$ ions, whereas weak acids only partly ionise forming an equilibrium mixture

explain why ethanoic acid reacts slower than hydrochloric acid of the same concentration and why it is less conductive

explain why gas volume made depends on the concentration, not the acid strength

explain why the pH and reactivity of a weak acid is lower than a strong acid at the same concentration

explain the difference between strength and concentration

construct symbol equations for ionisation of weak acids

explain why ethanoic acid is less conductive in terms of carrying charge

---

describe precipitation reactions as ions of one solution react with ions from another

describe the precipitate colours when lead nitrate reacts with halides

describe how barium chloride tests for sulfates

recognise state symbols – (aq), (s), (l) and (g)

label the apparatus used to make an insoluble salt by precipitation

understand that the ions in solids are in fixed positions, but they break up and move in solution

interpret data about testing with barium chloride for sulfates and lead nitrate for halides

construct word equations for precipitation reactions

describe the stages in preparing a dry sample of an insoluble salt

explain in terms of collisions between ions why precipitation reactions are very fast

construct ionic equations, with state symbols, for precipitation reactions

explain the role of 'spectator ions'

C5 Exam-style questions

Foundation Tier

AO1 **1 (a)** Which answer shows the unit used to measure how much substance?

gram   litre   mole   ratio   [1]

AO2 **(b)** Calculate the molar mass of sulfur dioxide gas, SO_2. [S=32, O=16]   [1]

AO2 **(c)** 2.4 g of magnesium forms 4.0 g of magnesium oxide. How much oxygen reacted?   [1]

[Total: 3]

2 Baby milk needs to be at the right concentration.

AO1 **(a)** Describe how to dilute baby milk at 100 g/dm³ down to 50 g/dm³.   [1]

AO1 **(b)** Why does baby milk need to be at the right concentration?   [1]

[Total: 2]

AO1 **3 (a)** Using acid in a burette, pipette, flask and litmus indicator, describe how to carry out an acid–base titration.   [6]

AO1 **(b)** What colour is litmus indicator in an acid?   [1]

[Total: 7]

AO2 **4 (a)** This table shows the gas collected during a reaction.

time in min	volume in cm³
0	0
1	30
2	40
3	42
4	42

AO2 **(i)** Plot a line graph to show the results.   [4]

AO2 **(ii)** At what time did the reaction finish?   [1]

AO2 **(iii)** When is the reaction fastest?

0–1 min   1–2 min   2–3 min   3–4 min   [1]

AO3 **(iv)** Gail writes a statement about the results.

The volume increases proportionally.

Is Gail's statement correct? Explain your answer.   [2]

[Total: 8]

5 Look at the table about changing temperature.

Temperature in °C	% of reactants at equilibrium	% of products at equilibrium
100	21	79
200	35	65
300	50	50
400	58	

AO2 **(a)** Work out the missing number in the table.   [1]

AO2 **(b)** Describe the pattern shown by the table.   [1]

AO2 **(c)** Predict the likely % of product formed at 250°C.   [1]

[Total: 3]

AO1 **6 (a)** Sulfuric acid is a strong acid. Name another strong acid.   [1]

AO1 **(b) (i)** Name the gas made when sulfuric acid reacts with magnesium metal.   [1]

AO1 **(ii)** Describe a difference when magnesium reacts with a weak acid rather than sulfuric acid.   [1]

AO1 **(iii)** Describe one other difference between a weak and strong acid.   [1]

[Total: 4]

AO1 **7 (a)** Describe the test to find if a compound is a sulfate.   [3]

AO1 **(b)** Lead nitrate can be used to test for halide ions. What coloured precipitate are made with:

(i) chlorides

(ii) bromides

(iii) iodides?   [3]

[Total: 6]

✳ Worked Example – Foundation Tier

(a) Which process in used to make sulfuric acid? [1]

Contact   Haber   Fritz   Thermit

Haber.

How to raise your grade!
Take note of these comments –
they will help you to raise your grade.

(b) Air, water and sulfur are all used to manufacture sulfuric acid. Describe how. [3]

Sulfur is heated up. When it burns it makes sulfur dioxide. When this dissolves in water it makes sulfuric acid.

This is incorrect. Always make sure you learn the key ideas. The answer should be the Contact Process.   0/1

(c) Part of the process involves a reversible reaction. Draw the symbol used for this. [1]

⇌

The missing stage is reacting sulfur dioxide with oxygen to make sulfur trioxide.   2/3

(d) Reversible reactions in gases are affected by pressure.

   (i) What happens to the space between gas particles when the pressure is increased? [1]

They get closer.

This is correct.   1/1

This is correct.   1/1

   (ii) Explain why increasing pressure speeds up reactions. [2]

If the particles are closer they will be more likely to hit each other.

This is worthy of credit, but collide would be a better word to use. 'Explain' needs a reason – a good answer would be that successful collisions are needed for a reaction.   1/2

(e) Explain how Universal Indicator solution can be used to test the pH of a solution. [3]

Add the indicator.
Look at the colour.

The student has indicated that the colour should be observed. There is no indication of what the universal indicator should be added to, nor is there reference to comparing with a colour chart to find the pH.   1/3

This student has scored 6 marks out of a possible 11. This is below the standard of Grade C. With more care the student could have achieved a Grade C.

C5 Exam-style questions

Higher Tier

1 Glucose has the molecular formula $C_6H_{12}O_6$.

AO1 **(a)** How many moles are in 45 g of glucose? [2]

AO1 **(b)** What is the empirical formula for glucose? [1]

AO1 **(c)** What is the percentage mass of carbon in glucose? [2]

[Total: 5]

2 An isotonic sports drink can be made by adding 185 g of concentrated orange squash and 1 g of salt, then making up to 1 dm³ with water.

AO2 **(a)** What is the concentration of the solution in g/dm³? [2]

AO1 **(b)** What mass of orange squash would be needed to make up 500 cm³ of solution? [1]

AO2 **(c)** The vitamin C concentration in the drink is 240 mg/dm³. How many moles are in 500 cm³ of this solution? [2]

[Total: 5]

AO1 **3** Mg + 2HCl → MgCl² + H₂

Describe how to measure the mass of hydrogen produced during this reaction, and why the reaction stops.
The quality of written communication ✐ will be assessed in your answer to this question. [6]

AO1 **4 (a)** Write a balanced symbol equation for making sulfur trioxide. [2]

AO1 **(b)** Describe and explain the conditions needed for the contact process. [3]

[Total: 5]

5 Jayne finds this information about GDA for salt on a packet of crisps. The crisps contain 0.6 g of salt which is 10% of the GDA.

| salt |
| 0.6 g |
| (10%) |

1 g of sodium is present in every 2.5 g salt.

AO2 **(a)** Calculate the mass of sodium present in the crisps. [1]

AO1 **(b)** Suggest why the amount of sodium in the crisps might be higher. [1]

AO3 **(c)** Jane thinks she can eat 10 packets of crisps a day and still be healthy. Do you agree? Explain your answer. [2]

[Total: 4]

6 This table shows two results from a titration neutralising sulfuric acid with sodium hydroxide, using phenolphthalein indicator.

	Volume of acid used in cm³	
	1st Run	**2nd Run**
Final volume in cm³	33.4	66.6
Initial volume in cm³	0.0	33.4
Titre in cm³	33.4	

AO2 **(a)** Work out the mean value. [1]

AO1 **(b)** Explain why repeat readings are needed. [1]

AO1 **(c)** Describe what would be seen when using phenolphthalein indicator. [1]

AO2 **(d)** If 25 ml of 0.2 mol/dm³ alkali was used, calculate the concentration, in mol/dm³ of the acid (assume the ratio is 1:1). [1]

AO2 **(e)** A pH probe gave these results:

Volume of acid added in cm³	pH reading
0	11
18	10
21.1	9
21.2	7
21.3	5
32	3

Sketch a pH titration curve for this reaction. [3]

AO3 **(f)** Phenolphthalein changes colour between pH 10 and pH 8.3.

Gary writes

I will use this indicator to find the neutral point. It's close enough.

Do you agree? Explain your answer. [2]

[Total: 9]

AO1 recall the science   AO2 apply your knowledge   AO3 evaluate and analyse the evidence

✱ Worked Example – Higher Tier

Precipitates form when some soluble solutions come into contact.

(a) Describe, in terms of the movement of ions, how precipitates form. [2]

Ions move and collide.

(b) Explain why precipitation reactions are very fast. [2]

The solutions contain ions that collide together.

(c) This table shows data about testing solutions with lead nitrate solution.

solution tested	result
sodium chloride	white solid
potassium nitrate	no change
potassium iodide	yellow solid
sodium carbonate	no change

(i) Which ions must be present for a precipitate to be formed? [2]

Sodium, potassium, chloride and iodide.

(ii) Write a word equation to show how lead nitrate reacts with sodium chloride. [2]

Lead nitrate + sodium chloride
→ lead chlorine + sodium nitrate

(iii) Identify the spectator ions in reaction (ii). [2]

Sodium and nitrate ions.

(iv) Write a balanced symbol equation to show how lead iodide forms. [3]

$Pb(NO_3)_{2(aq)} + 2KI_{(aq)} \rightarrow PbI_{2(ppt)} + 2KNO_{3(aq)}$

How to raise your grade!

Take note of these comments – they will help you to raise your grade.

⬇

There is a mark for collision but the idea that a solid lattice with ions fixed in place is formed should be added. 1/2

Ions do collide, but for the second mark the fast reaction needs to be linked to a high number of collisions. 1/2

Only the chloride and iodide are needed, so answer is incorrect. 0/2

Should be lead chloride, not lead chlorine. All 2 element compounds end in 'ide'. 1/2

This is correct. 2/2

A good answer. You only need to give State symbols when asked for in the question. 3/3

This student has scored 8 marks out of a possible 13. This is below the standard of Grade A. With more care the student could have achieved a Grade A.

C6 Chemistry out there

Ideas you've met before

Electrolysis and energy transfers

Electric current can be used to break chemicals apart.

The electrolysis of salt solution produces hydrogen at one electrode and chlorine gas at the other electrode.

Hydrogen can burn in oxygen to produce water.

When fuels burn they release useful energy.

 Why are we looking for an alternative fuel to petrol?

Metals and organic compounds

Metals are shiny, conduct electricity, can be hammered flat and conduct heat.

Metals are found on the left-hand side of the periodic table.

Some transition metals make salts that are coloured in solution.

Fats, oils and grease do not dissolve in water.

 Which metals are best to make saucepans?

Air quality and water resources

Carbon dioxide levels are increasing, which may be linked to the greenhouse effect.

Increased carbon dioxide levels are due to more fossil fuels being burned.

Water is conserved in reservoirs and underground in aquifers.

Water vapour evaporated from the oceans condenses to form clouds.

 How does the greenhouse effect work?

Emulsifiers and evaporation

Oil and water do not mix and can be separated using a separation funnel.

Mayonnaise is an emulsion of oil and water held together by egg yolk.

Crude oil can be separated by fractional distillation, which involves evaporation then condensation.

Evaporation of water is a physical process because steam is turned back to liquid.

 Why is melting ice a physical process?

In C6 you will find out about...

> how hydrogen and oxygen are made in electrolysis

> how the amount of product of electrolysis varies

> how fuel cells are used as there are no moving parts

> how the car industry is developing fuel cells

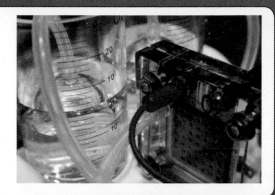

> how redox reactions involve reduction and oxidation

> how galvanising iron with zinc protects in two ways

> why the temperature for fermentation is in the range 25–50 °C

> how ethanol is both a renewable and non-renewable fuel

> how CFCs deplete the ozone layer by breaking in UV

> how highly reactive chlorine atoms react with ozone

> how temporary and permanent hardness in water occurs

> how to compare hardness in different samples of water

> how bromine water shows unsaturation in oils

> how immiscible liquids, like oil and water, form emulsions

> how there are advantages to low temperature washing

> how detergent molecules have a hydrophobic end

Electrolysis

You will find out:

> what happens in electrolysis

> about electrolysis of molten electrolytes

> about electrolysis of solutions

> why different substances appear at the anode and the cathode

Electrolysis and its uses

Electrolysis is the process of breaking down (decomposition of) compounds, when they are in solutions or when they are melted, by passing an electric current through the solution or melted compound to form two products. One product is formed at a negative electrode, and the other product forms at a positive electrode.

Sulfuric acid is decomposed and separated during electrolysis to make two gases, hydrogen and oxygen.

Minerals such as copper compounds can be broken down by electrolysis to make pure copper.

Conducting liquids

Electrolysis is the process of passing direct current (d.c.) through a solution or melted compound to break it down. It can be used to obtain useful products.

The solution or molten compound is called the **electrolyte**. An electrolyte is a liquid that conducts electricity and is decomposed during the electrolysis.

Two **electrodes** are used. They are called the **cathode** and the **anode**. These 'dip into' the solution or 'melt'. The cathode is the **negative electrode**. The anode is the **positive electrode**.

Electrolytes are made from ions. During electrolysis:

> **positive ions** are attracted to the cathode and are called **cations**

> **negative ions** are attracted to the anode and are called **anions**.

Cations have a formula with a positive charge, such as H^+ and K^+. Anions have a formula with a negative charge, such as OH^- and Cl^-.

Recognising formulae of ions

The table below shows ions that are usually moving in electrolysis cells.

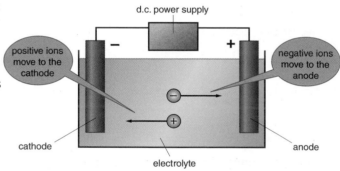

FIGURE 1: An electrolysis cell. What is the positive electrode called?

Anion		Cation	
OH^-	hydroxide	H^+	hydrogen
O^{2-}	oxide	Al^{3+}	aluminium
Cl^-	chloride	Na^+	sodium

Look at the table. What two products would form if molten sodium chloride was electrolysed? Other molten electrolytes make different products. What about aluminium oxide? Lead bromide would make lead and bromine. Potassium iodide would form potassium and iodine.

Questions

1 Write down the name of the process that breaks down compounds using electricity.

2 Which is the negative electrode?

3 What type of charge do anions have?

Why does an electrolyte have to be liquid?

Electric current is a **flow of charge**. Electrolytes are **ionic**, and the only way that the power supply can make charge move through the **electrolyte** is by making the ions move – if the electrolyte solidifies then the ions cannot move. They are held in fixed positions, so the current cannot flow and the electrolysis stops:

> ionic solids have ions in fixed positions that cannot move

> ions in molten liquid can move

> ions in solution can move.

What happens to the ions?

> Positive ions move toward the cathode.

> Negative ions move toward the anode.

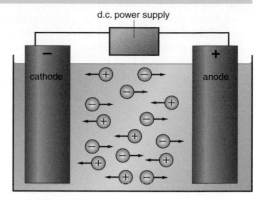

FIGURE 2: Why is an electrolyte a liquid?

When the ions get to the electrodes they are discharged – they turn into atoms or molecules.

Question

4 Explain why salt dissolved in water can be electrolysed.

What happens when a molten electrolyte decomposes?

The reactions that take place at the electrodes in the electrolysis of electrolytes, such as molten sodium chloride, can be written as half equations.

At the cathode:

$$Na^+ + e^- \rightarrow Na$$

The sodium chloride splits up into ions. The ions are free to move. The positive Na^+ ions migrate towards the negative cathode. Each Na^+ ion gains one extra electron from the cathode.

At the anode:

$$2Cl^- - 2e^- \rightarrow Cl_2$$

The negative Cl^- ions migrate to the positive anode. The Cl^- ions are discharged as chlorine gas. Two Cl^- ions each lose one electron and combine to form a chlorine molecule.

Other decompositions can be written as half equations if the formulae of the ions are known.

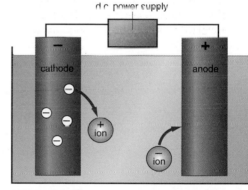

FIGURE 3: Half reactions occur at each electrode.

Electrolyte	Half equation at cathode	Half equation at anode
KCl	$2K^+ + 2e^- \rightarrow K$	$2Cl^- - 2e^- \rightarrow Cl_2$
$PbBr_2$	$Pb^{2+} + 2e^- \rightarrow Pb$	$2Br^- - 2e^- \rightarrow Br_2$
PbI_2	$Pb^{2+} + 2e^- \rightarrow Pb$	$2I^- - 2e^- \rightarrow I_2$
Al_2O_3	$2Al^{3+} + 6e^- \rightarrow 2Al$	$6O^{2-} - 12e^- \rightarrow 3O_2$

The reactions at the anode are more correctly written as $2Cl^- \rightarrow Cl_2 + 2e^-$.

Questions

5 Write a half equation for the formation of Li from its ions, Li^+.

6 Why do hydroxide ions move towards the anode?

Demonstrating electrolysis

Sulfuric acid solution can be broken down by electrolysis into hydrogen and oxygen. Figure 4 shows the apparatus that can be used in a school laboratory. It has a d.c. power supply, an anode and a cathode.

Testing for the gases given off during the electrolysis of sulfuric acid

To test for the two gases we:

> hold a lighted splint in hydrogen and it burns with a 'pop'

> hold a glowing splint in oxygen and it relights the splint.

The gases are made from the ions that migrate towards the two electrodes. The positive ions migrate towards the negative electrode. The negative ions migrate towards the positive electrode. When they reach the electrode they **discharge**.

Hydrogen discharges at the negative electrode. Oxygen discharges at the positive electrode.

The electrolysis of copper(II) sulfate using carbon electrodes

The electrolyte is a blue solution. As the electrolysis happens, the blue colour of the solution fades. The copper ions from the solution migrate to the cathode. The cathode becomes coated with copper. Bubbles can be seen at the anode.

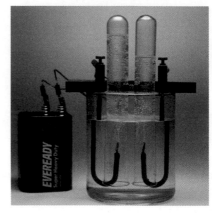

FIGURE 4: Electrolysis of sulfuric acid in the laboratory. Why is an electric current needed?

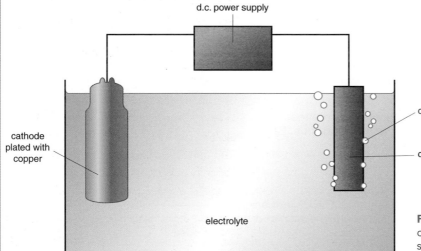

d.c. power supply

cathode plated with copper

oxygen bubbles

carbon anode

electrolyte

FIGURE 5: What happens at the cathode in the electrolysis of copper(II) sulfate?

What affects the amount of substance produced?

Only two things have any effect on electrolysis reactions:

> size of current

> time the current is passing through the electrolyte.

 Questions

7 What is the test for oxygen?

8 At which electrode does oxygen discharge?

Did you know?

Changing the voltage seems to have an effect, but all it is doing is changing the current.

electric current   cations

Products of electrolysis

If sodium hydroxide, NaOH, solution is electrolysed, hydrogen is formed at the cathode and oxygen is formed at the anode.

If sulfuric acid, H_2SO_4, solution is electrolysed hydrogen is formed at the cathode and oxygen is formed at the anode.

If copper sulfate, $CuSO_4$, solution is electrolysed with carbon electrodes copper is formed at the cathode and oxygen is formed at the anode.

Why is it only current and time?

The amount produced at each electrode increases with current and with time. The substances produced at the electrodes are **discharged ions**. The only thing that affects the number of ions discharged is the amount of charge transferred.

And what controls the amount of charge transferred?

current – the charge transferred every second

time – how long the current is passed through the electrolyte

Questions

9 What is formed at the cathode when a solution of NaOH is electrolysed?

10 What two factors affect the amount of substance produced during electrolysis?

Electrode reactions

The electrode reactions in the electrolysis of $NaOH_{(aq)}$ or $H_2SO_{4(aq)}$ are:

> at the cathode:

$2H^+ + 2e^- \rightarrow H_2$

> at the anode:

$4OH^- - 4e^- \rightarrow 2H_2O + O_2$

In the electrolysis of sodium hydroxide, hydrogen is made and not sodium as sodium is much higher up in the reactivity series, so hydrogen is discharged in preference.

The electrode reactions in the electrolysis of $CuSO_{4(aq)}$ with carbon electrodes are:

> at the cathode:

$Cu^{2+} + 2e^- \rightarrow Cu$

> at the anode:

$4OH^- - 4e^- \rightarrow 2H_2O + O_2$

The link between current and charge and mass

We can work out how much mass is increased by knowing that size of current changes the mass and that the time changes the mass.

Quantity	Symbol	Unit
time	t	second
current	I	amp
mass	*mass*	g

Example

A current of 0.1 A for 2 hours increased the mass of an anode by 0.24 g.

a What mass increase will be produced by a current of 0.2 A?

b What mass increase will 600 minutes produce?

Answer:

a If 0.1 A produces 0.24 g

Then 0.2 A produces 0.48 g

b 2 hours produces 0.24 g

120 minutes produces 0.24 g

600 minutes produces 1.2 g

So a mass increase of 1.2 g is produced.

Questions

11 Write the half equation of the reaction that produces oxygen at an anode.

12 If a current of 0.5 A is passed for 1 hour and deposits 2.4 g, how much mass would be deposited if the current was passed for 30 minutes?

Remember!

The symbol (aq) shows the compound is in the aqueous state.

Q anions   coulomb   negative electrode   positive electrode

Energy transfers – fuel cells

You will find out:
> what a fuel cell is
> where fuel cells are used
> about advantages of fuel cells

A better way to power your MP3

Have you ever wished that you did not have to keep replacing the batteries in your MP3 player? Or that you did not have to wait while your mobile phone charged?

Fuel cells are the answer. They make their electricity from fuel, so instead of changing the battery you just refill the tank.

FIGURE 1: Parts of the Space Shuttle are powered by fuel cells.

What are fuel cells?

Fuel cells are a special type of electric cell. They do not need replacing or recharging like ordinary batteries. Instead, fuel cells have a fuel tank that needs refilling every now and then.

Fuel cells are a very **efficient** way of producing electric current because they have no moving parts. Until recently their main use was in **spacecraft**. However, they are increasingly used in small vehicles for back-up power.

Fuel (hydrogen) in the fuel cell reacts with oxygen from the air. This is an **exothermic** reaction because heat energy is released. Fuel cells use exothermic reactions because the energy released can then be converted into **electrical energy**. This is a very efficient process.

Hydrogen is the fuel in a fuel cell. It makes a pollution-free fuel because when hydrogen reacts with oxygen, all it makes is water.

hydrogen + oxygen → water

FIGURE 2: This plane flaps its wings like an insect. It is powered by fuel cells.

Questions

1 What are fuel cells used for?

2 What two chemicals react inside most fuel cells?

3 The Space Shuttle carries tanks of oxygen to power its rocket motors and also the fuel cells. Suggest one other use for the oxygen.

4 Suggest one reason why a fuel cell is efficient.

Did you know?

Do you know the difference between an electric cell and a battery?

Cells make electricity.

The object you buy in the shop has several cells inside it, so it is called a battery.

Why are fuel cells used in spacecraft?

Fuel cells are used in **spacecraft** because:

> they are efficient – they waste very little energy

> the **water** produced is not wasted – the astronauts drink it

> they are **lightweight** – lighter than normal batteries, so the spacecraft can use the saving to carry a bigger payload

> they are **compact**

> there are **no moving parts**

> they can be used continuously – they do not need time out to be recharged

> they do not need a special fuel with its own separate storage system – the spacecraft has to carry hydrogen and oxygen anyway for its rocket engines.

Q fuel   fuel cell   oxygen   hydrogen   pollution-free fuel

Fuel cells and the car industry

Car makers are very interested in fuel cells. Not only are more laws being passed to reduce the pollution from vehicle exhausts, but the reserves of oil used to produce petrol will eventually run out. This is already leading to increasing prices on the garage forecourts. These fossil fuels are **non-renewable**. Manufacturers are researching alternatives that cause less **pollution**.

A normal car engine converts chemical energy into heat, and the heat is then converted into movement energy. Not only is this stage inefficient but, when fuels burn, the temperature gets hot enough to react the nitrogen and oxygen in the air to make nitrogen oxides. Nitrogen oxides in the air cause photochemical smog.

Fuel-cell powered cars make a very attractive option. The cells are more efficient than normal engines so they use less fuel. The fuel does not burn, so no high temperatures are involved and no oxides of nitrogen are produced.

The main product of a hydrogen-powered fuel cell is water, which is not a pollutant at all. There are **no carbon dioxide emissions** from the car. A problem with fuel-cell powered cars is that hydrogen is a gas. This is more difficult to store inside the car, and filling stations will need a totally different type of fuel pump. However, there is a large source of hydrogen available by **decomposing water**.

FIGURE 3: Instead of a diesel engine, this bus uses fuel cells and an electric motor. Guess how much noise it makes?

Did you know?

In a satellite weighing 4 tonnes, the batteries can weigh up to 300 kg. If the batteries are changed to fuel cells, that can be reduced to 75 kg.

Questions

5 Fuel cells work at low temperatures. Why does this create less pollution?

6 Suggest one other reason why a hydrogen–oxygen fuel cell creates less pollution.

7 Suggest one advantage of using fuel cells in an unmanned spacecraft.

8 If fuel-cell powered cars are to work, what must happen to petrol stations?

Fuel cell advantages and problems

Fuel cells are especially useful for mobile energy sources. Currently, mobile sources of electricity are the battery and the internal combustion engine.

Advantages of a fuel cell

> Direct energy transfer. Energy is converted directly from chemical energy in the fuel into electrical energy. The energy does not have to be converted into heat first.

> Fewer transfer stages. Every new stage in any process increases the energy losses.

> Within the fuel cell itself the energy conversion is all done in a single stage.

> More efficient. Not only are there fewer stages, but the energy-conversion stage is highly efficient. Almost all the energy is converted. Even if the hydrogen for the fuel cell is made from oil, it is more efficient to do this at a refinery than burn oil directly in a car engine.

> Less polluting. Not only is the main product of a hydrogen–oxygen fuel cell water, but no nitrogen oxides are produced.

> Conventional rechargeable batteries need replacing sooner. If battery electrodes are just thrown away, they will contaminate landfill sites.

> Fuel cells weigh less. Currently, batteries in electric vehicles are very heavy. Fuel cells would make cars lighter, reducing fuel consumption.

Pollution production

There will still be some pollution when using hydrogen–oxygen fuel cells. This will be because they:

> use poisonous catalysts that have to be disposed of at the end of the life-time of the cell

> involve the use of energy, which may have to come from the burning of fossil fuels, to produce the hydrogen and oxygen needed.

Questions

9 Fuel cells produce water. Suggest why this is useful on manned Space missions.

10 Which aspect of a petrol engine makes it so inefficient?

11 Use your notes on ammonia to suggest how hydrogen might be produced as a fuel.

12 Hydrogen is also produced near some hydroelectric power stations. How?

efficient fuel   internal combustion engine   fuel-cell technology

Why are fuel cells less polluting?

You will find out:

> about the chemical reaction that takes place in a fuel cell

> about exothermic reactions

> about energy-level diagrams

Petrol, used in a car engine, comes from a fossil fuel. It burns in oxygen to make carbon dioxide. Increased carbon dioxide levels have been linked with climate change. It is thought to contribute to global warming. Fuel cells use hydrogen as a fuel. Water is made as the only product. Hydrogen reacts with oxygen to make water.

In this reaction, the reactants are hydrogen and oxygen. When they react, the product is water. The word equation for this reaction is:

hydrogen + oxygen → water

The reaction between hydrogen and oxygen is exothermic. This means that energy is given out. The hydrogen–oxygen fuel cell uses the energy released from the reaction to produce electrical energy efficiently.

You may remember the reaction between hydrogen and oxygen that you have tested before. Look at Figure 4. Do you remember the squeaky pop?

In a fuel cell, this reaction takes place in a sealed chamber. The waste gas is water vapour.

lighted splint

'pop'

hydrogen

FIGURE 4: The exothermic reaction between hydrogen and oxygen.

Questions

13 Why is oxygen gas needed in the fuel cell?

14 What type of reaction happens when hydrogen gas and oxygen react.

15 In a fuel cell, what type of energy is transferred out of the cell?

16 Why is a fuel cell less polluting than an engine burning petrol?

Getting energy from fuel

A fuel on its own will not make a fuel cell work. The energy is only released when the fuel reacts with oxygen in air. A reaction which gives out energy is exothermic.

If the fuel cell uses hydrogen, the reaction is:

hydrogen + oxygen → water

$2H_2 + O_2 \rightarrow 2H_2O$

When hydrogen reacts with oxygen by burning, the chemical energy is given out as heat. A fuel cell converts chemical energy directly into electrical energy – there is no heat.

The overall reaction is:

$2H_2 + O_2 \rightarrow 2H_2O$ with electrons being exchanged from the cathode to the anode.

Other fuels, such as methanol, are also being tried.

FIGURE 5: Fuel cells come in different sizes. Some are small, some much bigger.

🔍 direct energy transfer   chemical energy   electrical energy   energy losses

Questions

17 What does the fuel react with in a fuel cell?

18 What is the name for chemical reactions that give out energy?

19 Methanol is a liquid. Suggest one advantage of using liquid fuels to recharge the batteries in laptop computers.

20 Methanol contains carbon. Suggest one disadvantage of using methanol as a large-scale fuel.

Energy-level diagrams

The fuel and oxygen have chemical energy inside them.

When the fuels react, some of the chemical energy is given out. This leaves less chemical energy inside the chemicals.

This is shown in an **energy-level diagram**.

> The hydrogen forms an H^+ ion at the catalyst releasing an electron at the **negative electrode**. This is the reaction:

$2H_2 \rightarrow 4H^+ + 4e^-$

This is an **oxidation** reaction.

The electrons move through the wire, while the ions travel through the electrolyte.

> The ions are reduced while reacting with oxygen, the **positive electrode** in the fuel cell takes in the electrons:

$4H^+ + O_2 + 4e^- \rightarrow 2H_2O$

This is a **reduction** reaction.

A reaction where electrons are gained and lost is called a **redox** reaction.

The overall reaction is:

$2H_2 + O_2 \rightarrow 2H_2O$

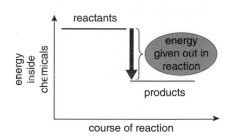

FIGURE 6: A simple energy-level diagram. What type of energy change is shown in this diagram?

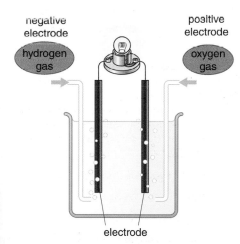

FIGURE 7: Diagram of a fuel cell. What happens at the negative and positive electrodes?

Questions

21 Draw a labelled energy-level diagram for the combustion of hydrogen.

22 What happens to the amount of chemical energy in a reaction that gives out heat?

23 Why should the electrodes be made from metal?

24 The fuel in a fuel cell is oxidised. Use one of the half equations to explain why.

Redox reactions

You will find out:

> what makes iron rust and some ways to stop it

> what a redox reaction is

> why rusting is a redox reaction

Rusting iron

When metals corrode they are taking part in a **redox** reaction. Rusting iron is just one example of this.

Redox is a word made up from two processes that work together. The two processes are **oxidation** and **reduction**.

Oxidation is the addition of oxygen to a substance. Reduction is the removal of oxygen from a substance.

FIGURE 1: What is needed for objects to rust?

What makes iron rust?

Iron and steel only rust when oxygen (in the air) and water are touching their surfaces.

> Iron needs oxygen to rust.

> Iron needs water to rust.

Rust is not just iron oxide. Its scientific name is:

> hydrated iron(III) oxide

The 'hydrated' part tells us that water is needed as well as oxygen.

Six ways of preventing rust

FIGURE 2: The Titanic sank almost 100 years ago. Why is it rusting only slowly?

1 cover the iron with oil or grease

2 cover the iron with paint

3 cover the iron with **tin plate**

FIGURE 3: Six ways to prevent rust.

4 **galvanising:** cover the iron with a layer of zinc

5 **sacrificial protection:** connect the iron to a reactive metal that corrodes instead of iron

6 **alloying:** mix other elements in with the iron when it is made

To stop iron from rusting, just remove either water or oxygen:

> stop water touching the iron – iron does not rust in air if the air is very dry

> stop oxygen from touching the iron.

Covering the iron with a layer of oil, grease or paint stops the iron rusting for both of these reasons.

If a ship sinks to the bottom of the sea, air can still get to it because air dissolves in water. As both water and air are touching the ship, the ship rusts. If the water is deep there is very little dissolved air. The Titanic sank in water 2.5 miles deep, so it is rusting very slowly.

Questions

1 What is the chemical name for rust?

2 State two things needed for iron to rust.

3 Name five ways to prevent iron from rusting.

4 Suggest why painting the chain in Figure 3 above does not stop it rusting for long.

Q redox reaction   redox   oxidation   reduction   galvanising

Rust and redox

Rusting is a redox reaction.

In a redox reaction, both **red**uction and **ox**idation take place.

iron + oxygen + water → hydrated iron(III) oxide

How to prevent iron from rusting by *galvanising*

Galvanising protects iron from rusting when it is covered by a layer of zinc. The layer of zinc stops water and oxygen from reaching the surface of the iron. It also acts as a sacrificial metal. This means that it reacts with the water or oxygen instead of the iron, as it is a more reactive metal.

> **Did you know?**
>
> Only iron can rust. Most metals corrode, but when iron corrodes we give it a special name – rusting.

> **Did you know?**
>
> Salt acts as a catalyst for rusting iron. Salt doesn't make iron rust in the first place.

Questions

5 A car left in the Sahara desert does not rust as quickly as a car in the UK. Suggest why.

6 Explain why ships that sink in shallow water rust quite quickly.

7 What happens in a redox reaction?

8 Write a word equation for the rusting of iron.

More on redox reactions

Oxidation and reduction can also be described in terms of electrons.

> A substance is **oxidised** if it **loses electrons**. An **oxidising agent** takes electrons away from a substance.
> A substance is **reduced** if it **gains electrons**. A **reducing agent** pushes electrons onto another substance so that it gains them.

Rust and redox

When iron rusts, the oxygen is the oxidising agent; it takes electrons from the iron. The electrons go onto the oxygen, so the oxygen itself has been reduced.

> Iron **loses electrons** – it is oxidised.

> Oxygen **gains electrons** – it is reduced.

Reversing the reaction

Redox reactions can be forced in either direction. One direction is oxidation; the other direction is reduction.

> Iron is normally oxidised, but it can be pushed back (Figure 4).

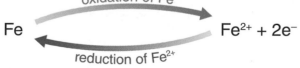

$$Fe \qquad\qquad Fe^{2+} + 2e^-$$

oxidation of Fe

reduction of Fe^{2+}

FIGURE 4: Oxidation/reduction of iron.

> Iron ions can be forced from one type to another (Figure 5).

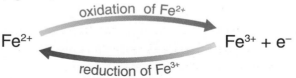

$$Fe^{2+} \qquad\qquad Fe^{3+} + e^-$$

oxidation of Fe^{2+}

reduction of Fe^{3+}

FIGURE 5: Oxidation/reduction of iron ions.

> Chlorine is like oxygen, it is normally reduced (Figure 6).

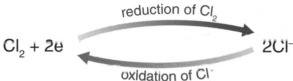

reduction of Cl_2

$$Cl_2 + 2e \qquad\qquad 2Cl^-$$

oxidation of Cl^-

FIGURE 6: Oxidation/reduction of chlorine.

When these reactions take place, an electrode potential is set up. Each reaction at the electrode has a different electrode potential. These can be used in calculations. When a magnesium ion gains two electrons to become a magnesium atom, the voltage measured is -2.37 V. When a zinc ion gains two electrons to become a zinc atom, the voltage measured is -0.76 V. This is because zinc is a less reactive metal than magnesium.

Questions

9 Chlorine accepts electrons. Is it an oxidising agent or a reducing agent?

10 Use the above equations to suggest what the '(III)' in iron(III) oxide tells you.

11 When sulfur reacts with hydrogen it gains electrons from hydrogen.

a What has been oxidised?

b What has been reduced?

Displacement reactions

If magnesium is put into iron(II) sulfate solution, the magnesium pushes the iron out of the iron(II) sulfate. Solid iron and a solution of magnesium sulfate are made. This is called a **displacement reaction**.

It happens because magnesium is more reactive than iron. The iron has been pushed out or 'displaced'. It forms a coating on the rest of the magnesium.

You can see this taking place. The magnesium turns darker where it dips into the iron(II) sulfate solution and has become coated with iron. If one metal is much more reactive than another, the solution can get quite warm.

The table shows what happens when four different metals are tried out.

Solution used	Metal being added			
	magnesium	zinc	iron	tin
magnesium sulfate	✗	✗	✗	✗
zinc sulfate	✓	✗	✗	✗
iron(II) sulfate	✓	✓	✗	✗
tin sulfate	✓	✓	✓	✗

Key:

✗ means that nothing happens
✓ means that metal gets coated

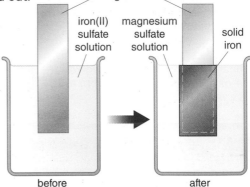

FIGURE 7: A displacement experiment. Describe what happens in this reaction.

FIGURE 8: Here is a displacement experiment using zinc and copper sulfate. What is the dark coating on the zinc?

Questions

12 Look at the table above. Which metals does magnesium displace?

13 a When magnesium is put into zinc sulfate, the magnesium becomes coated. What with?

b What does the magnesium turn into?

14 If zinc is put into iron(II) nitrate solution, what does the zinc form?

Displacement reactions

When a word equation for a displacement reaction is written, it does not matter whether the compound is a chloride, a nitrate or a sulfate, it is the metals that are important. In every case, the more reactive metal swaps places with the less reactive metal.

> Magnesium is more reactive than zinc (Figure 9).

magnesium + zinc sulfate → magnesium sulfate + zinc

more reactive metal

less reactive metal

FIGURE 9: Zinc is less reactive than magnesium.

If we compare zinc with iron, zinc is now the more reactive metal (Figure 10).

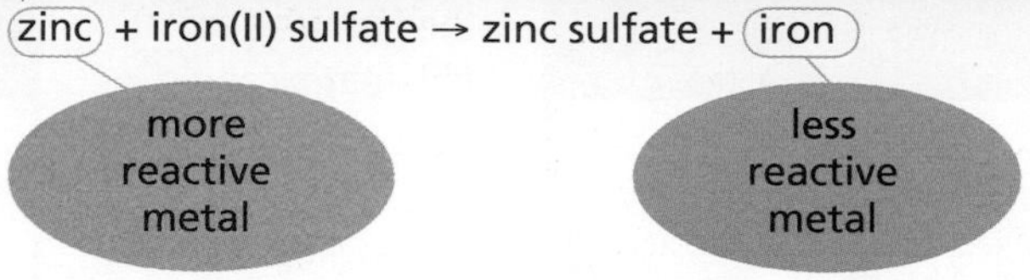

zinc + iron(II) sulfate → zinc sulfate + iron

more reactive metal

less reactive metal

FIGURE 10: Iron is less reactive than zinc.

Here are some more:

zinc + iron(II) chloride → zinc chloride + iron
iron + tin chloride → iron(II) chloride + tin

To work out what happens in a displacement reaction you do not need to remember every reaction. You need to know the order of reactivity: magnesium, zinc, iron, tin.

This means that:

> magnesium metal displaces zinc, iron and tin

> zinc displaces iron and tin

> iron displaces tin.

Questions

15 Write the word equation for the reaction between zinc metal and tin sulfate.

16 Write the word equation for the reaction between zinc metal and iron(II) sulfate.

17 Write the word equation for the reaction between magnesium metal and tin chloride.

18 In the experiment shown in Figure 8, when zinc is dipped into copper sulfate, the copper sulfate solution turns paler during the reaction. Suggest why.

Displacement – what is going on?

Displacement reactions depend on having a reactive metal element and the compound of a less reactive metal. All metals react by pushing off electrons and turning into ions. This is **oxidation**. The more reactive the metal, the harder it pushes off electrons. These electrons have to go somewhere; they are forced onto the ions of other metals that are not so reactive. Metal atoms that gain these electrons are **reduced** (Figure 11).

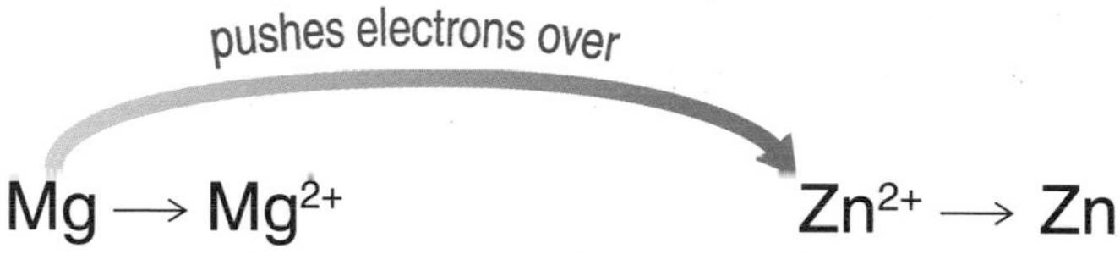

pushes electrons over

$$Mg \longrightarrow Mg^{2+} \qquad Zn^{2+} \longrightarrow Zn$$

FIGURE 11: Displacement reaction of magnesium by zinc.

Displacement reactions all follow the pattern in Figure 12.

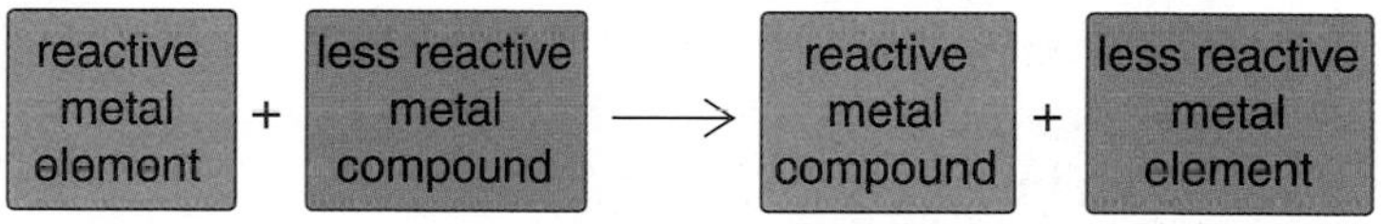

Examples, and also the ionic equations to show what really happens, are:

| reactive metal element | + | less reactive metal compound | → | reactive metal compound | + | less reactive metal element |

FIGURE 12: Pattern of displacement reactions.

$Mg + ZnSO_4 \rightarrow MgSO_4 + Zn$ $\quad$ $Mg + Zn^{2+} \rightarrow Mg^{2+} + Zn$
$Zn + FeSO_4 \rightarrow ZnSO_4 + Fe$ $\quad$ $Zn + Fe^{2+} \rightarrow Zn^{2+} + Fe$
$Zn + FeCl_2 \rightarrow ZnCl_2 + Fe$ $\quad$ $Zn + Fe^{2+} \rightarrow Zn^{2+} + Fe$
$Fe + SnCl_2 \rightarrow FeCl_2 + Sn$ $\quad$ $Fe + Sn^{2+} \rightarrow Fe^{2+} + Sn$

Tinning, galvanising and sacrificial protection

Iron can be protected by coating it with another metal. The effect of this metal depends on how reactive it is.

> Iron coated in tin is 'tinned steel'. The tin acts as a barrier. It is less reactive than iron, so when the tin layer is scratched the iron loses electrons in preference to the tin and the iron rusts even faster than on its own.

> Iron coated in a thin layer of zinc is 'galvanised iron'. If the zinc layer is scratched the iron does not rust. Zinc is more reactive than iron, so it loses electrons in preference to iron. If the iron cannot give off electrons, it cannot react. The zinc has sacrificed itself and the process is called sacrificial protection.

> Coating the whole of the iron with the sacrificial metal is not necessary. It is enough to connect blocks of the sacrificial metal at intervals. Underground pipelines and the metal legs of seaside piers often have lumps of magnesium attached at intervals.

Questions

19 Tin cans attract magnets. What does this tell you?

20 Explain why magnesium is more effective than zinc at protecting iron.

21 Write a balanced equation for the reaction between magnesium and zinc nitrate, $Zn(NO_3)_2$.

22 Write a balanced equation for the reaction between magnesium and tin sulfate, $SnSO_4$.

🔍 tinning   galvanising   sacrificial protection

Alcohols

You will find out:
> the formulae for different alcohols
> how ethanol is made
> what ethanol is used for

Drink or other uses?

The alcohols are a complete group of compounds that have a wide range of uses. The most common alcohol is ethanol, and it is the only one we can drink.

FIGURE 1: Ethanol is in alcoholic drinks.

Ethanol

Ethanol is not a hydrocarbon as it has an oxygen atom in its formula, C_2H_5OH. Huge amounts of ethanol are made every year. Apart from being used to make other chemical compounds, the three main uses of ethanol are:

> in petrol replacements
> as a **solvent**, such as industrial methylated spirits
> in alcoholic drinks.

Most **ethanol** is made using plants in a process called **fermentation**. Plants are **renewable**, so ethanol made for fuel in this way is a renewable fuel. The process of fermentation uses **yeast**. The fermenting is done in enclosed tanks. Theses tanks are temperature-controlled. Fermentation needs:

> sugars from plants
> **water**
> enzymes from **yeast**
> a temperature between **25 °C** and **50 °C**
> an **absence of oxygen**.

FIGURE 2: Fermentation tanks. Why must air not be allowed to enter these tanks?

FIGURE 3: 'Alcool' is made from ethanol.

FIGURE 4: Perfumes are dissolved in ethanol.

Questions

1 Give three uses of ethanol.

2 Give three conditions needed for fermentation.

3 Why can ethanol be called renewable?

4 Ethanol is one of a group of compounds. What is the name of the group?

The formula of ethanol

The **molecular formula** for ethanol is:
C_2H_5OH

It is easy to see what a molecular formula means if the displayed formula is used, as in Figure 5.

Making ethanol by fermentation

glucose → ethanol + carbon dioxide
The balanced chemical equation for fermentation is:
$C_6H_{12}O_6 \rightarrow 2C_2H_5OH + 2CO_2$
Fermentation has been used for thousands of years.

Q alcohol and ethanol   hydrocarbon   fermentation   fractional distillation

The first fermentation happened naturally when wild **yeasts** blown by the wind landed on ripe fruit. People slowly worked out what was happening – that fermentation needed the yeast as well as the glucose solution, and that there was an **optimum temperature** for carrying out the process successfully.

Ethanol is still made by fermentation of **glucose solution**. The reaction is catalysed by the **enzymes** in yeast. This will only happen if there is an **absence of air**. The ethanol produced is dilute. There needs to be **fractional distillation** of the dilute liquid to produce almost pure ethanol.

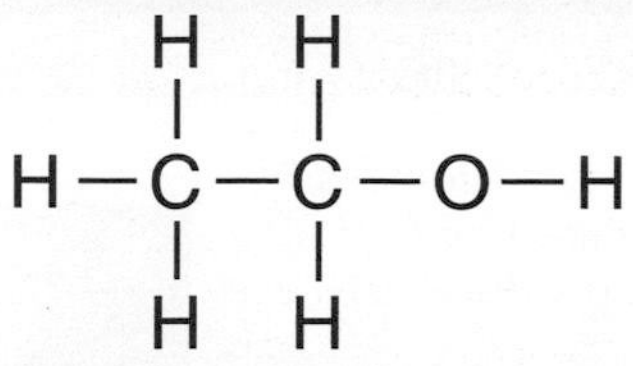

FIGURE 5: Displayed formula of ethanol.

Questions

5 What does the term 'optimum temperature' mean?

6 How many 'C–H' bonds are there in a molecule of ethanol?

7 How many 'C–O' bonds are there in a molecule of ethanol?

8 How many 'O–H' bonds are there in a molecule of ethanol?

There is more than one alcohol

The alcohols are a group of compounds.

An alcohol has the **general formula** $C_nH_{2n+1}OH$

The formulae of alcohols can be worked out using this general formula. Both the molecular formulae and the displayed formulae of alcohols with one to four carbon atoms are shown in the table. There are many more alcohols than this.

	Molecular formula	Displayed formula
methanol	CH_3OH	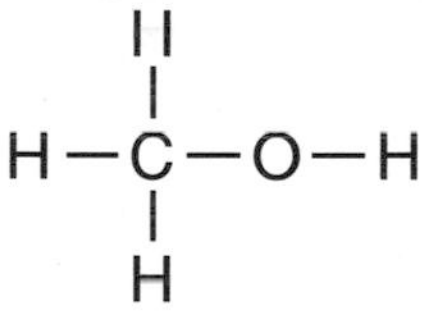
ethanol	C_2H_5OH	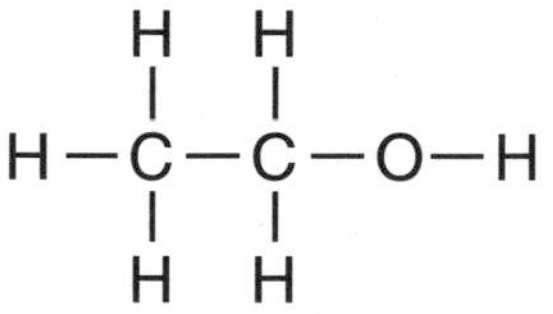
propanol	C_3H_7OH	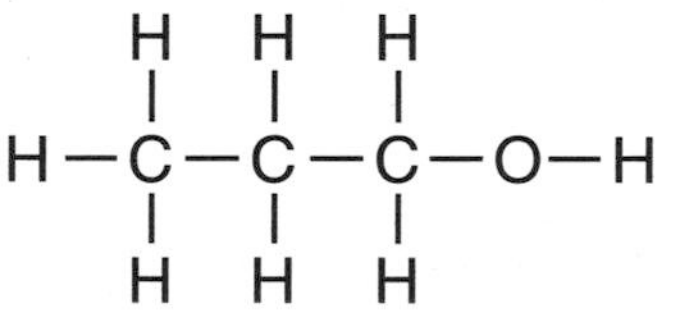
butanol	C_4H_9OH	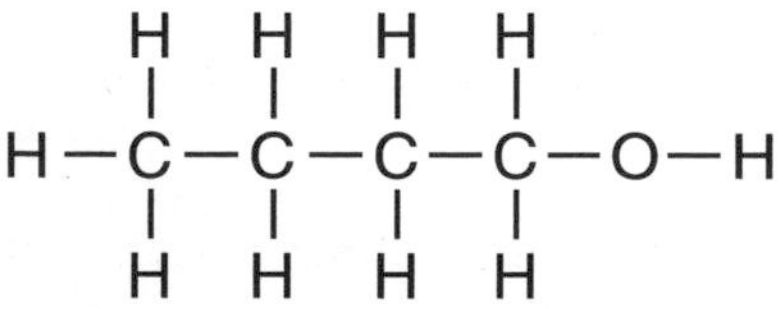

Fermentation conditions

The balanced chemical equation for fermentation is:

$$C_6H_{12}O_6 \rightarrow 2C_2H_5OH + 2CO_2$$

Fermentation is carried out under carefully controlled conditions:

> if the temperature is too cold the enzymes in yeast are inactive

> if the temperature is too hot the enzymes in yeast are denatured

> if air is present there is a different reaction, producing ethanoic acid instead of ethanol.

Did you know?

Pure alcohol is very poisonous. The strongest whisky is about 50% water – any stronger and it would do people serious harm and even kill them.

Questions

9 Write down the molecular formula for pentanol, an alcohol with five carbons.

10 Suggest the displayed formula for pentanol.

11 What is a common feature of all alcohols?

12 Fermentation produces ethanol and also a second chemical that can be sold. What is it?

Making ethanol from ethene

Most of the world's ethanol is made from **biomass**, but there is another way: it can be made from **ethene**. The UK is the world's largest producer of ethanol in this way.

Ethene is made from crude oil or from natural gas. Fresh supplies of oil cannot be grown, so if ethanol is made from oil, the ethanol is non-renewable.

The ethanol is made by reacting ethene and water. The reaction is called a hydration reaction.

Ethanol is important because it can be used to make other compounds.

Did you know?

A catalyst is needed to make this reaction viable.

FIGURE 6: Ethene being made at a refinery.

Questions

13 a What chemical from crude oil is used to make ethanol?

b What is this chemical reacted with to make ethanol?

c What is the type of reaction called that produces ethanol from this chemical?

14 Why is most of the world's ethanol 'renewable'?

Making ethanol from ethene

Ethanol that is used for drinking is always made by **fermentation**. Ethanol made by fermentation can also be used as a fuel. It is a renewable fuel because the plants that make the sugar for the process can be grown very quickly.

Ethanol that is used in industry can also be made from **ethene**. Ethanol made in this way is a **non-renewable fuel** because the ethene it is made from comes from fossil fuels. These **fossil fuels** cannot be replaced and are a **finite resource**.

This reaction to make ethanol from ethene is known as a **hydration** reaction because water is added to the ethene molecule:

ethene + water → ethanol

> The water is heated to make steam.

> Ethene and steam are then passed over a **hot phosphoric acid catalyst**.

> The percentage yield of this reaction is not very high, so the unreacted ethene is recycled back through the catalyst.

The balanced symbol equation for the hydration reaction is:

$C_2H_4 + H_2O \rightarrow C_2H_5OH$

Questions

15 What catalyst is used during the hydration of ethene?

16 What happens to the ethene that does not react?

Advantages and disadvantages of ethanol production

Making ethanol by fermentation has many advantages.

> It is a renewable resource.

> It is carbon neutral. The carbon dioxide released when it burns is balanced by the carbon dioxide taken in by photosynthesising plants.

> Household waste can be fermented to make ethanol, reducing the need for landfill sites.

However, crops are often grown specially to make the ethanol. This works best in hot countries and needs large areas of land.

> This land might be better used for growing food for the local population rather than cash crops.

> Large areas of natural forest may be cut down to make room for the crops.

> It is not an efficient process. There are high transport costs in getting the crops to the fermentation plants and there are also energy costs in producing the fertiliser and pesticides that the plants need to grow well.

In the UK, industrial ethanol is made from the hydration of ethene that comes from oil and natural gas.

> Advantages. This is a much cheaper route. The climate and the amount of land available in the UK make it less economic to grow crops for fermentation.

> Disadvantages. It is a non-renewable method. Once the oil has run out an alternative source of ethene will have to be found.

Remember!

Atom economy is the

$$\frac{M_r \text{ of desired products}}{\text{sum of all } M_r \text{ of all products}} \times 100$$

the higher the atom economy the less waste there is.

Evaluating the methods of production

	by fermentation	by hydration
conditions	35 °C, less energy needed	hot catalyst, more energy needed
processing	batch	continuous
sustainability	sustainable	from a finite resource
purification	fractional distillation needed	relatively pure on production
percentage yield and atom economy	low percentage yield and atom economy	high percentage yield and atom economy

Questions

17 List the advantages of producing ethanol by fermentation.

18 Suggest why cutting down forests may not be a good idea.

19 Explain how the climate in the UK is not ideal for growing crops to make ethanol.

Q finite resource   hot phosphoric acid catalyst   industrial ethanol

Preparing for assessment: Applying your knowledge

To achieve a good grade in science, you not only have to know and understand scientific ideas, you also need to be able to apply them to other situations and investigations. These tasks will support you in developing these skills.

✳ Beer making

Richard works for the local authority in a city in the north of England. He has lived there ever since he was a student and still enjoys the taste of beer. These days he's sometimes disappointed by the flavour of what he calls 'keg beers' – he's actually happier drinking real ale.

For some years now he's brewed his own beer as well and has become something of an expert. Some of his first attempts were pretty grim – they had certainly fermented, but the cloudiness was a bit off-putting, and so was the taste. A bit of experimenting and picking the brains of other people soon got things sorted out.

In the living room of his house are a row of demijohns, the large glass bottles he uses for the brew.

At first they make little popping noises; nothing very loud, but quite persistent.

"It's the air lock," he explains. "What you can hear is the gas escaping. As the brew ferments the gas builds up and it has to be let out."

The airlock consists of a tube with two 180° bends in it. Water is put in the U-shaped bend and sits there. Gas can push through it, but nothing can pass through the other way.

It seemed a bit odd to have these demijohns all lined up in the living room though, but apparently the fermentation works better if they're warm. It takes several weeks and some careful filtering, but Richard swears by the outcome.

"I could win prizes with this," he says. "It's like nectar."

Task 1

What is the gas that is being formed?

Write down the word equation that shows its formation.

Task 2

What are the conditions needed for the fermentation to work effectively?

How has Richard set things up to provide these?

Task 3

Write down the word equation that describes the process of fermentation.

What is the function of yeast in this process? How is it being fed?

Task 4

However long the brew is left, the alcohol concentration never rises above a certain proportion.

Why is this?

Task 5

Explain why it is important that the gas is allowed to escape.

Explain why it is important that air from the outside is not allowed to enter.

Task 6

Now convert the word equation for fermentation into chemical symbols.

What happens to this process if the temperature is not carefully controlled?

Maximise your grade

Answer includes showing that you can...	
F	Describe some of the conditions needed for fermentation.
	State the two products of fermentation.
	Write down the word equation for fermentation.
C	Describe all the conditions needed for fermentation and how more concentrated alcohol is obtained.
	Construct the balanced symbol equation for the process of fermentation knowing that $C_6H_{12}O_6$ is glucose and C_2H_5OH is ethanol.
	Explain the role that temperature plays in this process.
A	Explain why air has to be kept out of the fermentation process.
	Construct the balanced equation without being given any of the formulae.
	As above, but with particular clarity and detail.

159

Depletion of the ozone layer

You will find out:
> how the ozone layer protects us
> how chlorofluorocarbons (CFCs) are damaging the ozone layer
> why the UK has banned CFCs

Getting a tan

Between 20 km and 40 km (12–25 miles) up in the atmosphere there is a very thin layer of ozone.

This ozone layer reduces the risk of us dying from skin cancer by preventing the Sun's harmful rays from reaching Earth. CFCs are contributing to the destruction of this layer.

FIGURE 1: What is she at risk of?

What is ozone?

Oxygen atoms can combine to form two types of molecule.

> Molecules with two oxygen atoms make oxygen gas.

> Molecules with three oxygen atoms make a different gas, called ozone, O_3.

Ozone is a good thing and a bad thing. It is poisonous, so any ozone at ground level is a bad thing. Ozone high in the atmosphere is a good thing, as it blocks out most of the Sun's fierce **ultraviolet (UV) radiation**.

Why is UV light dangerous?

UV light is the part of sunlight that gives us a tan. You may not mind getting **sunburnt**, or even the way that a lot of Sun makes the **skin 'age'** and go leathery, but **skin cancer** and **cataracts** of the eyes are much more dangerous.

ozone molecule

oxygen molecule

FIGURE 2: How many atoms are there in a molecule of oxygen and in a molecule of ozone?

Questions

1 What is the formula of oxygen gas?

2 What is the formula of ozone gas?

3 Give four problems caused by UV radiation.

4 Where is the ozone layer?

The problem with the ozone layer

The ozone layer is in the stratosphere. There are only tiny amounts of ozone in this layer, but it still absorbs most of the UV radiation from the Sun. In 1985 scientists discovered that the amount of ozone high over the South Pole was much less than it should be, and they called this the 'hole' in the ozone layer.

They have now also found a huge hole over the North Pole, and ozone levels over the Northern hemisphere are lower in general.

The more depleted the layer of ozone gets, the more UV light can get through to the Earth's surface.

Chlorofluorocarbons are now banned

When chlorofluorocarbons (CFCs) were first discovered, they were thought to be totally safe.

Scientists now know that CFC molecules slowly move upwards into the stratosphere where they attack the ozone layer. Society agreed with the scientists' views that it is CFCs that deplete the ozone layer. This is why the use of CFCs has been banned in the UK.

The ban on the use of CFCs should slowly allow the amount of ozone to return to its original level.

CFCs can be replaced with alkanes or hydrofluorocarbons (HFCs) which will not damage the ozone layer.

chlorofluorocarbons   skin cancer   cataracts   ozone layer

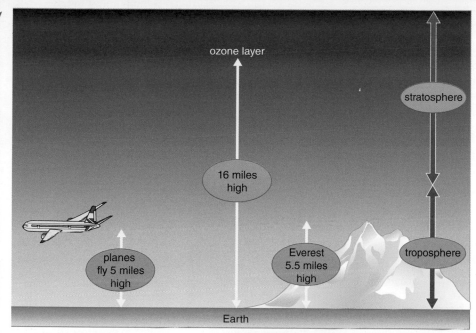

ozone layer

stratosphere

16 miles
high

planes
fly 5 miles
high

Everest
5.5 miles
high

troposphere

Earth

FIGURE 3: Which part of the atmosphere do planes fly in?

Questions

5 What part of the atmosphere contains the ozone layer?

6 What do CFCs do to the ozone layer?

7 What does ozone do to UV radiation?

8 What action has the UK and many other countries taken on CFCs?

How does ozone work?

UV radiation is part of the electromagnetic spectrum. Visible light can pass through the ozone layer easily, but UV radiation is absorbed by it.

The UV part of the electromagnetic spectrum has exactly the right frequency to make ozone molecules vibrate. The energy of the UV radiation is converted into movement energy inside each molecule. The thicker the ozone layer, the more UV radiation is absorbed.

Attitudes to CFCs

> CFCs were discovered in the 1930s, and they started out as wonder chemicals that were going to make life a lot better. This is because they are inert. Scientists were very enthusiastic in their attitude towards their use.

> In the 1970s, scientists began to suspect that CFCs might affect the ozone layer. They linked the depletion of the layer with the presence of CFCs. Their advantage of inertness is the problem. They do not react with anything in the lower atmosphere, so they are able to diffuse up to the stratosphere – the one place where they do damage. Ozone molecules are broken down into oxygen molecules and oxygen atoms.

> Once the first ozone hole was discovered, CFCs were soon identified as the cause. CFCs do not stay above

the country where they are released; they spread around the whole of the Earth. Scientists and the rest of the world community accepted the evidence. This is a global problem and most countries, including the UK and EU, agreed to a worldwide ban on CFCs.

> Unfortunately, not all countries have agreed to the ban. Some countries ban only some CFCs and other countries cannot afford to make the change from using CFCs.

Questions

9 When UV light is absorbed by ozone, where does its energy go?

10 UV and visible radiation are both electromagnetic. Find the major difference between them.

11 Why does visible light not get absorbed by the ozone layer?

12 Why is it important for all countries to ban CFCs?

Q ultraviolet radiation   stratosphere   ozone depletion

What are CFCs?

You will find out:
> more about CFCs
> how CFCs make radicals
> what are used as alternatives to CFCs

CFC stands for **chlorofluorocarbon**. CFCs are organic molecules made from carbon, chlorine and fluorine atoms and nothing else. CFCs are part of a group of compounds called freons.

CFCs have many uses.

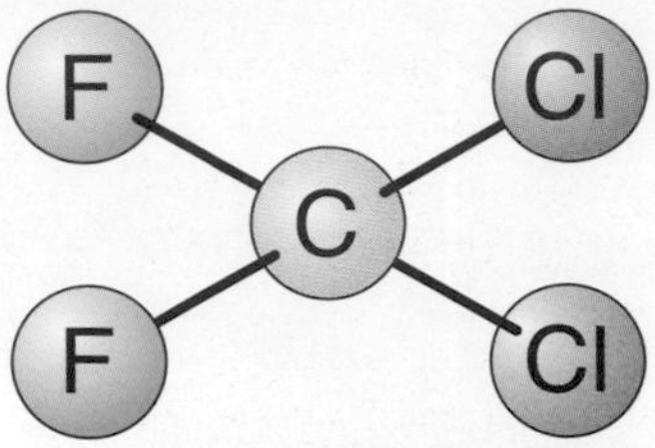

Why are CFCs so good at their job?

All CFCs:

> have **low boiling** points

> are **chemically inert**, which made them especially useful for aerosols – if chemicals do not react, they cannot poison you

> are **insoluble in water**.

What is the problem with CFCs?

CFCs are chemicals that have been destroying the **ozone layer**, so letting more UV rays through to reach Earth. They are now banned in the UK and the European Union (EU).

Alternatives to CFCs

In the countries where CFCs are banned, people have switched to **hydrocarbons**.

Hydrocarbons can provide a safe alternative, although they are not quite as good.

FIGURE 8: Propane is a safe alternative to CFCs in some cases, such as refrigerants, but it is not quite as good.

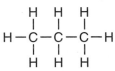

propane

FIGURE 4: What atoms are in chlorofluorocarbons?

FIGURE 5: As **refrigerants**, a liquid that is pumped through pipes at the backs of fridges.

FIGURE 6: As **aerosol propellants**, they push the liquid out of aerosol cans. They should not be used.

Questions

13 What does CFC stand for?

14 Give two uses of CFCs.

15 Give three properties of CFCs.

16 What are used now instead of CFCs?

CFCs, ozone and radicals

In the stratosphere, the UV radiation from the Sun is strong enough to break single chlorine atoms off the CFC molecule. These chlorine atoms are highly reactive. A single chlorine atom is called a **chlorine radical**. These chlorine atoms attack ozone molecules, turning the ozone back into oxygen gas and depleting the ozone layer. More chlorine atoms are regenerated so they can react with more ozone molecules.

The CFCs are only slowly broken down by UV light so they last a long time. As CFCs are removed from the stratosphere only very slowly, each CFC molecule has time to do a lot of damage.

$$CF_2Cl_2 \xrightarrow{UV162} CF_2Cl + Cl\bullet$$

Alternatives to CFCs

The main alternatives are alkanes and HFCs. HFCs do not contain chlorine, so they cannot make chlorine radicals and so are safer.

FIGURE 7: Fridges waiting to have CFCs removed from them. How do CFCs from old fridges get into the atmosphere?

refrigerant   aerosol propellants   chlorine radical

Questions

17 What particles attack the ozone in the ozone layer?

18 Why do CFCs do so much damage?

19 What are the alternatives to CFCs?

20 Explain why HFCs are safer than CFCs.

How radicals form

A **covalent bond** is made of two **shared electrons**. When a covalent bond breaks, either:

> both electrons go to one end to form ions

> or the bond splits into equal halves to make radicals.

Ultraviolet radiation causes **radicals**. Each radical sets off a chain reaction. Each reaction produces another chlorine radical, which reacts further. One chlorine radical causes the breakdown of more than 100 000 ozone molecules.

The chain reaction has three stages:

1 Ultraviolet light breaks a bond in CFC molecules to form chlorine radicals.

2 Chlorine radicals react with ozone molecules, creating more chlorine radicals to continue the chain. Because the Cl• radical is regenerated by this process, only a small number of chlorine atoms is needed to destroy a large number of ozone molecules.

$$Cl\bullet + O_3 \rightarrow OCl\bullet + O_2$$
$$OCl\bullet + O_3 \rightarrow Cl\bullet + 2O_2$$

Combining these two equations gives:

$$2O_3 \rightarrow 3O_2$$

3 Eventually two radicals collide, removing both of them. One possible termination reaction is:

$$Cl\bullet + Cl\bullet \rightarrow Cl_2$$

The problem with CFCs

Different CFCs have different lifetimes in the stratosphere, but lifetimes of 20–50 years before the CFCs are broken down by UV light are common. CFCs will continue to have an effect long after they have been banned.

However, there is evidence that the ozone holes may be getting smaller, and it is hoped that they may be gone in 50 years time.

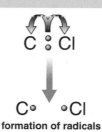

FIGURE 8: Radicals from the C–Cl bond are very reactive.

formation of radicals

Questions

21 How do the equations (left) show that chlorine radicals act as a catalyst?

22 Oxygen atoms exist in the stratosphere as well as oxygen molecules. Suggest how the oxygen atoms might have been formed.

23 The first ozone hole was discovered over the South Pole. CFCs are mainly produced in the northern hemisphere. What does this tell you about the atmosphere?

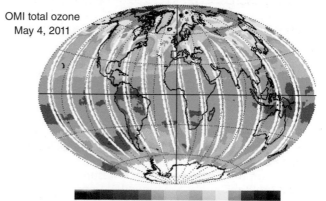

OMI total ozone May 4, 2011

Dobson units: dark gray < 100 and > 500 DU

FIGURE 9: Using the scale, approximately how many Dobson units of ozone were there in May 2011 over the Magnetic North Pole? In what region was there the least ozone?

Hardness of water

You will find out:

> how to tell if water is hard

> what causes hardness in water

> about temporary and permanent hardness

> how to compare the hardness of different water samples

Do you have hard water?

A quick look inside an electric kettle will tell you. If your kettle is full of white limescale, you have hard water. If it isn't, count yourself lucky!

You can find out from a map or an internet search which areas of the UK have the hardest water.

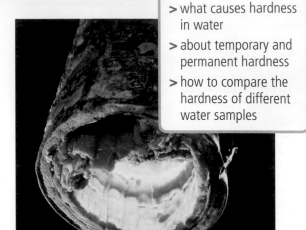

FIGURE 1: Limescale can block pipes. Why does this matter?

Hard water

Hard water clogs up water pipes. It also makes it very difficult for **soap** to make bubbles – it will not **lather**. Hardness is caused by calcium or magnesium **ions** dissolved in the water. The soap reacts with the ions to make a solid scum. Once the soap has reacted with all the ions, the rest of the soap works for washing. If you live in a hard-water area, most of the scum round the edge of your bath is not dirt, it is reacted soap! Soapless **detergents** do the same job as soap, but they are not affected by hardness in water.

Two types of hard water

> Sometimes hardness in water is destroyed when the water is boiled, so it is called **temporary hardness**.

> If boiling the water has no effect on the hardness, it is called **permanent hardness**.

 Questions

1 Name the ions that cause hardness.

2 What do we call the two types of hardness in water?

3 How can you tell the difference between the two types of hardness?

4 Why are soapless detergents so useful?

Look at how much soap was needed to make a lather in four water samples.

water sample	A	B	C	D
soap used in cold water in cm³	10	1	10	10
soap used in boiled water in cm³	2	1	10	8

You can tell that sample A contains temporary hardness as the hardness is less after boiling. Which sample has permanent hardness? Which is the softest water? Which has mostly permanent hardness?

How can you compare the hardness of different water supplies?

Although there are special water-testing kits that show water hardness, a simpler way is to measure how much soap is needed to produce a lather. When a soap solution is shaken in a water sample, calcium ions in the water react with the soap and turn it into scum. Eventually, as more solution is added and the mixture is shaken, the soap reacts with all of the calcium ions. After that, any more soap produces a stable lather for the first time. The volume of soap solution needed is a measure of water hardness.

Q hard and soft water   detergents   temporary hardness

How does hardness get into water?

Hard water is formed when rainwater dissolves some of the rock that it flows over on the way to a reservoir.

Permanent hardness is produced when calcium sulfate rock dissolves.

Rainwater may contain dissolved carbon dioxide. This makes it slightly acidic. Rocks such as chalk, limestone and marble are forms of calcium carbonate. These react with water and carbon dioxide to form calcium hydrogencarbonate. Calcium hydrogencarbonate dissolves to form temporary hard water.

The equation for the formation of temporary hard water is:

calcium carbonate + carbon dioxide + water → calcium hydrogencarbonate

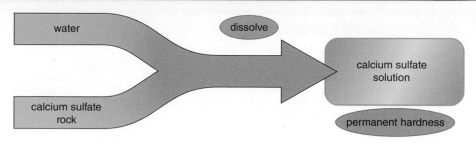

FIGURE 2: How does permanent hard water form?

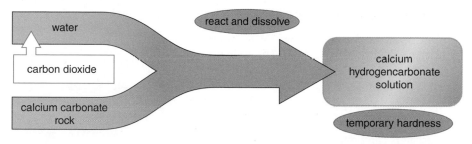

FIGURE 3: How does temporary hard water form?

Questions

5 How does rainwater pick up hardness?

6 Dolomite rock is magnesium sulfate. What sort of water hardness does it cause?

7 Suggest a pH for a solution of carbon dioxide in water.

8 What particles do all samples of temporary hard water contain?

What causes hard water?

As calcium sulfate rock dissolves in water its calcium ions go into solution.

Rocks based on calcium carbonate undergo a different process. Calcium carbonate does not dissolve in pure water, but rainwater is not pure. Rainwater contains some carbon dioxide dissolved in it from air, which makes it slightly acidic. There is then a chemical reaction between all three substances to produce a compound that is soluble, calcium hydrogencarbonate.

$$CaCO_{3(s)} + CO_{2(g)} + H_2O_{(l)} \rightarrow \qquad Ca(HCO_3)_{2(aq)}$$
$$\text{calcium carbonate} \qquad \text{calcium hydrogencarbonate}$$

It is calcium hydrogencarbonate that provides calcium ions in temporary hard water. Magnesium ions behave the same way as calcium ions:

calcium sulfate	magnesium sulfate	calcium hydrogen-carbonate	magnesium hydrogen-carbonate
↓	↓	↓	↓
permanent hardness	permanent hardness	temporary hardness	temporary hardness

Questions

9 When magnesium carbonate rock makes hard water, what is the compound in the water that produces the hardness?

10 Write an equation for the formation of the compound in question 9 from magnesium carbonate.

11 What dissolved ions make water hard?

12 What dissolved ions make hardness temporary?

Removing hardness

You will find out:
> how hardness can be removed
> about limescale and limescale removers

Water softeners

Some homes have water-softening machines to remove hardness. They often fit under a kitchen sink. This sort of water softener has a device called an **ion-exchange column** inside it.

Washing soda

One of the oldest water softeners of all is in the soap powder. It is called **washing soda**. The chemical name for washing soda is **sodium carbonate**. It **precipitates** out the ions that cause hardness before they can react with the soap. The solid precipitate is washed out of the clothes along with the dirt.

Removing limescale

If water is not softened before it is heated, kettles and hot-water pipes clog up with solid **limescale**. Descalers can be bought that get the limescale out of kettles. Limescale is made from calcium carbonate, so descalers contain acids that react with carbonate.

FIGURE 4: What do water-softening machines do?

FIGURE 5: Your grandparents used washing soda. How did it help them?

FIGURE 6: This is a descaler. What is it used for?

Questions

13 What is inside a water-softening machine?

14 A kettle full of hard water is boiled. What is seen inside the kettle?

15 What is limescale made from?

16 What removes limescale?

Getting into 'hot water'!

Temporary hardness is a big problem. The calcium hydrogencarbonate decomposes easily in hot water to form calcium carbonate, water and carbon dioxide.

Calcium carbonate does not dissolve – it is a solid deposit inside hot-water pipes. This is limescale, and eventually it clogs up the pipes totally.

Heating removes the calcium ions from the water, and so removes temporary hardness – but at the cost of the limescale produced.

Permanent hardness is not affected by heating – calcium sulfate is too stable.

Removing hardness using ion-exchange resins

Ion-exchange resins are a better way of removing hardness. The water flows over beads of solid **resin**, which trap the calcium and magnesium ions on its surface, taking them out of the water. Ion-exchange resins remove both temporary and permanent hardness.

Removing limescale

Strong acids, such as hydrochloric acid, not only remove limescale, they might also react with the metal of the tap. Descalers contain weak acids that are less likely to damage anything else.

acid + carbonate
↓
salt + carbon dioxide + water

The costs of having limescale in washing machines, hot water tanks and dishwashers is high, so it is better to remove the hardness first rather than have a build-up of limescale.

🔍 ion exchange column   washing soda uses   limescale

FIGURE 7: This hardness was removed by boiling.

Question

17 Write the word equation for the thermal decomposition of calcium hydrogencarbonate.

Softening water

All methods of softening water remove the ions that react with soap. This is usually done by turning the ions into a compound that does not dissolve.

Using thermal decomposition

Thermal decomposition softens temporary hardness only. It 'locks' the calcium ions up as insoluble calcium carbonate. The equation is the reverse of that for the formation of hard water.

$$Ca(HCO_3)_{2(aq)} \rightarrow CaCO_{3(s)} + CO_{2(g)} + H_2O_{(l)}$$

Using washing soda

Washing soda is sodium carbonate, Na_2CO_3, and it can soften both temporary and permanent hard water.

When washing soda dissolves it puts large numbers of carbonate ions into solution. Each time a carbonate ion collides with a calcium ion in the water, the two ions react to form insoluble calcium carbonate, 'locking up' calcium ions.

$$Ca^{2+}_{(aq)} + CO_3^{2-}_{(aq)} \rightarrow CaCO_{3(s)}$$
from hard water · · · · · · · · · from washing soda

Using ion-exchange resins

The resin in an ion-exchange column starts with sodium ions on its surface. As calcium ions flow past the resin they are attracted to it and stick to it, pushing sodium ions off into the water.

The calcium ions are now 'locked' onto the resin surface. The sodium ions in the water do not affect soap, so the water is now soft.

Once the resin surface is completely coated with calcium ions it can be discharged using concentrated salt solution. This knocks the calcium ions back off the resin and replaces them with fresh sodium ions.

Using limescale removers

The action of limescale remover is that of any acid on a carbonate. Hydrochloric acid may be aggressive on the metal of the tap, but it works well on the limescale.

$$2HCl + CaCO_3 \rightarrow CaCl_2 + CO_2 + H_2O$$

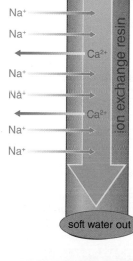

hard water in

Ca^{2+}
Na⁺
Na⁺
Ca^{2+}
Na⁺
Na⁺
Ca^{2+}
Na⁺
Na⁺

soft water out

FIGURE 8: How is an ion-exchange column discharged?

Questions

18 Suggest one disadvantage of washing soda as a softening agent.

19 Water containing calcium and sulfate ions flows into an ion-exchange column. What ions come out?

20 Suggest why weak acids are less aggressive than strong acids.

21 Explain why turning calcium ions into a solid softens a sample of water.

Natural fats and oils

You will find out:

> the difference between fats and oils

> what fats and oils are used for

> how to test for unsaturated fats and oils

Living off the fat of the land

Natural fats and oils come from animals and plants. They were being used long before crude oil was discovered.

FIGURE 1: **a** Cow's milk gives us butter.   **b** Palm trees give us palm oil.

What is the difference between fats and oils?

Fats and **oils** are all the same type of chemical, but:

> oils are liquid at room temperature

> fats are solid at room temperature.

Fats and oils are used for much more than cooking

Natural fats and oils are important raw materials in the chemical industry. They are used to make **margarine** and soap. They can also be turned into alternative fuels, such as **biodiesel**, that could become very useful when crude oil runs out.

FIGURE 2: Bacon gives us lard and olives give us olive oil. Which is a solid and which is a liquid at room temperature?

Did you know?

Margarine was invented in 1869 as a cheap replacement for butter. There wasn't enough real butter to go round.

Did you know?

Look carefully at the list of ingredients on a margarine pack.

If it says partially **hydrogenated** or partially polyunsaturated then some of the double bonds are still there.

Questions

1 What is the difference between fats and oils?

2 Name three substances that can be made from vegetable oil.

3 Why will biodiesel become more and more important?

4 Which industry uses vegetable oils as a raw material?

🔍 fats and oils   margarine   biodiesel   esters   hydrogenated

What are fats and oils?

Fats and oils are all part of a group of compounds called **esters**. They all have chains of carbon atoms.

Testing for unsaturation

Bromine water is orange. If you shake it with an unsaturated compound it loses its colour – it is decolourised. When shaken with a saturated compound, bromine remains orange.

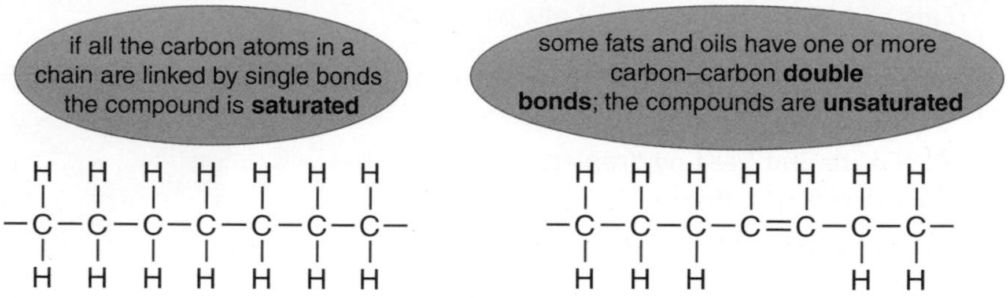

if all the carbon atoms in a chain are linked by single bonds the compound is **saturated**

some fats and oils have one or more carbon–carbon **double bonds**; the compounds are **unsaturated**

FIGURE 3: How can you tell if a hydrocarbon compound is saturated or unsaturated?

One industrial use of vegetable oils is to make margarine. Vegetable oils are **unsaturated**, which makes most of them runny. The first stage is to 'harden' them – to turn them into **saturated** compounds. Hydrogen is bubbled through the oil at about 200 °C using a **nickel catalyst**. The hydrogen reacts with the carbon–carbon **double bonds** and turns them into **single bonds**. If only some of the carbon–carbon double bonds are allowed to react, the oil only partially solidifies. Partially solidified oils are used to make soft margarines that are easy to spread. If all of the double bonds react, the margarine is relatively hard. Runny vegetable oils, like olive oil, are thought to be better for your health.

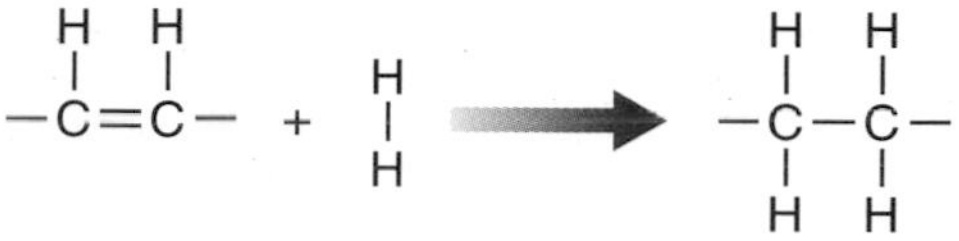

FIGURE 4: Hydrogen turns carbon–carbon double bonds into single bonds.

Questions

5 What group of chemical compounds do vegetable oils belong to?

6 What makes a molecule unsaturated?

7 What is the test for an unsaturated compound?

8 How can an unsaturated compound be turned into a saturated compound?

FIGURE 5: Some of the bromine water from this bottle has just been shaken with the oil in the gas jar. Is the oil sample unsaturated or saturated?

BROMINE WATER

Fats, oils and health

Saturated fats and oils usually come from animals and unsaturated fats and oils come from plants. 'Polyunsaturated' means the compound contains more than one carbon–carbon double bond. People whose diet has a higher percentage of unsaturated fats and oils have lower levels of the type of **cholesterol** that causes **heart disease**.

Why does the bromine test work?

Bromine water is a solution of bromine molecules in water. It is the bromine molecules that make the liquid orange. Bromine reacts, in an **addition reaction**, with the carbon–carbon double bonds in the carbon chain. The reaction uses up the bromine molecules, so the colour disappears. The colourless compound is a **dibromo** compound.

Saturated compounds cannot react with bromine since they have no carbon–carbon double bond.

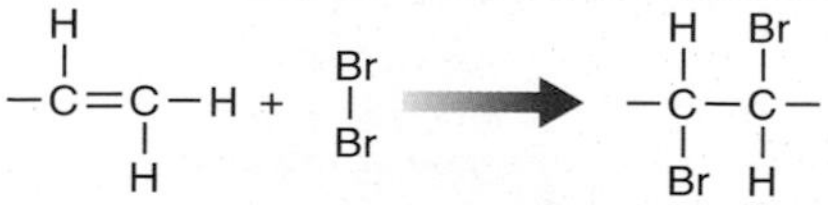

FIGURE 6: What compound is made when bromine water reacts with a double carbon–carbon bond in an unsaturated molecule?

Questions

9 a Describe what happens as increasing amounts of bromine water are added to a small amount of unsaturated oil.

b Explain why this happens.

10 Suggest what 'polyunsaturated' means.

11 Some vegetable oils react with more hydrogen than do others. What does this tell you?

🔍 nickel catalyst   double covalent bonds   single covalent bonds

You will find out:
> what an emulsion is made from
> how soap is made

 ## Mixing oil and water

Fats and oils do not normally mix with water. However, if some oil is shaken in water, tiny droplets of oil spread throughout the water. The oil drops are **dispersed** through the water.

A mixture that has tiny droplets of one liquid dispersed inside another is called an **emulsion**.

Getting help with mixing

Fats and oils mix with water much more easily if soap is added – and soap is made from fats and oils!

Soap is made by reacting vegetable oils with sodium hydroxide.

FIGURE 7: Milk is made from tiny droplets of oil dispersed through water. It is an **oil-in-water emulsion**. Butter is made from tiny droplets of water dispersed through oil or fat. It is a **water-in-oil emulsion**.

Questions

12 Guess which well-known soap brand is made from a mixture of palm oil and olive oil.

13 If you cut a pack of butter when it is cold there is water on the butter where the knife went through. Where has the water come from?

14 What kind of emulsion is milk?

15 What is soap made from?

 ## Getting fats into water

Oil and water are **immiscible** liquids. They usually do not mix. They do not dissolve in each other, but it is possible to disperse tiny droplets of one liquid inside the other. A **suspension** of one substance inside another is a **colloid**. If the suspension is of droplets of one liquid in another liquid, it is a special type of colloid called an **emulsion**.

Milk is a good example of an oil-in-water emulsion. Cold cream and margarine are water-in-oil emulsions.

 water-in-oil emulsion contains droplets of water spread through oil

 oil-in-water emulsion contains droplets of oil spread through water

FIGURE 8: Water and oil can make two types of emulsion. What is an example of each?

Making soap

Fats and oils are difficult to wash from clothes because they do not dissolve in water. However, fats and oils are the starting point for making soap.

fat + sodium hydroxide → soap + glycerol

In the manufacture of soap:

> vegetable oils are heated in large vats with sodium hydroxide solution

> salt is added at the end of the reaction to make the soap precipitate out

> the solid soap is then removed

> colouring and perfume may then be added.

The reaction between an oil and an alkali is sometimes called **saponification**.

The Latin word for soap is 'sapo'.

How was soap discovered?

One theory is that ancient tribes daubed themselves with ash from their cooking fires before battle. They realised that the ash made them cleaner. Fat which dripped from cooking meat had reacted with alkalis in the ash to make soap.

Questions

16 What is the process of saponification?

17 Suggest what happens when water is added to an oil-in-water emulsion.

18 Suggest what happens when a water-soluble dye is added to an oil-in-water emulsion.

19 Explain why an emulsion is a special type of colloid.

Saponification

The esters that make up fats and oils have the structure shown in Figure 9.

When an ester reacts with sodium hydroxide, the ester forms one **glycerol** and three soap molecules, as in Figure 10.

This reaction is really the reaction of alkaline water; it is a **hydrolysis** reaction. This hydrolysis reaction is a saponification reaction.

The reaction is:

fat + sodium hydroxide → soap + glycerol

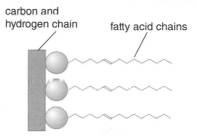

FIGURE 9: Structure of esters in fats and oils.

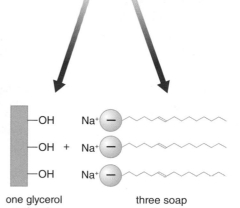

FIGURE 10: Reaction of an ester with sodium hydroxide.

Questions

20 What type of reaction is the reaction between oils and sodium hydroxide?

21 Suggest what happens when oil is added to an oil-in-water emulsion.

22 Suggest what happens when an oil-soluble dye is added to an oil-in-water emulsion and stirred vigorously.

23 The pH changes during a saponification reaction. What change in pH would you expect if all the reagents were totally used up?

Detergents

You will find out:
> what is in washing powders
> why clothes are washed at low temperatures
> how soap gets dirt off clothes

How clean can we go?

Manufacturers tell us that their washing powders wash 'whiter than white'. This is not just hype – they really do. There is a huge amount of science inside a box of washing powder.

 ## What is in a box of washing powder?

There is more in washing powder than just soap. Figure 1 shows some things that might be in it.

The **active detergent** does the cleaning.

There are different types of **detergent**; soap is just one of them. Many detergents are also **salts**.

Did you know?

An average load of washing has 40 g of dirt. That is about three large spoonfuls. Very dirty washing has a lot more!

Did you know?

Bleach doesn't remove stains; it just turns them colourless so that they can't be seen.

 ### Questions

1 List five things in a soap powder.

2 Washing powders that have enzymes in them must only be used at low temperatures. Find out why.

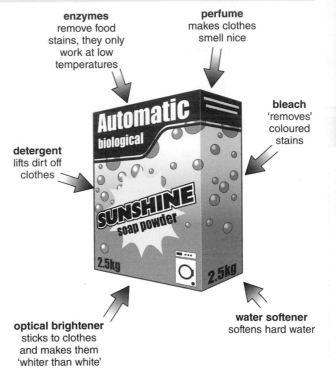

enzymes remove food stains, they only work at low temperatures

perfume makes clothes smell nice

bleach 'removes' coloured stains

detergent lifts dirt off clothes

water softener softens hard water

optical brightener sticks to clothes and makes them 'whiter than white'

FIGURE 1: Some of the things in washing powder. What do enzymes do?

 ## What are detergents?

Detergents can be made by **neutralising** some organic **acids** with alkali.

acid + alkali → salt + water

The salt is the detergent. It is suitable for cleaning uses because:

> it dissolves grease stains

> it dissolves in water.

Once the detergent has dissolved the grease, it lifts the grease stain off into the water.

Why can a detergent molecule lift off the grease?

A detergent molecule has two parts. It has a **hydrophilic** head and a **hydrophobic** tail.

Why are clothes washed at low temperatures?

It is good for the environment to wash clothes at 40 °C instead of at high temperatures. Why is this?

Washing machines have to heat up a lot of water.

This needs energy, so the lower the temperature of the water the less energy is used and the less of the **greenhouse gases** are put into the atmosphere.

Washing clothes at low temperatures is also good for coloured clothes – many dyes are easily damaged by high temperatures.

Did you know?

'Hydrophilic' means water loving. 'Hydrophobic' means water fearing.
The words come from the Greek:
- philia = love
- phobia = fear
- hydro = water.

Questions

3 Explain why lower wash temperatures result in lower emissions of greenhouse gases.

4 Which end of the detergent molecule dissolves in grease?

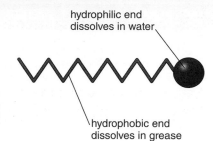

hydrophilic end dissolves in water

hydrophobic end dissolves in grease

FIGURE 2: A detergent molecule. How do the two ends react with water?

How does a detergent work?

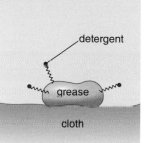

detergent

grease

cloth

1 The hydrophobic (water hating) ends of detergent molecules forms strong intermolecular forces with the molecules of grease. This leaves the hydrophilic (water liking) ends on the outside.

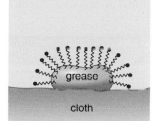

grease

cloth

2 Eventually, so many detergent molecules have formed strong intermolecular forces with the grease that the outside of the grease is covered in hydrophilic ends.

grease

cloth

3 The grease is now surrounded by hydrophilic ends that form strong intermolecular forces with the water molecules, and so it lifts off the cloth into the water.

Did you know?

When a washing powder contains an enzyme it works well at low temperatures. Enzymes are denatured at high temperatures.

FIGURE 3: How a detergent works.

A detergent molecule is made of two parts: one part dissolves in oil and grease (the **hydrophobic** part) and the other part dissolves in water (the **hydrophilic** part). The hydrophobic end of the molecule forms strong intermolecular forces with the molecules of oil or fat, as in the first picture of Figure 3.

The hydrophilic end of the detergent molecule forms strong **intermolecular** forces with water molecules, as in the third picture of Figure 3.

'When clothes are washed, oil or grease will not dissolve in water. Detergents form strong intermolecular forces with the hydrophobic end of the molecule. The water makes strong intermolecular forces with the water, so the grease spot surrounded by detergent is 'lifted' off the clothes into the water.'

Questions

5 Give two reasons why clothes should be rinsed after they have been washed.

6 Explain how the intermolecular forces between detergent molecules and water or grease help to clean clothes.

🔍 low-temperature washes   greenhouse gases

Solvents and solutes

Fresh coffee stains dissolve in water but Biro marks do not. However, Biro marks will often dissolve in methylated spirit. Methylated spirit and water are both **solvents** – they dissolve other, different, substances. A substance that dissolves in a solvent is a **solute**. Biro ink dissolved in a solvent makes a **solution**. Different solvents dissolve different substances. If a substance dissolves, it is **soluble**. If it does not dissolve, it is **insoluble**.

What is in a bottle of washing-up liquid?

The **active detergent** in washing-up liquid does the cleaning. Water is often used as the **thinning agent** so that it can be **dispensed** more easily.

Comparison of washing-up liquids

Comparing washing-up liquids can be done by seeing how many plates can be washed by each. To compare the liquids you would have to make fair comparisons. What would you need to keep the same in the experiment?

An example of the results you might get is given in the table.

Washing-up liquid	A	B	C	D
Number of plates washed	23	32	12	18

You may say that washing-up liquid C was the least-effective liquid. Would your results support this conclusion? What evidence do you have about the most-effective liquid?

perfume and colouring make the cleaner attractive to use

thinning agent makes the detergent runny and easier to squeeze out of the bottle

detergent lifts dirt off crockery

rinse agent helps water run off crockery

FIGURE 4: Some of the things in washing-up liquid. What does rinse agent do?

Questions

7 David makes a cup of tea and adds some sugar to it.

a What is the solute that he has used?

b What is the solvent?

8 Write down the names of three different solvents.

9 Name one substance that is in washing-up liquid but not in soap powder.

Dry cleaning

Some fabrics will be damaged if they are washed in water – they must be **dry cleaned**. A dry cleaning machine washes clothes in an **organic solvent**. The 'dry' does not mean that no liquids are used, just that the liquid is not water.

Most of the stains on clothing contain grease from the skin or from food.

Grease-based stains do not dissolve in water, but they do dissolve easily in a dry cleaning solvent.

Questions

10 What does the 'dry' in dry cleaning mean?

11 Dry cleaning removes grease stains better than water. Explain why.

🔍 dry cleaning   organic solvent   solute   insoluble

How does dry cleaning work?

Molecules stick to each other. They are held together by weak forces of attraction. Forces between molecules are called **intermolecular forces**. They are different from the chemical bonds that hold the atoms inside a molecule together.

Molecules of grease are held together by **weak intermolecular forces**. The same types of forces hold molecules of dry cleaning solvent together (weak intermolecular forces).

The dry cleaning solvent molecules link with molecules of the grease by intermolecular forces. The solvent molecules can then surround the molecules of grease, so the grease then dissolves in the solvent.

Molecules of water are held together by stronger intermolecular forces, called hydrogen bonds.

The water molecules cannot stick to the grease because they are sticking to each other much too strongly.

Look at the percentage of stains which were cleaned by different cleaning agents. Which stain was muddy grass stains, which was grease and which was blood?

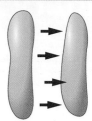

grease sticks
to grease

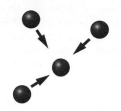

molecules of dry cleaning
solvent stick to other molecules
of dry cleaning solvent

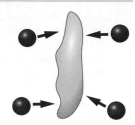

molecules of dry cleaning
solvent stick to grease

FIGURE 5: Dry cleaning solvent molecules stick to grease as happily as they stick to each other. How does this make the solvent a good cleaning agent?

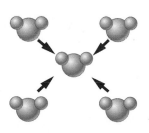

water sticks strongly
to water

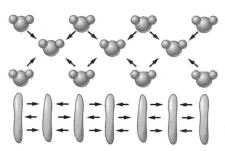

water sticks to water too
strongly to stick to grease

FIGURE 6: Water molecules are held tightly together and do not stick to grease molecules.

Powder sample	A	B	C
Percentage cleaned in dry cleaning fluid	100%	10%	15%
Percentage cleaned in water at 35 °C	65%	100%	70%
Percentage cleaned in water at 60 °C	5%	15%	98%

Look at the percentage of stains which were cleaned by different powders at different wash temperatures.

Which powders work best in a hot wash? Can you explain how you know which powder sample contains enzymes?

Powder sample	A	B	C	D
Percentage cleaned in water at 15 °C	10%	10%	10%	10%
Percentage cleaned in water at 35 °C	65%	100%	98%	45%
Percentage cleaned in water at 60 °C	100%	15%	20%	82%

Question

12 When a washing machine washes clothes in water it spins the clothes to remove most of the water. The clothes are then hung out to dry to let the remaining water evaporate. Dry cleaning uses a more complicated spin cycle because the clothes are not hung out.

Suggest why clothes are not hung out to dry in dry cleaning.

hydrophobic   hydrophilic

Preparing for assessment: Research and collecting secondary data

To achieve a good grade in science, you not only have to know and understand scientific ideas, but you need to be able to apply them to other situations and investigations. These tasks will support you in developing these skills.

✳ Tasks

> Recommend two ways that water from different areas can be used more effectively for washing.

✳ Context

Many areas of the country have hard water. There are some disadvantages to having water that is too hard. Limescale can result which builds up inside of water cylinders and washing machines.

This limescale is a result of one type of hardness. The limescale can be removed by descaler but it can also be prevented using a water softener.

The hardness of water can be measured by experiment both before and after using a water softener.

✳ General rules

1 You may work with other students but your written work should be done on your own.

2 You cannot get detailed help from your teacher.

3 You are not allowed to redraft your work.

4 Your work can be handwritten or word processed.

5 It is expected that you complete this task in two hours.

6 You are allowed to do this research outside the laboratory.

What do you need to research?

✺ Research and collecting secondary data

To research:

- what type of hardness causes this damage in hot cylinder or hot washing machines
- where in the country this type of hardness causes a problem
- how water softeners work
- how you could use experiments to investigate how hard water is in different areas and
- how much difference using a water softener makes.

Where would you find this information?

Information sources:

- Science textbooks in the school library
- The internet. The search term 'hard water areas' will give you sites that are government information sites and sales sites. Which will you choose?

  'Water softeners' search gives lots of sales sites only, so try 'ion exchange resin softeners'.

What do you find out about?

Research found:

Maps showing areas of different degrees of hardness. Broadly, the further south and east in the UK the harder the water.

'Ion-exchange resin softeners' gives sites with information on how ion-exchange systems work. The best sites are those showing images of the movement of ions.

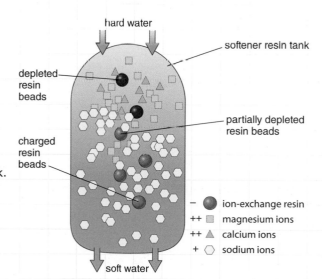

How do you record all this information?

Record:

- Use a notebook to write down the information and where it came from.
- Use a computer to record images, conversations and diagrams.

Put together or print off a final account of your research and hand it to your teacher.

C6 Checklist

To achieve your forecast grade in the exam you'll need to revise

Use this checklist to see what you can do now. It gives you many of the important points you will need to know. Refer back to the relevant pages in this book if you're not sure and to see if there is anything else you need to know. Look across the three columns to see how you can progress.

Remember you'll need to be able to use these ideas in various ways, such as:

> interpreting pictures, diagrams and graphs
> applying ideas to new situations
> explaining ethical implications

> suggesting some benefits and risks to society
> drawing conclusions from evidence you've been given.

Look at pages 272–294 for more information about exams and how you'll be assessed.

To aim for a grade E

know that electrolysis is liquid decomposition by passing electric current through it

know the anode is the positive electrode, cathode is negative

recognise anions and cations from their formulae

know that the test for hydrogen is that it 'pops' when lit with a splint

know the cathode gets plated with copper in $CuSO_4$ electrolysis

---

know why fuel cells use exothermic reactions

know the reaction of hydrogen and oxygen is exothermic

know a fuel cell releases energy to produce electrical energy

know one use of fuel cells is for the electrical power in spacecraft

explain why a hydrogen-oxygen fuel cell has no polluting product

---

describe oxidation as the addition of oxygen

describe reduction as the removal of oxygen

know that the rusting of iron needs both water and oxygen

list methods of preventing rust and know how oil prevents it

know the order of reactivity of metals and interpret observations

To aim for a grade C

know electrolysis is the flow of charge by ions to electrodes

know hydrogen and oxygen are made in electrolysis of H_2SO_4

know that H_2 is made at the cathode and O_2 at the anode

know how the amount of product varies with time and current

know that molten liquids can be electrolysed as the ions can move

---

construct the balanced symbol equation between H_2 and O_2

construct the balanced symbol equation for reaction in a fuel cell

know the advantages of using fuel cells in spacecraft

explain why the car industry is developing fuel cells

---

know that redox reactions involve oxidation and reduction

know rusting is a redox reaction and construct the word equation

explain that galvanising iron with zinc acts as a barrier to rusting

explain that when galvanising zinc acts as a sacrificial metal

construct word equations for displacement reactions

To aim for a grade A

construct half equations for cathode reactions of NaOH

construct half equations for anode reactions of H_2SO_4

construct the cathode half equation $Cu^{2+} + 2e^- \rightarrow Cu$

calculate amounts made in electrolysis by time and current

construct half equations for electrode reactions of electrolytes

---

draw and interpret energy-level diagrams for H_2 and O_2 reaction

construct equations for electrode reactions of fuel cells, given ions

explain the change at electrodes in terms of redox reactions

explain the advantages of fuel cells in electricity generation

explain why the use of fuel cells will still produce pollution

---

know that oxidation involves the loss of electrons

know that reduction involves the gain of electrons

explain why rusting is a redox reaction with electron loss/gain

explain how sacrificial protection stops iron rusting by electron loss

construct balanced symbol equations for displacement reactions

To aim for a grade E

explain why alcohols are not hydrocarbons

know the conditions needed for fermentation

know fermentation needs yeast and a temperature of 25–50 °C

know the main uses of ethanol – beverages, solvents and fuel

know that the hydration of ethene produces ethanol

---

know that a chlorofluorocarbon (CFC) has Cl, F and C atoms

know that ozone has three oxygen atoms in the formula O_3

know that CFCs are chemically inert and have low boiling points

know that increased levels of UV light can cause skin cancer

know that hydrocarbons can give safer alternatives to CFCs

---

know that hard water does not lather well with soap, soft does

know that hard and soft water both lather well with detergents

know water hardness is caused by calcium and magnesium ions

know how hardness in water can be removed by ion exchange

interpret data on water hardness and removal by washing soda

---

know that natural fats and oil are important industrial raw materials

know that vegetable oils can be used to make biodiesel

know that at room temperature oils are liquids and fats are solids

describe an emulsion and that milk is an oil-in-water emulsion

know that vegetable oils reacting with sodium hydroxide give soap

---

relate each ingredient in a washing powder to its function

understand the words solvent, solute, solution and insoluble

know that the active detergent does the cleaning

know the rinse agent helps the water drain off the crockery

interpret data from experiments on washing powders and liquids

To aim for a grade C

know the molecular formula and displayed formula of ethanol

know the word and balanced symbol equations for fermentation

explain why ethanol made by fermentation is a renewable fuel

explain why ethanol made from ethene is a non-renewable fuel

know that hydration of ethene needs a phosphoric acid catalyst

---

explain why the use of CFCs has been banned in the UK

describe how CFCs deplete the ozone layer by breaking in UV

know that highly reactive chlorine atoms react with ozone

know how depletion of the ozone layer lets more UV reach Earth

know that CFCs can be replaced with alkanes or HFCs

---

describe the origin of temporary hardness in water from rocks

know that temporary hardness is from calcium hydrogencarbonate

know that permanent hardness is caused by calcium sulfate

describe how boiling removes temporary hardness to limescale

plan experiments to compare hardness in different water

---

know that animal and vegetable fats and oils are esters

describe how bromine water is used to show unsaturation in oils

know how margarine is manufactured from vegetable oils

know how immiscible liquids, like oil and water, form emulsions

know that the splitting up of fats using hydroxides is saponification

---

explain the advantages of using low-temperature washing

describe detergent molecules as hydrophilic and hydrophobic ends

describe dry cleaning as using solvents that are not water

know that dry cleaning solvents act on stains insoluble in water

make simple conclusions from data on washing powders

To aim for a grade A

know the general formula of an alcohol is $C_nH_{2n+1}OH$

draw the displayed formula of alcohols with up to five C atoms

explain the effect of low or high temperature in fermentation

evaluate the two methods of producing ethanol

evaluate continuous and batch processes and atom economy

---

explain how the attitude of scientists to CFCs has changed

explain how a carbon–chlorine bond breaks

explain why small numbers of chlorine atoms destroy much O_3

explain why CFCs will continue to deplete ozone for a long time

explain how ozone absorbs UV light in the stratosphere

---

know the formula for calcium hydrogencarbonate is $Ca(HCO_3)_2$

know that $Ca(HCO_3)_2$ breaks to form $CaCO_3$, H_2O and CO_2

know that washing soda is sodium carbonate

explain how sodium carbonate can soften hard water

---

explain why unsaturated fats are healthier in a balanced diet

explain why bromine tests for unsaturation in fats and oils

explain it is an addition reaction at carbon–carbon double bonds

explain that a colourless dibromo compound forms

explain that fat + sodium hydroxide makes soap + glycerol

---

explain how a hydrophilic end has intermolecular forces with water

know a hydrophobic end has intermolecular forces with oil

explain dry cleaning solvents using intermolecular forces

know intermolecular forces of solvents form with grease

interpret experimental data to see if detergents have enzymes

Foundation Tier

1 Surnathi has used an iron padlock to secure the gate to her garden. She stops it rusting by placing a few drops of oil on to the padlock.

AO1 **(a)** Write down why the oil stops the iron from rusting. [2]

AO1 **(b)** Iron objects can also be protected by placing a more reactive metal such as magnesium in contact with the iron object. Explain how this prevents the iron object from rusting. [1]

[Total: 3]

2 Matt has carried out a series of reactions between metals and solutions of metal salts. The table below shows a summary of his results. A tick (✓) shows that when a metal was put in the metal salt solution, a reaction took place. A cross (✗) shows that when a metal was put in the metal salt solution, no reaction took place.

metal salt solution/ metal	magnesium Mg	copper Cu	iron Fe
magnesium sulfate $MgSO_4$		✗	✗
copper sulfate $CuSO_4$	✓		✓
iron sulfate $FeSO_4$	✓	✗	

AO1 **(a)** From the table which metal is most reactive? [1]

AO2 **(b)** Write a word equation for the reaction between magnesium and copper sulfate solution. [2]

AO2 **(c)** If a new metal was put into these solutions in turn and reacted with all three solutions is it likely to be nickel (a transition metal) or lithium (a group 1 metal)? [1]

[Total: 4]

3 Ethanol can be used in alcoholic drinks.

AO1 **(a)** Write down **two other** uses of ethanol. [2]

AO1 **(b)** Ethanol can be produced by fermentation using a yeast catalyst.

Copy and complete this word equation for fermentation.

glucose → ethanol + ... [1]

[Total: 3]

4 Fats and oils are important raw materials.

AO1 **(a)** Write down the state of fats and oils at room temperature. [2]

AO1 **(b)** Animal fats are often saturated. Vegetable oils are often unsaturated. Describe, in terms of bonding, the difference between a saturated fat and an unsaturated fat. [2]

AO2 **(c)** Sam has 3 test tubes A, B and C. He has a different fat or oil in each tube. He adds bromine water to each. Test tube A looks orange/brown. Test tube B looks colourless. Test tube C looks orange/brown. Which test tube has a saturated fat and which has a unsaturated oil? [2]

[Total: 6]

5 (a) Write down the names of the metal ions which dissolve in water to make it hard. [2]

AO1 **(b)** Write down two ways that hard water can be softened. [2]

AO3 **(c)** Jenny tests four water samples with soap. She uses 50 cm³ of water each time. She wants to know which water sample is distilled water and which water samples contain some kind of hardness.

water sample	A	B	C	D
amount of soap used in water before boiling in cm³	1	10	10	10
a mount of soap used in water after boiling in cm³	1	1	5	10

What do you think Jenny finds out? Justify your answer with the evidence. [6]

The quality of written communication ✎ will be assessed in your answer to this question. [Total: 10]

AO1 recall the science   AO2 apply your knowledge   AO3 evaluate and analyse the evidence

✳ Worked Example – Foundation Tier

Leo and Jia are experimenting with electrolysis.
They start with a solution of sulfuric acid.

(a) Which two gases will they see bubbling at the electrodes?

[2]

cathode	anode
oxygen	hydrogen

(b) How will they test for one of these gases? [1]

Hydrogen 'pops' with a lighted splint

Next they want to electrolyse copper sulfate.

(c) Jia says that they cannot use solid copper sulfate to electrolyse.

Explain why. [1]

The ions in a solid are not free to move.

They choose to use a solution. They pass a current through the solution of copper sulfate using carbon electrodes.

(d) What will happen at the cathode? [1]

It will get plated with pinkish copper.

(e) What will they see at the anode? [1]

It will dissolve.

(f) What will they see happening to the copper sulfate solution? [1]

It will stay blue.

(g) What charge does the cathode have? [1]

negative

(h) What kind of charge do cations have? [1]

positive

(i) Leo and Jia now try to see if they produce more of the substance on the cathode. Leo suggests that they should leave the electrolysis working for a longer time. Jia suggests that they should increase the current they use.

Who is correct and why? [3]

Leo is correct as the longer they leave it the more they will get.

How to raise your grade!

Take note of these comments –
they will help you to raise your grade.

↓

The gases are correct but at the wrong electrodes. **0/2**

This is correct. **1/1**

This is correct. **1/1**

This is correct. **1/1**

The student is confusing this process with the use of impure copper in the purification process. **0/1**

The colour disappears. **0/1**

This is correct. **1/1**

This is correct. **1/1**

Time will increase product, but both Leo and Jia are correct as current increase will also produce more product. **1/3**

This student has scored 6 marks out of a possible 12. This is below the standard of Grade C. With more care the student could have achieved a Grade C.

Higher Tier

1 Surnathi has used an iron padlock to secure the gate to her garden. She stops it rusting by putting drops of oil on to the padlock.

AO1 **(a)** Write down the word equation for the rusting of iron. [2]

AO1 **(b)** She could also protect the padlock by placing a more reactive metal such as magnesium in contact with the iron object. Explain how this prevents the iron object from rusting. [1]

[Total: 3]

2 Matt has carried out a series of reactions between metals and solutions of metal salts. The table below shows a summary of his results. A tick (✓) shows that when a metal was put in the metal salt solution, a reaction took place. A cross (✗) shows that when a metal was put in the metal salt solution, no reaction took place.

metal salt solution/ metal	magnesium Mg	copper Cu	iron Fe
magnesium sulfate $MgSO_4$		✓	✗
copper sulfate $CuSO_4$	✓		✓
iron sulfate $FeSO_4$	✓	✗	

AO2 **(a)** Write a word equation for the reaction between magnesium and copper sulfate solution. [2]

AO2 **(b)** Silver is less reactive than copper. If copper was put into a solution of silver nitrate write the word equation of what would happen. [1]

[Total: 3]

3 Ethanol can be used in alcoholic drinks.

AO1 **(a)** What is the molecular formula of ethanol? [1]

Ethanol can be produced by fermentation using a yeast catalyst.

AO1 **(b)** Construct the balanced symbol equation for fermentation. [2]

AO1 **(c)** There is an optimum temperature for fermentation to occur. If the temperature for fermentation is too low the rate of reaction slows down. Explain why the reaction slows down at low temperatures. [1]

AO1 **(d)** If the temperature is too high the reaction does not work as well. Explain why. [1]

[Total: 5]

4 Fats and oils are important raw materials.

AO1 **(a)** Animal fats are often saturated. Vegetable oils are often unsaturated. Describe, in terms of bonding, the difference between a saturated fat and an unsaturated fat. [2]

AO1 **(b)** Sam has 3 test tubes A, B and C. He has a different fat or oil in each tube. He adds bromine water to each. Test tube A looks orange/brown. Test tube B looks colourless. Test tube C looks orange/brown. Which test tube has a saturated fat and which has a unsaturated oil? [2]

[Total: 4]

5 You have been given two samples of water and some soap. One sample is of hard water, the other is of soft water.

AO1 **(a)** Describe how you could tell them apart. [1]

AO1 **(b)** Write down the name of a chemical that causes permanent hardness in water. [1]

AO1 **(c)** Explain how an ion-exchange resin can soften water. [2]

[Total: 4]

AO3 **6** Jenny tests four water samples with soap. She uses 50 cm^3 of water each time. She wants to know which water sample is distilled water and which water samples contain some kind of hardness.

water sample	A	B	C	D
amount of soap used in water before boiling in cm^3	1	10	10	10
amount of soap used in water after boiling In cm^3	1	1	5	10

What do you think Jenny finds out? Justify your answer with the evidence. [6]

The quality of written communication ✏ will be assessed in your answer to this question.

AO1 recall the science   AO2 apply your knowledge   AO3 evaluate and analyse the evidence

✳ Worked Example – Higher Tier

Leo and Jia are experimenting with electrolysis.
They start with a solution of sulfuric acid.

(a) Which gas will they see bubbling at the anode? [1]

Oxygen

(b) Construct the half equation for the process of hydrogen released at the cathode? [2]

$H^+ + e^- \rightarrow H_2$

Next they want to electrolyse copper sulfate.

(c) Jia says that they cannot use **solid** copper sulfate to electrolyse. Explain why. [1]

The ions in a solid are not free to move.

They choose to use a solution. They pass a current through the solution of copper sulfate using carbon electrodes.

(d) What will happen at the cathode? [1]

It will get plated with pinkish copper.

(e) What will happen at the anode? [1]

It will dissolve.

(f) Leo and Jia now try to see if they produce more of the substance on the cathode.
Leo suggests that they should leave the electrolysis working for a longer time.
Jia suggests that they should increase the current they use.
Who is correct and why? [3]

Leo is correct as the longer they leave it the more they will get.

(g) They find that after 30 minutes, using a current of 0.5 amps the mass of the cathode increases by 0.25 g.
If they needed to increase the mass by 1 g using the same current, how long would it take? [2]

To produce 1g would need four times the length of time. It would take 2 hours.

How to raise your grade!

Take note of these comments – they will help you to raise your grade.

⬇

This is correct. 1/1

This is not balanced. 1/2

This is correct. 1/1

This is correct. 1/1

The student is confusing this process with the use of impure copper in the purification process. 0/1

Time will increase product, but both Leo and Jia are correct as current increase will also produce more product. 1/3

This is correct. 2/2

This student has scored 7 marks out of a possible 11. This is below the standard of Grade A. With more care the student could have achieved a Grade A.

P5 Space for reflection

Ideas you've met before

Motion

Average speed = $\dfrac{\text{distance travelled}}{\text{time taken}}$; acceleration = $\dfrac{\text{change in velocity}}{\text{time taken}}$

Speed is measured in m/s and acceleration in m/s^2.

The velocity of an object is its speed combined with its direction.

Objects accelerate as they fall vertically due to gravity.

● The speed of a car increases from 10 m/s to 20 m/s in 2.5 s. What is its acceleration?

Momentum and force

Momentum = mass × velocity. The unit of momentum is kg m/s.

Force = $\dfrac{\text{change in momentum}}{\text{time}}$ and is measured in newtons (N).

In a vehicle accident, the force on occupants can be reduced by increasing the stopping or collision time. Crumple zones, airbags and seatbelts all do this.

 Why is a high-speed crash likely to cause more serious injuries than one at a lower speed?

Optical fibres

Laser light is used to transmit digital signals along an optical fibre.

Lasers produce an intense coherent beam of monochromatic light (of a single colour).

Light is totally internally reflected from the sides of the very thin glass fibre as the angle of incidence is greater than the critical angle.

Infrared radiation can also be transmitted along an optical fibre.

 What is the critical angle?

Wireless communication

All electromagnetic waves can be reflected, refracted and diffracted.

For long-distance communication:

> some radio waves are refracted and subsequently reflected by the ionosphere in a similar way to total internal reflection for light

> microwaves pass through the ionosphere and are then received and re-transmitted by satellites.

DAB provides more stations and there is less interference with other broadcasts.

 How can radio waves be received out of line of sight?

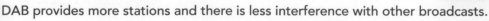

In P5 you will find out about...

> natural and artificial satellites

> geostationary artificial satellites and their uses

> circular motion and centripetal force

> relative speed of two cars travelling on a straight road

> scalar and vector quantities

> using equations for uniformly accelerated motion

> projectile motion

> calculating the resultant velocity of a projectile

> action and reaction (Newton's third law)

> conservation of momentum

> using a particle model to explain pressure

> using simple kinetic theory to explain rocket propulsion

> the use of radio waves with different frequencies for different forms of communication

> satellite transmitting and receiving dishes

> constructive and destructive interference of waves

> diffraction of light for both a single slit and double slits

> polarisation of light

> wave and particle theories of light

> refraction of light and refractive index

> conditions for total internal reflection and its uses

> explaining dispersion in terms of spectral colours having different speeds in glass

> a convex lens and its use as a magnifying glass

> the magnification produced by a convex lens

Satellites, gravity and circular motion

You will find out:
> about artificial and natural satellites
> about gravity as a force of attraction between all masses
> about the varying speeds of comets
> how the orbit period of a planet depends on its distance from the Sun

Communicating

Satellites have played a major part in the development of world communications. Look at the photograph of a communications satellite orbiting Earth. The satellite has two long solar-cell arrays (blue) to produce power, and a number of round-dish antennae to receive and transmit communication signals. Communications satellites usually have a high geostationary orbit so that they maintain the same position above the Earth's surface. Satellites like this relay telephone and television signals around the planet.

FIGURE 1: Why have satellites revolutionised communications?

Satellites

A **satellite** is an object that orbits a larger object in space. Satellites stay in **orbit** because of the **gravitational attraction** between the satellite and the object it orbits.

We call this **gravity**.

Many planets, such as Earth and Jupiter, have natural satellites. For example, the Moon is a natural satellite of Earth.

During the last 50 years, many artificial satellites have been placed in orbit around Earth. Those in a low orbit take less time for one orbit than satellites in high orbits. The height of an artificial satellite above Earth affects its orbit, which affects what it is used for.

Did you know?

The world's first artificial satellite, Sputnik I, was launched by the Soviet Union in 1957 and took about 98 minutes to orbit Earth.

Questions

1 In what direction is the force on the Moon that keeps it in orbit around Earth?

2 What is the difference between a natural and an artificial satellite?

Gravity

Every object in the Universe attracts every other object. There is a force of gravitational attraction between you and all the objects around you, but (except for the pull of Earth) this force is so small you are not aware of it. The attractive force only becomes significant on an astronomical scale. The planets, including Earth, stay in orbit around the Sun due to the gravitational attraction between them.

Any object moving in a circle needs a force towards the centre of the circle to maintain its circular path. This is called a **centripetal force**. The pull of gravity, g, is 10 N/kg on Earth but is different elsewhere in the Universe. On Jupiter, the largest planet, it is 25 N/kg.

Lee has a mass, m, of 70 kg. His weight, W, is given by $W = mg$. On Earth his weight is $70 \times 10 = 700$ N. On Jupiter he would weigh $70 \times 25 = 1750$ N.

FIGURE 2: Why does the Moon remain in orbit around Earth?

🔍 artificial satellites   geostationary satellites

Geostationary artificial satellites

A **geostationary** satellite is placed in an orbit above the equator and takes exactly 24 hours to orbit Earth. This is the same time Earth takes to rotate once on its axis, so the satellite stays over the same point on Earth. It is ideal for **communications** as it provides a constant link.

The time for one orbit is called the **orbital period**. It increases with height above Earth's surface. All geostationary satellites orbit Earth at a height of 36 000 km above the equator.

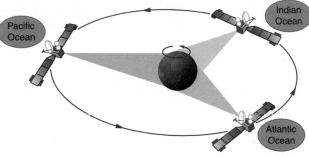

FIGURE 3: Just three geostationary satellites can cover Earth. What height above Earth are these satellites?

● Questions

3 How far does a geostationary satellite travel in 1 day? (The radius of Earth is 6400 km.)

4 An artificial satellite takes 11 hours to orbit Earth. Is its height above the surface of the Earth greater or less than 36 000 km? Explain your answer.

5 Jupiter has *many* moons. What force keeps them in orbit around Jupiter?

Gravitational force

> There is a gravitational force between all masses.

> Larger masses have a larger gravitational force between them.

> The further apart the masses, the smaller the gravitational force, F, between them. If their distance apart is *doubled*, the force between them, F, is *one quarter* of its original value. It obeys an **inverse square law**. If d = distance apart: F is proportional to $1/d^2$.

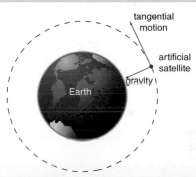

FIGURE 4: Why does a satellite accelerate towards Earth?

FIGURE 5: Why does a comet's speed vary?

Orbit periods for planets

Planets closer to the Sun have shorter orbit times because:

> they do not travel as far in one orbit

> they travel faster because there is a bigger gravitational pull on them.

Artificial satellites

Artificial satellites in low orbits around Earth have short orbit times. Gravity makes a satellite **accelerate** towards Earth but, because it moves at a tangent and Earth is curved, it maintains a near-circular orbit.

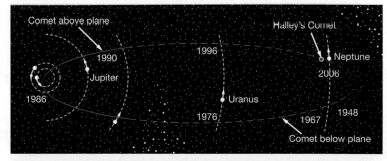

Comets

A **comet** has a very elongated **elliptical** orbit so its speed varies considerably. Both the gravitational force and the comet's speed decrease as its distance from the Sun increases.

> When it is close to the Sun, a comet travels very fast.

> When it is far away from the Sun, a comet travels more slowly.

Planets have nearly circular orbits. As the orbital period depends on their distance from the Sun, their speed is almost constant.

● Questions

6 Use the data in the table to calculate the orbital speeds of Mercury and Earth.

Planet	Distance from Sun (million km)	Time to orbit Sun (years)
Mercury	58	0.24
Earth	150	1.00

7 Why does the speed of a comet vary while the speed of the planets around the Sun is almost constant?

You will find out:

> about some uses of artificial satellites

> more about low polar orbit satellites

> more about geostationary satellites

Some uses of artificial satellites

> Communications – Satellite Earth Stations are used to relay telephone messages and computer data, and to broadcast television signals. We are all familiar with satellite television. We can telephone friends and family around the world as easily as speaking to someone nearby. If you are telephoning a friend in America, the 'dish' **aerial** transmits a signal to a satellite. The satellite then **amplifies** the signal and **re-transmits** it down to another dish aerial in America.

> Weather forecasting – images received from satellites orbiting Earth have made weather forecasts more reliable in recent years.

> Military – uses including spying.

> Scientific research – satellites orbiting Earth can take clearer pictures of objects in space as they are above Earth's atmosphere.

> **Global Positioning Systems (GPS)** – are used by ships, aircraft and cars to locate their position very accurately.

FIGURE 6: Telecommunications satellite dishes. Why are they pointing upwards?

FIGURE 7: Satellite image of the Japanese earthquake and tsunami in 2011 – areas coloured red show plants, bare earth is tan coloured and the city shows as silver. How are such images helpful?

Did you know?

24 GPS satellites orbit Earth roughly twice a day at a height of 19 000 km travelling at about 10 000 km/h.

Questions

8 Suggest why satellites have made weather forecasting more reliable.

9 Why are cars fitted with GPS?

Low polar orbit and geostationary satellites

Low **polar orbit** satellites orbit Earth over the poles at a height of 100 km to 200 km. They have a typical orbital period of about 90 minutes. As Earth rotates on its axis, their trajectory passes over a different area on each orbit.

Uses of low polar orbit satellites

> Weather forecasting – satellites in low polar orbit are used for short-range forecasts. They produce visible and infrared pictures. Infrared pictures show hot and cold areas in different colours so are good for forecasting temperatures.

> Imaging the Earth's surface – low polar orbit satellites give a detailed picture of Earth's surface.

> Military uses – buildings and moving things can be observed.

Uses of geostationary satellites

> Communications – geostationary satellites have made high-quality worldwide telephone communication possible.

low polar orbit

equatorial orbit

FIGURE 8: What are low polar orbit satellites used for?

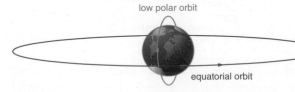

> TV and radio broadcasts – geostationary satellites allow events to be heard and seen in real time, all over Earth.

> Weather forecasting – geostationary meteorological satellites are used for longer term weather predictions.

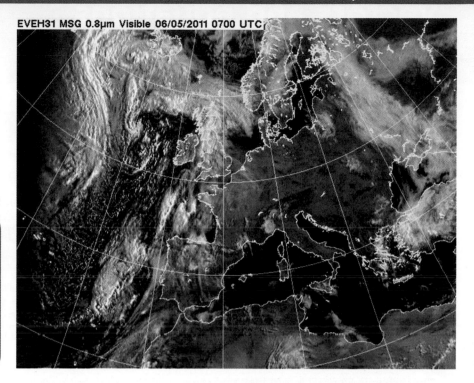

FIGURE 9: Infrared picture from a geostationary weather satellite. (The outline of all the land masses visible has been added.) Which parts of Great Britain may be sunny?

 Questions

10 Give **two** advantages of using infrared pictures for weather forecasting.

11 Explain why a geostationary satellite *must* orbit above the equator.

More on satellites

Geostationary satellites compared with low orbit satellites

A low polar orbit satellite has a short orbital period.

> This is because it is close to Earth so the gravitational attraction is strong.

> This gives a large acceleration towards Earth – a centripetal acceleration.

> So, the speed of the satellite is high.

> The distance travelled in one orbit is small (= $2\pi r$), where the radius of the orbit, r, is only a little greater than the radius of Earth.

> time = $\frac{distance}{speed}$, so orbital period is small.

The orbital period is typically about 2 hours.

A geostationary satellite has a longer orbital period.

> This is because it is a large distance above Earth so the gravitational force is much weaker.

> This gives a smaller centripetal acceleration.

> So, the speed of the satellite is lower.

> The distance travelled in one orbit is large (= $2\pi r$), where r is much greater than the radius of Earth.

> time = $\frac{distance}{speed}$, so orbital period is larger.

The height of a geostationary satellite is calculated so that its orbit time is exactly 24 hours.

All geostationary satellites must be in the same orbit. This makes it rather crowded. The satellites cannot be too close together as their signals would overlap due to diffraction.

FIGURE 10: Part of a satellite communications dish antenna in London's Docklands. In what orbit is the satellite in space that it sends signals to, or receives signals from?

Questions

Take Earth's radius, r = 6400 km.

12 All geostationary satellites are 36 000 km above Earth.

How fast are they moving?

13 Calculate the orbit time of a polar satellite 150 km above Earth that is travelling at 27 400 km/h.

Vectors and equations of motion

You will find out:
> about relative motion
> about the difference between scalar and vector quantities
> how to calculate the resultant of two vectors

Relative motion

If you were on one of the trains shown here, travelling in the same direction on parallel lines, you would appear to be moving very slowly and take a long time to pass. Indeed, if both trains had the same speed you would appear to be stationary, relative to the other train. But if the trains were going in opposite directions they would pass each other very quickly.

FIGURE 1: If both trains were travelling at 40 m/s and you were on one of the trains, how fast would you appear to be going when seen from the other train?

Direction of motion

The walker in Figure 2 is told that a refuge is only 3 km away but he is not likely to find it unless he also knows in which direction to walk. Direction is important when describing motion. '3 km due North' is in the opposite direction to '3 km due South'!

Relative motion

People often do not realise how fast they are travelling on a motorway because all the other cars are moving in the same direction and at similar speeds. Their **relative speed** is low. The cars in the outside lanes on each carriageway are moving past each other in opposite directions. They pass each other very quickly, giving an impression of very high speed. Their relative speed is high.

Speed tells us how fast an object is moving but not the direction in which it is moving. Speed is a **scalar** quantity.

FIGURE 2: Why can't the walker find the refuge?

Questions

1 Bristol is 200 km from London. What else do you need to know to locate it?

2 What is the relative speed of two cars both moving at 100 km/h in the same direction?

3 Two trains travelling in opposite directions at 160 km/h pass each other. What is their relative speed?

FIGURE 3: Motorway traffic. Why may you be moving faster than you think?

Scalar and vector quantities

A scalar quantity has **magnitude** (size) only. For example, a train travelling at 20 m/s. Speed is a scalar quantity; direction is not relevant.

A **vector** quantity has magnitude and direction. For example, a force of 5 N acting vertically downwards.

Velocity is speed in a named direction. For example, a car travelling at 20 m/s due North. Velocity is a vector. A vector is drawn on a diagram as a straight line with an arrow, or a '+' or '–' sign, to indicate direction.

+5 m/s →   ← –5 m/s

5 m/s to the right   5 m/s to the left

Q relative motion for kids

Scalar	Vector
speed	velocity
energy	force
mass	weight
time	acceleration

Resultant of two parallel vectors

The resultant of two parallel vectors is found by adding them algebraically.

Example [1]:

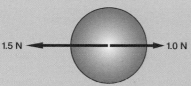

What is the **resultant** of the two vectors shown in red?

1.5 N ← → 1.0 N

The resultant force is 0.5 N to the left.

Example [2]:

a   +3 N   +   +2 N   =   +5 N

b   +3 m/s   +   −2 m/s   =   +1 m/s

FIGURE 4: Which side pulled harder in this tug-of-war?

Questions

4 Sort the following list into scalars and vectors.
force   distance   energy   velocity   acceleration

5 Find the sum of the following vectors.

3 N ←——○——→ 8 N

○ ——→ 20 m/s

○ ——→ 10 m/s

Resultant of two non-parallel vectors

When two forces are not in the same straight line the resultant is found using the **parallelogram of forces**. If the forces are perpendicular, the parallelogram becomes a rectangle, as shown in the example.

Example [3]:

Find the resultant of two perpendicular forces of magnitude 3 N and 4 N acting at a point.

3 N   resultant   θ   4 N

The resultant, R, is given by the diagonal of the rectangle.

This can be found
> by a scale drawing
> by using Pythagoras's theorem.

$R^2 = 3^2 + 4^2 = 9 + 16 = 25$ $R = 5\text{ N}$

Force is a vector so the direction of R must also be found. This can be expressed as the angle the resultant makes with one of the forces. The angle can be measured on a scale drawing or calculated using trigonometry.

So the angle, θ, in the rectangle shown above is:

$\tan\theta = \dfrac{3}{4} = 0.75$ So, $\theta = 36.9°$

The resultant force is 5 N making an angle of 36.9° with the 4 N force.

> If the vectors are perpendicular, it is easier to find the resultant by calculation.

> If the vectors are not perpendicular, it is easier to find the resultant by scale drawing.

FIGURE 5: The two small tugboats pull the much larger liner. In which direction does the liner move?

Questions

6 Tom can row at 4 m/s in still water. What is his resultant velocity if:

a he rows upstream against a current of 1 m/s?

b he rows straight across the stream?

7 Find the resultant of two perpendicular forces of 5 N and 12 N.

Average speed

Speed changes during a journey so it is usually more useful to calculate average speed.

average speed = total distance / total time

Example [1]:

> Concorde, the world's only supersonic passenger aircraft, travelled from London to New York in about 3.5 hours as opposed to about 8 hours for a subsonic flight. Its average speed was 1596 km/h. How far did it travel?
>
> Distance = average speed × time
> = 1596 × 3.5 = 5586 km.

Travelling westwards, the 5-hour time difference meant Concorde effectively arrived before she left. She travelled 'faster than the Sun'. Concorde was taken out of service in 2003. No more supersonic passenger aircraft have been built.

Equations of motion

When an object has a constant acceleration in a straight line, the equations of motion are used to find out more about how it moves. The equations use five special symbols:

s = distance travelled, in m
u = initial velocity, in m/s
v = final velocity, in m/s
a = acceleration, in m/s^2
t = time taken, in seconds

Distance travelled = average speed × time

$$s = \frac{(u + v)}{2} \times t$$

Example [2]:

> A car accelerates uniformly from 10 m/s to 20 m/s in 12 s. How far does it travel?
>
> $s = ?$ $u = 10$ m/s $v = 20$ m/s $t = 12$ s
>
> $$s = \frac{(u + v)}{2} \times t = \frac{(10 + 20)}{2} \times 12 = 180 \text{ m}$$

Final speed = initial speed + (acceleration × time) $v = u + at$

Example [3]:

> A car moving at 15 m/s starts to accelerate uniformly at 2 m/s^2.
> How fast is it moving after 6 s?
> $u = 15$ m/s $v = ?$ $a = 2$ m/s^2 $t = 6$ s
> $v = u + at = 15 + (2 \times 6) = 15 + 12 = 27$ m/s

Remember!
average speed = distance / time

FIGURE 6: Boarding one of the world's fastest scheduled trains at Guangzhou, China. It covers 922 km in 2.95 hours. What is its average speed?

FIGURE 7: Concorde. Suggest why no more supersonic passenger aircraft have been built.

Did you know?

Shanghai Maglev (magnetic levitation) train, is an airport rail link that reaches a top speed of 431 km/h. It has no wheels and 'floats' above a special track. How does this help it to go very fast?

Questions

8 Give **three** reasons why the speed of a car may change during a journey.

9 What does it mean when people say that Concorde travelled 'faster than the Sun'?

10 A car started to move with a constant acceleration of 5 m/s^2.

a How fast was it going after 4 s?

b How far did it go in this time?

More about the equations of motion for uniform acceleration

acceleration $= \dfrac{\text{change in velocity}}{\text{time taken}}$ $\quad a = \dfrac{v - u}{t} \quad v = u + at$ **①**

average speed $= \dfrac{\text{distance travelled}}{\text{time taken}}$ $\quad \dfrac{u + v}{2} = \dfrac{s}{t} \quad s = \dfrac{(u + v)t}{2}$ **②**

Example:

> A car reached a speed of 144 km/h after accelerating uniformly at 2 m/s² for 10 s. How fast was it going before accelerating?
>
> $v = 144 \text{ km/h} = \dfrac{144 \times 1000}{3600} = 40 \text{ m/s}$
>
> Write down 'suvat' and fill in what you know.
>
> s (not needed) $u = ?$ $v = 40$ m/s $a = 2$ m/s² $t = 10$ s
> $v = u + at$ $u = v - at = 40 - (2 \times 10) = 20$ m/s (72 km/h)

Example:

> A cyclist accelerated uniformly at 3 m/s² from rest for 8 s, travelling a distance of 48 m. What speed did he reach?
>
> $s = 48$ m $u = 0$ (at rest) $v = ?$ a (not needed) $t = 8$ s
> $s = \dfrac{(u + v)t}{2}$ $u + v = \dfrac{2s}{t} = \dfrac{(2 \times 48)}{8} = 12$ $v = 12 - 0 = 12$ m/s

Example:

> QRIO ('Quest for Curiosity') is a humanoid autonomous robot, standing 58 cm high.

FIGURE 8: QRIO. What is special about it?

> It is the world's first two-legged robot capable of running (i.e. moving while both legs are off the ground at the same time). How long does it take to reach its maximum speed of 0.23 m/s, starting from rest if it travels 0.58 m in this time?
>
> $s = 0.58$ m $u = 0$ (starts from rest)
> $v = 0.23$ m/s a (not needed)
> $t = ?$
> $s = \dfrac{(u + v)t}{2}$
> $t = \dfrac{2s}{(u + v)} = \dfrac{(2 \times 0.58)}{(0 + 0.23)} = \dfrac{1.16}{0.23} = 5.04 \text{ seconds}$

Questions

11 A sports car accelerates at 5 m/s² for 4 s reaching a speed of 108 km/h.

a How fast was it moving before accelerating **i** in m/s **ii** in km/h?

b How far did it travel in this time?

12 A cyclist decelerated from 12 m/s to 6 m/s over a distance of 45 m. How long did this take?

Remember!

When using motion equations always work in the correct units: s in m, u and v in m/s, a in m/s² and t in seconds.

More on equations of motion for uniform acceleration

Using equation **①** to replace v in equation **②**:
$$s = \frac{[u + (u + at)]\,t}{2} \qquad s = ut + \tfrac{1}{2}at^2 \text{ ③}$$

Using equation **①** to replace t in equation **②**:
$$s = \frac{(u + v)}{2}\frac{(v - u)}{a} = \frac{v^2 - u^2}{2a} \qquad v^2 = u^2 + 2as \text{ ④}$$

If three of the 'suvat' quantities are known then the other two can be found.

Example:

> A sprinter reaches a speed of 12 m/s from rest after running 20 m. Find her acceleration.
> $s = 20$ m $u = 0$
> $v = 12$ m/s $a = ?$ t (not needed)
> Using $v^2 = u^2 + 2as$
> $(12)^2 = 0 + (2 \times a \times 20)$
> $(12)^2 = 40a$ $a = 3.6$ m/s²

Falling under gravity

> Ignoring air resistance all objects accelerate towards the Earth at 10 m/s² due to gravity.

> An object thrown up from the Earth **decelerates** at 10 m/s². Its acceleration, $a = -10$ m/s² and at its highest point its velocity, $v = 0$.

Questions

13 Brian hit a ball vertically upwards at 18 m/s. How high did it go?

14 Ian dropped a stone from the top of a cliff 45 m high. With what speed did it hit the water?

Projectile motion

You will find out:
> about the path taken by a ball after it is hit or thrown
> how to apply the equations of motion to projectiles
> how to calculate the resultant velocity of a projectile (H)

Ball skills

Many sports involve throwing, hitting or kicking a ball.

We all develop the ability to control a ball's speed and direction as well as to judge its path so that we can receive it.

We cannot all be as talented as the professionals but we can still enjoy participating and following our favourite team or players.

FIGURE 1: What path will the football take when kicked?

Projectiles

A **projectile** is the name given to any object that moves in Earth's gravitational field, such as:

> balls

> darts

> long-jumpers.

The path of a projectile is called its **trajectory**.

All projectiles that are thrown horizontally or at an angle to the ground follow a curved path.

FIGURE 2: What is the path of a projectile called?

Questions

1 Sketch the trajectory of the tennis ball shown in Figure 3. Why do you think the ball in the photo is blurred?

2 Suggest why projectiles follow a curved path.

FIGURE 3: The golf ball and tennis ball are projectiles. Which path, A or B, is more likely to be followed by the golf ball?

Projecting horizontally

If two similar balls are released from the top of a tower simultaneously, a red one by dropping and a black one by throwing horizontally, they reach the ground at the same time. The graph (Figure 4) shows the paths of the two balls at one second intervals. The trajectory of the black ball is **parabolic**. Note that, if air resistance is ignored:

> there is no acceleration in the horizontal direction; the black ball has a constant horizontal velocity

> the acceleration due to gravity acts equally on both balls in the vertical direction.

> This is true no matter how hard the ball is thrown horizontally.

Remember!
Velocity is a vector.

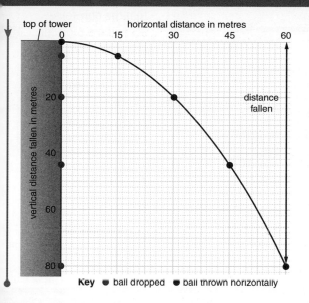

FIGURE 4: Graph to show how the two balls behave. Successive positions are shown at 1 s intervals. How would the trajectory of the black ball change if it was thrown with a greater force?

Key ● ball dropped ● ball thrown horizontally

Questions

3 Use the graph in Figure 4 to help you answer this question.

a What horizontal velocity was given to the projectile?

b How far has the projectile fallen after: **i** 2 seconds? **ii** 4 seconds? Comment on your answers.

c Copy the graph and add a curve for a ball thrown at half the speed of the one shown.

Resultant velocity

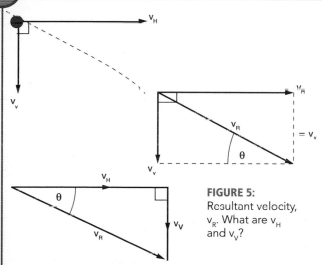

FIGURE 5: Resultant velocity, v_R. What are v_H and v_V?

The resultant velocity of a projectile at a point on its trajectory is the vector sum of the horizontal (v_H) and vertical (v_V) velocities at that point.

By definition these are perpendicular vectors. The resultant velocity, v_R, is represented by the diagonal of the vector rectangle shown. The value of v_R is found using Pythagoras' theorem: $v_R^2 = v_H^2 + v_V^2$ and its direction with the horizontal by: $\tan \theta = \dfrac{v_V}{v_H}$

Example:

Find the resultant velocity of a ball at a point in its trajectory where $v_H = 12$ m/s, $v_V = 9$ m/s.
$v^2 = 12^2 + 9^2 = 144 + 81 = 225$ giving $v = 15$ m/s

$$\tan \theta = \frac{9}{12} = 0.75; \quad \theta = 36.9°$$

The resultant velocity at this point is 15 m/s at 36.9° to the horizontal.

Calculations on projectiles

The equations for uniformly accelerated motion on page 193 can be applied to the motion of projectiles.

Example:

A darts player stands 3 m from a darts board. The dart leaves her hand horizontally and takes 0.2 seconds to hit the board.

Calculate:

a the initial velocity of the dart

b the vertical height the dart falls in flight

Answer:

a The initial velocity is in a horizontal direction. Horizontally there is no acceleration, so initial velocity =

$$\frac{\text{distance}}{\text{time}} = \frac{3.0}{0.2} = 15 \text{ m/s}$$

$s = ?$ $u = 0$ v (not needed) $a = 10$ m/s² $t = 0.2$ s

Using $\qquad s = ut + \frac{1}{2} at^2$

$s = (0 \times 0.2) + (\frac{1}{2} \times 10 \times 0.2^2)$

$s = 0 + 0.20 = 0.2$ m

Questions

4 Sam kicked a football horizontally off the edge of a wall 0.8 m high. The ball travelled a horizontal distance of 6 m before hitting the ground. At what speed did it leave the wall?

5 In the example above, calculate the velocity with which the dart hits the board.

🔍 physics of darts   adding vectors for KS4

Range of a projectile

> The horizontal distance a projectile travels is called its **range**.
> The range depends on the **launch angle**.
> For maximum range the launch angle must be 45°.

 Questions

6 Sketch the trajectory for a ball thrown horizontally from the top of a cliff. Mark the range on the diagram.

7 Louis is playing golf. What **two** things can he do to make the ball go as far as possible?

FIGURE 6: Trajectories for large and small launch angles. What launch angle gives the greatest range?

Forces on projectiles

If air resistance is ignored, once a projectile has been released the only **force** acting on it is gravity. Gravity is a force that always acts vertically downwards so, since $F = ma$, the projectile has a downwards acceleration. This only affects the vertical velocity, decelerating the projectile at 10 m/s^2 as it moves up and accelerating it at 10 m/s^2 as it moves down.

There is no force in a horizontal direction so horizontal velocity is constant.

The sequence of images revealed by a strobe light shows that a ball's trajectory is a parabola. The delay between the images is the same, so the changing speed of the ball can be seen. The ball moves fastest near the ground and slowest at the top of the arc. The image also shows that the horizontal speed is approximately constant, and that only the vertical speed is changing.

Range of a projectile

The range of a projectile depends on its:

> initial velocity – the greater the initial velocity the greater the range.

> launch angle – a large launch angle gives it a large vertical velocity but it does not travel far horizontally; a small launch angle gives it a large horizontal velocity but it does not stay long in the air, quickly hitting the ground.

FIGURE 7: A bouncing ball shown as a **stroboscopic** image. A light flashing at regular intervals allows successive positions of the ball to be seen. How do you know the ball moves slowest at the top of the arc?

 Questions

8 Why does a bouncing ball rise to a lower height on each successive bounce?

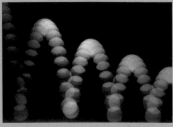

9 How does the angle at which a tennis ball is hit affect its range?

10 Why is the horizontal velocity of a projectile constant?

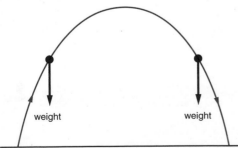

FIGURE 8: Why does a projectile have a downwards acceleration?

weight   weight

FIGURE 9: A long-jumper. What does he do, that a ball cannot, to maximise his range?

Q looking at projectile motion   long jump techniques

Projecting an object horizontally

> The horizontal motion is not affected by gravity. The object travels the same distance horizontally each second. It has a constant horizontal velocity – no acceleration.

> The object falls further vertically in each successive second. It has a steadily increasing vertical velocity. The acceleration due to gravity acts in the vertical direction.

Newton's projectile

Newton suggested that a projectile fired from a very high peak (V) on Earth would never hit the ground if it was launched with enough initial speed. Instead, it would remain in orbit around Earth, prevented from flying away by the force of gravity and never getting closer to the surface due to Earth's curvature.

This is the principle used to put satellites into orbit today. Newton did not have the technology to test his theory.

FIGURE 10: A javelin thrower. What angle of throw will give the maximum range? The world record, set in 1996 by Jan Zelezny, is 98.48 m.

FIGURE 11: The trajectory of a long-jumper is a parabola but he has to work hard to keep his body in the air for the maximum amount of time. The world record, set by Mike Powell in 1991, is 8.95 m.

FIGURE 12: Part of one of Newton's drawings taken from his 'Principia' written in 1687, showing his theory of projectiles. Why does it need a very large initial speed to orbit Earth?

Questions

11 A new javelin that had its centre of gravity moved forward was introduced in 1999. Suggest why this led to shorter distances being achieved. How may this have affected the world record?

12 A bullet is fired horizontally at a speed of 200 m/s at a target 100 m away.

a Calculate how far the bullet has fallen when it hits the target.

b How far would the bullet have fallen if it had been fired at half the speed?

c Calculate the velocity with which the bullet hits the target.

13 In the cartoon on the right, the monkey lets go at the same time as the hunter fires horizontally at it. Does the monkey escape? Explain your answer.

Remember!

• The horizontal velocity does not change as there is no acceleration in a horizontal direction.
• The vertical velocity decreases as the projectile rises and increases as it falls due to gravity.

Action and reaction

You will find out:

> about collisions
> about equal and opposite reactions
> that momentum is always conserved in collisions
> how to use ideas about momentum to explain explosions and recoil

Collisions

We tend to think of **collisions** as unpleasant accidental events when cars, trains or aircraft collide, but there are all sorts of collisions, from snooker balls to cars, as well as rockets and fireworks.

FIGURE 1: What collision takes place?

Pairs of forces

Forces always occur in pairs.

Sarah and Ben are about to start an ice dance routine. If Sarah pushes Ben with a force of 50 N, Ben pushes Sarah with a force of 50 N in the opposite direction. This means they move away from each other.

FIGURE 2: Sarah and Ben's ice dance. What is the action force?

> Sarah pushing Ben is the **action** force

> Ben pushing Sarah is the **reaction** force.

(Note that we get the reaction force by turning the names around.) In general, the force on an object A due to an object B is equal and opposite to the force on B due to A.

Remember!

Every action has an equal and opposite reaction.

Questions

1 The tyres of a car push on the road. This is an action force. What is the reaction force?

2 Two cars collide head on, as shown.

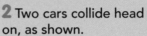

a What are the opposite reactions in this collision?

b Copy and complete the following sentence: The force on the red car due to the orange car is…

More on pairs of forces

A collision occurs every time a ball is struck when playing sport.

The action and reaction pairs of forces:

> are equal in magnitude (size)

> are opposite in direction

> act on different objects.

This is Newton's third law.

FIGURE 3: Pool ball and cue – there will be several collisions after the ball is hit. What objects will collide then?

The boy has weight because he is attracted to Earth due to gravity.
The Earth is also attracted to the boy; this is the reaction force.

> Action – downward force on boy due to gravitational attraction of Earth.

> Reaction – upward force on Earth due to gravitational attraction of boy.

FIGURE 4: Describe the action and reaction forces on the boy and Earth

These pairs of forces are equal in magnitude (size) and opposite in direction. If the boy jumps in the air he is attracted to Earth and moves down to it. Earth is also attracted to the boy with an equal force but its mass is so great that it only moves a very small amount, much too small for us to notice.

Recoil

On firing the gun, there is an action force on the bullet by the gun; the bullet moves forwards.

There is also an equal force on the gun from the bullet but in the opposite direction. The gun moves backwards – it **recoils**.

FIGURE 5: The Moon orbits Earth due to gravitational attraction. Are the forces on the Moon and Earth equal?

FIGURE 6: Why doesn't the gun recoil as fast as the bullet moves forwards?

Questions

3 Santosh kicks a football. What is the reaction force to the push of his boot on the football in an easterly direction? What can you say about the size of these forces?

4 If you jump off a stool, you move towards Earth due to gravitational attraction. Why does Earth not move up towards you?

5 When swimming, you push the water backwards. What is the reaction force?

Conservation of momentum

The **momentum** of an object increases if there is an increase in either its mass or velocity.

momentum = mass × velocity   Unit: kg m/s

In any collision, if no external forces act, the total momentum before the collision is equal to the total momentum after the collision. We say that momentum is **conserved**. If the colliding objects stick together (**coalesce**) we can write:

$m_1 u_1 + m_2 u_2 = (m_1 + m_2) v$

Example:

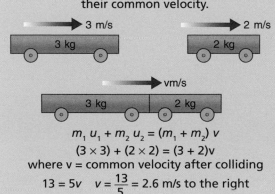

Two trolleys collide and stick together. Find their common velocity.

3 m/s   2 m/s
3 kg   2 kg

v m/s
3 kg 2 kg

$m_1 u_1 + m_2 u_2 = (m_1 + m_2) v$
$(3 \times 3) + (2 \times 2) = (3 + 2)v$
where v = common velocity after colliding
$13 = 5v$ $v = \frac{13}{5} = 2.6$ m/s to the right

Firing a gun

> Before firing a shell, the total momentum is zero.

> After firing the shell, the total momentum is again zero so the momentum of the shell forwards equals the momentum of the gun backwards and the gun recoils.

FIGURE 7: Why are no soldiers standing behind the gun?

Explosions

> An explosion is the opposite of a collision. In an explosion, objects move apart instead of colliding.

> Momentum is still conserved.

Questions

6 If you push on a wall you feel a force against your hand. Why doesn't your hand accelerate through the wall?

7 A car of mass 1200 kg travelling at 30 m/s runs into the back of a stationary lorry.

Find the mass of the lorry if the car and the lorry move at 4 m/s after impact.

8 In the cartoon, explain why Harry falls into the water when he tries to step ashore.

How a gas exerts a pressure

> the particles in a gas are moving **randomly**
> they collide with each other as well as with the walls of their container
> when they hit a wall there is a force on the wall
> **pressure** = force / area so the gas exerts a pressure on the wall.

Launching a rocket

Have you noticed what happens if you let the air out of a balloon? The air goes one way and the balloon moves in the opposite direction. **Rockets** work in much the same way.

In a rocket:

> gas is burned inside the rocket
> the rocket pushes some particles of hot gas out downwards – action force
> some particles push the rocket upwards – reaction force.

Jet engines work in a similar way.

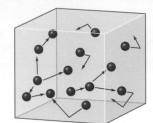

FIGURE 8: How does the gas exert a pressure?

FIGURE 9: Why does the balloon move in the opposite direction to the air escaping from the balloon?

air escapes this way

balloon moves this way

Questions

9 The rocket shown in Figure 10 rises slowly from the launch pad. Why must the hot gases move very quickly?

10 A space rocket, unlike a jet aircraft, must carry its own oxygen supply. **a** Why is oxygen needed? **b** Why doesn't a jet engine need to carry oxygen while a space rocket does?

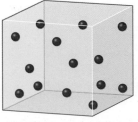

FIGURE 10: Why does the rocket rise up?

FIGURE 11: Jet aircraft. In which direction must the exhaust gases move?

Kinetic theory of gases

> Every material is made of tiny moving particles; they have kinetic energy.

> The higher the temperature, the more kinetic energy the particles have and the faster they move.

> In a gas, the particles are far apart and free to move.

> The fast-moving gas particles collide with the walls of their container, creating a force, and hence a pressure, on the walls.

> Changing volume – if the same number of particles, at the same temperature, are in a smaller container they hit the walls more frequently, increasing the pressure.

> Changing temperature – if the same number of particles in the same container are heated (so the volume is constant) the particles move faster and hit the walls more frequently and with greater force, increasing the pressure.

FIGURE 12: Why is the gas pressure greater in the smaller box?

Q how rockets work for kids   uses of space rockets   kinetic theory of gases

More about rockets

Look at the rocket launch in Figure 10.

> Inside a rocket some of the fast-moving gas particles colliding with the walls create a force.

> In addition, some particles are forced down out of the base of the rocket at high speed.

> This results in an equal and opposite reaction force upwards, launching the rocket.

Questions

11 Use the kinetic theory of gases to explain why the particles of hot gases move very fast.

12 Explain how a rocket works.

More about gas pressure

$F = ma$ and $a = (v - u) / t$

$F = (mv - mu) / t$ = change of momentum / time

Force = rate of change of momentum

Consider a moving particle in a sealed box:

> A particle of mass m moving with velocity $-v$ has momentum $-mv$.

> Each time the particle collides perpendicularly with a wall it rebounds with the same speed but in the opposite direction. Its velocity is now $+v$ and its momentum is $+mv$.

> This means the change of momentum of the particle is $mv - (-mv) = 2mv$.

> Force on wall, $F = 2mv / t$ where t = time between collisions with wall.

Example:

> Explain why the pressure of a gas increases if there is a greater mass of gas present in the same volume and at the same temperature.
> There will be more gas particles in the box so there will be:
> > more collisions with the walls of the box
> > so a greater rate of change of momentum
> > so a greater force on the wall
> > so a greater pressure.

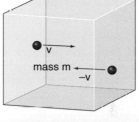

FIGURE 13: What is the change in momentum of the particle when it hits the wall?

mass m

Rocket propulsion

> The rocket moves up while the hot gases move down.

> Momentum is conserved. This means that the momentum of the rocket upwards is equal to the momentum of the hot gases downwards.

> Large rockets, used to lift satellites into Earth orbit, have a very large mass.

> This means the upward momentum (= mass × velocity) required is large.

> The gases ejected downwards have the same momentum.

> So there must be a large number of exhaust gas particles moving at very high speeds.

FIGURE 14: Saturn V, NASA's largest rocket, carried the crew of Apollo 11 to the Moon in 1969. Why did it need to be so large?

Questions

13 Use ideas about momentum to explain what happens to the pressure of a gas when (a) its volume increases (b) its temperature increases.

14 Use ideas about momentum to explain how a rocket is able to lift off.

15 Explain why a rocket can travel in space without using its motors.

Remember!
The action and reaction forces are the same size.

Preparing for assessment: Applying your knowledge

To achieve a good grade in science, you not only have to know and understand scientific ideas, you also need to be able to apply them to other situations and investigations. These tasks will support you in developing these skills.

✳ Under siege!

Nisha is 12 years old. In her history class she is learning about siege weapons. The class has been given the job, in small groups, of researching one of the weapons and seeing if they can build something to show how it worked. Nisha's group has chosen a trebuchet as they think it will be great fun to lob things across the room.

A trebuchet, they soon realise, is a bit like a see-saw, but with one side much longer than the other. The short side has a weight hanging from the end, and starts off up in the air. When it drops, the other, much longer, side flicks up. On the end of this side is a long cord, with a sling on the end. In the sling goes the missile, and Nisha's group is told that in this case it is going to be a golf ball. The target will be a Lego castle.

They get some pictures and a video clip from the internet and then get to work with old ice-lolly sticks and differently sized pieces of dowelling. There's quite a bit of fiddling to get it right and they decide not to bother with wheels, but soon the golf balls are flying across the room.

The accuracy of their trebuchet is not great but soon they get the hang of how to make it work. A couple of good shots land on the castle and Lego bricks fly everywhere to loud cheers from the rest of the groups, and the teacher, who are watching.

Task 1

Have a look at some more pictures of trebuchets on the internet. Nisha's group found the video clip really useful – see if you can find one.

✹ Task 2

Draw a picture showing the trebuchet at the instant the missile is about to leave it. Label the weight, the arm, the sling and the missile.

✹ Task 3

Draw another picture, this time to a smaller scale, showing the trebuchet at one side of the room and the Lego castle at the other. Now draw the path of the golf ball through the air.

✹ Task 4

Now draw a picture of the golf ball the instant it has left the sling and is in flight. Discuss what forces are acting upon it, and show these forces with arrows in the correct directions.

✹ Task 5

Nisha's teacher didn't want them to use anything more potent than a golf ball. Imagine a streamlined missile being launched in this way. Describe the horizontal component of its velocity during the flight. How might the golf ball vary from this?

✹ Task 6

Nisha is directing the launch of the golf ball using the trebuchet and is safely behind it. Her job is to make sure it is in line with the castle. The golf ball travels away from her and hits the castle. Describe the vertical component of the golf ball's journey as she sees it. What force or forces are responsible for this?

✹ Maximise your grade

	Answer includes showing that you can…
	Label the parts of the trebuchet.
F	Explain how the trebuchet works.
	Correctly draw the path of the projectile.
	Explain how gravity affects the motion of the projectile.
C	Explain why and how the projectile is accelerating.
	Compare the horizontal and vertical components of the projectile's motion.
A	Explain fully the motion of the projectile in terms of horizontal and vertical components.
	As above, but with particular clarity and detail.

Satellite communication

Mobile phones

Communications technology has developed at an incredible rate since the launch of the first communications **satellite** in 1962. For many young people the mobile phone is the most amazing invention of the last century. The latest mobile phones are mini-computers, including cameras, internet and e-mail access. Worldwide communication at the speed of light!

You will find out:
> about radio waves of different frequencies
> about use of satellites in communication systems
> how microwaves are used for communications

FIGURE 1: What do you think was the most amazing invention of the last century?

Radio waves

Radio waves have **wavelengths** varying from a fraction of a millimetre up to 10 km; much longer than the wavelengths of other **electromagnetic** waves. The highest **frequency** radio waves are called **microwaves**. The higher the frequency of a radio wave, the more penetrating it is. This means that:

> high frequency radio waves pass through Earth's atmosphere

> low frequency radio waves are stopped by Earth's atmosphere.

Some radio waves are reflected by part of Earth's atmosphere.

Geostationary satellites stay above the same point on Earth. They orbit above the equator within a period of 24 hours at a height of 36 000 km. They can only be reached by high frequency waves such as microwaves. Low-orbit satellites can be reached by low frequency radio waves.

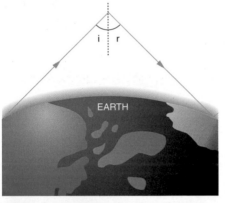

FIGURE 2: Reflection of radio waves by Earth's atmosphere. Are they high or low frequency radio waves?

FIGURE 3: Large parabolic antenna used for sending and receiving signals to and from satellites orbiting Earth. What does the dish do?

Questions

1 What is the orbit time of a geostationary satellite?

2 Suggest one use for a low-orbit satellite.

Did you know?

Radio waves are given off by stars, sparks and lightning. This is why you hear interference on your radio in a thunderstorm.

Using microwaves

Microwave signals are:

> sent into Space from a parabolic **transmitter**

> received and amplified by a geostationary satellite

> re-transmitted back to Earth or to another satellite

> picked up by a parabolic **receiver**.

Digital signals are used for satellite communication. The signals travel a long way to and from a geostationary satellite.

> Digital signals do not **attenuate** as rapidly.

> There is less **noise** with digital signals.

Choosing the best frequency for satellite communication

> The **ionosphere** reflects radio waves with frequencies below 30 MHz (30 million hertz), so higher frequencies are needed for satellite communication.

> Radio waves with frequencies above 30 GHz (30 000 million hertz) are easily **absorbed** and **scattered** by rain, dust and other atmospheric effects reducing signal strength.

> The frequency range used is between 3 GHz and 30 GHz. This corresponds to microwaves with a wavelength between 10 cm and 1cm.

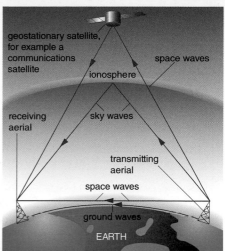

FIGURE 4: High frequency microwaves can travel into Space to reach satellites. What is meant by a geostationary satellite?

geostationary satellite, for example a communications satellite

space waves

ionosphere

receiving aerial

sky waves

transmitting aerial

space waves

ground waves

EARTH

Remember!

1 kHz = 1000 Hz (or 10^3 Hz)
1 MHz = 1 000 000 Hz (or 10^6 Hz)
1 GHz = 1 000 000 000 Hz (or 10^9 Hz)

FIGURE 5: The dishes on this mast receive, amplify and relay mobile phone signals. What frequency range is used?

Questions

3 All electromagnetic waves travel at a speed of 3×10^8 m/s in a vacuum (or air). Use the equation velocity = frequency × wavelength ($v = f\lambda$) to find the wavelength of radio waves having a frequency of: **a** 3 GHz **b** 30 GHz.

4 Why does satellite communication use digital signals?

Sending microwaves to a geostationary satellite

Geostationary satellites orbit Earth about 36 000 km above the equator.

> The microwave frequency range 3 GHz to 30 GHz is used for satellite communication. These microwaves have a short wavelength of a few centimetres.

> The size of the aerial dish is many times the microwave wavelength so there is very little **diffraction**. It produces a narrow beam that does not spread out.

> This means the receiving dish and satellite dish need exact alignment to ensure that the signals do not 'miss' the geostationary satellite.

There is only one possible geostationary orbit so it is very crowded. The satellites cannot be too close together or they would receive unwanted signals but, as microwaves are sent as a narrow beam, this is not a major problem.

The ionosphere

The ionosphere is a region of the atmosphere, between 100 km and 500 km above Earth, where molecules have been ionised by radiation from the Sun. Radio waves with frequency less than 30 MHz undergo a series of refractions as they enter the different layers and speed up, until total internal reflection occurs.

Waves reflected off the ionosphere can also be reflected from Earth's surface, especially from water, allowing the waves to travel to the other side of Earth.

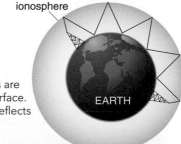

ionosphere

EARTH

FIGURE 7: Radio waves are reflected off Earth's surface. What type of surface reflects them best?

ionosphere

EARTH

FIGURE 6: Total internal reflection of radio waves at the ionosphere. What can you say about their frequency?

Questions

5 Why do satellite transmitting and receiving dishes need very careful alignment?

6 Suggest why water reflects microwaves better than land.

Q geostationary satellites ionosphere microwaves

Radio waves

You will find out:

> about diffraction of radio waves

> how the amount of diffraction depends on wavelength

> how radio waves of different frequencies behave

Compared with other parts of the electromagnetic spectrum, radio waves have a very long wavelength. But radio waves, including microwaves, cover a wide range of wavelengths. Their wavelength affects the way they behave and therefore their use.

Type of radio wave	frequency	wavelength	Examples of use
Long wave	30 kHz–300 kHz	10 km–1 km	Long wave AM radio such as Radio 4
Medium wave	300 kHz–3 MHz	1 km–100 m	Medium wave AM radio such as Radio 1
Short wave	3 MHz–30 MHz	100 m–10 m	International and amateur radio
VHF (very high frequency)	30 MHz–300 MHz	10 m–1 m	FM radio, shipping and police communications
UHF (ultra high frequency)	300 MHz–3 GHz (3000 MHz)	1 m–10 cm	Television, Bluetooth, GPS
microwaves	3 GHz–30 GHz	10 cm–10 mm	Satellite communications, mobile phones

Diffraction

Diffraction is the spreading out of waves when they pass through a gap or around a large object.

Receiving radio and television (TV) programmes

> An aerial is required to detect waves in the air for radio and **terrestrial** TV signals.

> Satellite TV signals require a 'dish' to pick up weak microwave signals from a geostationary satellite.

FIGURE 9: What type of signal is this satellite TV dish picking up?

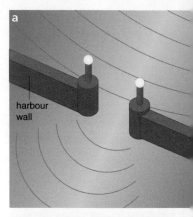

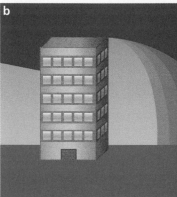

FIGURE 8: a Water waves spread out as they pass through a harbour wall; b radio waves spread out around a large building. What do we call this effect?

Questions

7 Which radio waves will diffract most, long waves or microwaves?

8 When you speak, sound is heard all around, not just in front of your mouth. What happens to the sound waves when they leave your mouth?

Diffraction of waves

The smaller the size of the gap, the greater the diffraction.

> Long wavelength radio waves have a very long range because they diffract around hills and buildings, as well as over the horizon, following the curvature of the Earth.

> Waves used for TV signals have shorter wavelengths that do not diffract very much so can only be received along a **line of sight**.

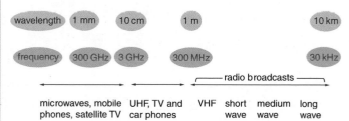

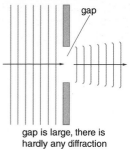
gap is large, there is hardly any diffraction

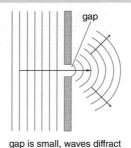
gap is small, waves diffract

FIGURE 10: Diffraction of waves. What is meant by diffraction?

FIGURE 11: Uses of radio waves depend on wavelength. What are radio waves of wavelength between 1 m and 10 km used for?

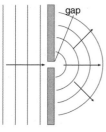

FIGURE 12: The house shown here has good radio reception but very poor television reception. Why is television reception poor?

Questions

9 Explain why radio waves are easily diffracted.

10 City Radio broadcasts on a wavelength of 194 m; and Radio X (long wave) uses a wavelength of 1500 m. Which station would be likely to have better reception in the house shown in Figure 12? Explain your answer.

Factors affecting diffraction

The amount of diffraction through a gap depends on the:

> size of the gap (See Figure 10.)

> wavelength.

Extensive diffraction occurs when the wavelength is equal to, or smaller than the size of the gap.

Long wavelength radio waves

They have very long wavelengths, comparable to the gaps between hills and large buildings, so diffract easily. They can also diffract over the horizon.

the smaller the gap, the larger the wavelength, the more diffraction

FIGURE 13: What is the amount of diffraction through a gap dependent on?

Questions

11 Sketch diagrams to show water waves of wavelength 3 cm passing through gaps of:

a 8 cm

b 4 cm.

12 Explain why long wavelength radio waves may give better reception than short wavelength radio waves in a hilly area.

Nature of waves

You will find out:
> about the interference effect of two sets of overlapping waves
> how to explain constructive and destructive interference

Interfering waves

When we talk about interference we usually mean 'noise' or 'crackle' in a radio receiver. In Physics, interference means the effect produced when two or more sets of waves overlap.

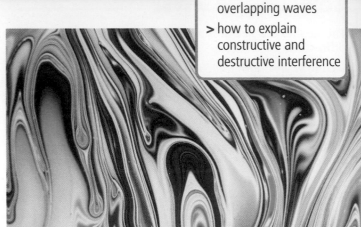

FIGURE 1: The colourful interference pattern seen here is caused by a thin film of oil floating on water. Two sets of overlapping waves are produced by reflection and refraction of light by oil and water. Suggest how this produces two sets of overlapping waves.

Interference of waves

A **ripple tank** can be used to create two overlapping sets of water waves. Two small dippers are attached to a strip of wood that moves up and down. Each dipper produces a circular wave pattern and these circular wave patterns overlap to create an **interference** pattern. It shows areas where the waves:

> add together giving a big disturbance
> subtract from each other giving calm water.

Overlapping sound and light waves also interfere.

	sound	light	water
waves adding	louder area	brighter area	higher crests, deeper troughs
waves subtracting	quieter area	darker area	calm water

FIGURE 2: Interference pattern in water waves in a ripple tank. Where do the waves overlap?

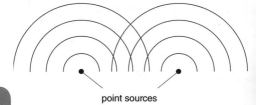

point sources

FIGURE 3: The interference pattern in water waves can be drawn. Red lines show the crests (tops) of the waves. What happens where two crests meet?

Questions

1 Suggest why it is easier to study interference in water waves, rather than in sound or light.

2 Give **two** differences between sound waves and light waves.

Overlapping waves

Whenever two sets of waves overlap:

> output is a maximum when the waves meet 'in step' (two **crests** or two **troughs** arrive together) – **reinforcement**

> output is a minimum when the waves meet 'out of step' (crest and trough arrive together) – **cancellation**.

To produce a stable interference pattern the two sources need to be **coherent**, having the same frequency and therefore wavelength. This can be achieved by using a single source and dividing the waves coming from it.

Using interference of sound waves

The destructive interference of sound waves is used in noise reduction systems. Special earphones are used by factory and construction workers. Such earphones take in sound from the environment and use computer technology to produce a second sound wave that is out of step with it. The combination of these two sound waves results in cancellation, reducing a worker's exposure to loud noise.

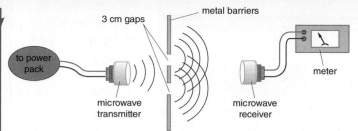

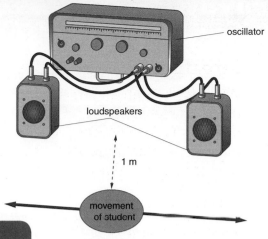
oscillator

loudspeakers

1 m

movement of student

FIGURE 4: Microwaves with a wavelength of about 3 cm pass through two 3 cm gaps, formed by three vertical metal plates. The receiver, connected to a meter, is moved along a line parallel to the plates. Why does the meter reading increase and decrease regularly?

FIGURE 5: Two loudspeakers connected to the same oscillator produce the same frequency sound. If you walk in front of the loudspeakers, about 1 m away, you hear alternate loud and quiet sounds.

Questions

3 a When demonstrating interference of microwaves, why should the gaps between the metal plates be about 3 cm or less?

b What would be the effect, if any, of making the gaps smaller?

4 Why is it difficult to demonstrate interference with light?

Explaining interference patterns

When two waves from the same source meet:

> **constructive** interference (reinforcement) – waves meet 'in step' (in **phase**)

> **destructive** interference (cancellation) – waves meet 'out of step' (in **antiphase**).

To produce a regular interference pattern, the two wave sources must be **coherent**. This means they have exactly the same frequency, and are either in phase or have a constant phase difference.

> Constructive interference occurs when the **path difference** from the two sources is an *even* number of half wavelengths.

> Path difference = $n\lambda$, n = 0, 1, 2, etc.

> Destructive interference occurs when the path difference from the two sources is an *odd* number of half wavelengths.

> Path difference = $(n + \frac{1}{2})\lambda$, n = 0, 1, 2, etc.

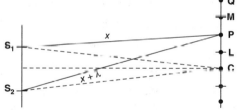

constructive interference:
• the central point **C** is a point of constructive interference as $S_1C = S_2C$ by symmetry (no path difference).
• If **P** is the next point of constructive interference, $S_2P - S_1P = \lambda$.
• If **Q** is the next point of constructive interference, $S_2Q - S_1Q = 2\lambda$.

destructive interference:
• If **L** is a point of destructive interference, $S_2L - S_1L = \frac{1}{2}\lambda$.
• If **M** is the next point of destructive interference, $S_2M - S_1M = \frac{3}{2}\lambda$.

FIGURE 7: When does constructive interference occur and when does destructive interference occur?

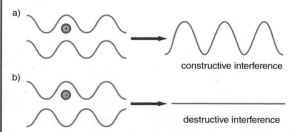

a) constructive interference

b) destructive interference

FIGURE 6: Constructive and destructive interference. Describe each type of interference.

Questions

5 When demonstrating interference of sound, the frequency of the sound was increased. What would you notice?

6 Tal set up an interference pattern in water waves in a ripple tank. She increased the wave frequency. Use Figure 7 to explain how the interference pattern changed.

Light travels in straight lines

We cannot normally see light travelling through air. However, **laser** light can be seen and it follows a straight path.

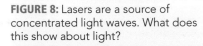

FIGURE 9: A bar code and a bar code scanner. Suggest why a laser is used to 'read' the bar code.

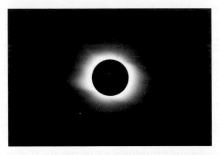

FIGURE 8: Lasers are a source of concentrated light waves. What does this show about light?

FIGURE 10: (left) Total eclipse of the Sun. How does this show that light travels in straight lines?

Laser pointers are often used when giving a presentation or teaching.

Sometimes light appears to 'bend' but there is always a reason for the change in direction.

What is light?

Scientists have argued about the nature of light for hundreds of years. In the seventeenth century:

> Newton described light as particles called **corpuscles**. Different colours were due to corpuscles of different sizes. He explained reflection in terms of the corpuscles bouncing off the reflecting surface.

> Huygens described light as waves that spread out in all directions. Different colours were due to different wavelengths.

Questions

7 How does the picture of a palm tree show that light travels in straight lines?

8 What was the main difference between the two theories of light that scientists disagreed about in the seventeenth century?

Wave nature of light

Two overlapping sets of light waves from the same source interfere to produce a series of bright and dark bands, or **fringes**.

> Bright bands are constructive interference.

> Dark bands are destructive interference.

To get an interference pattern between light waves from the two slits, the slits must be

> very narrow so that the light diffracts

> close together so that the waves overlap.

This was first done by Young in 1801. It is much easier to do using laser light as a larger and brighter fringe pattern can be produced. Interference gives strong evidence for the wave nature of light. Thinking of light as a particle cannot account for the cancelling effect seen in Figures 11 and 13; it can only be explained by a wave model.

Q shadows and eclipses for KS4   how bar code scanning works

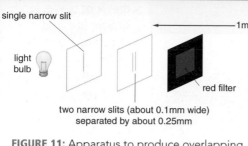

FIGURE 11: Apparatus to produce overlapping light waves. Why is a red filter used?

Did you know?

Electrons have wave properties. They can be diffracted, but their wavelength is much smaller than that of light.

Light diffracting through a single slit can also produce an interference pattern. (See Figure 13.)

Polarisation

All electromagnetic waves are transverse waves. Oscillations occur at right angles to the wave direction.

> In ordinary light, these oscillations occur in every direction perpendicular to the wave direction.

> **Polarised** light has oscillations in only one plane.

> Longitudinal waves cannot be polarised.

many planes of oscillation (unpolarised)   one plane of oscillation (polarised)

FIGURE 12: What is polarised light?

Questions

9 Explain why the particle theory of light is not universally accepted.

10 Can sound waves be plane polarised? Explain your answer.

Remember!
Only transverse waves can be plane polarised.

More about light

Single slit diffraction

The size of the gap must be of the order of the wavelength of light for diffraction to occur. The wavelength of light is between 400 and 700 nanometres (0.000 000 4 m and 0.000 000 6 m). Light diffracting from a single slit creates a diffraction pattern due to interference between waves from different parts of the slit. The centre band is wider and much brighter than the other fringes. (See Figure 13.)

Polaroid filters and sunglasses

Polaroid stops all oscillations except those in one plane. This cuts down the amount of light getting through. Light reflected from surfaces such as asphalt roadways and water is partly plane polarised. This reflected light can cause a glare.

The use of Polaroid sunglasses blocks this partially polarised light, removing reflected light causing the glare.

Wave or particle theory?

> Newton's corpuscular theory predicted light would travel faster in a denser medium.

> Huygen's wave theory predicted light would slow down in a denser medium.

The speed of light could not be measured at that time. Newton's influence was so strong that his particle theory was believed to be correct. When the speed of light could be measured it was found that light travels more slowly in glass than in air. For once, Newton was wrong!

FIGURE 13: Light diffracting from a single slit also creates a **diffraction** pattern. What are these light and dark bands called?

Questions

11 What causes a single slit diffraction pattern?

12 Why do fishermen wear Polaroid sunglasses?

13 Why did the wave theory of light become generally accepted?

Refraction of waves

You will find out:
> about refraction
> how to calculate refractive index
> about dispersion

Shimmering water or just a mirage?

A mirage may be seen when the layer of air closest to the ground is warmer than the air immediately above it. Typically, such a layer is above a sunlit road or sandy area such as a desert. The temperature difference causes a gradual change in air density and makes the light bend, or refract, so that it looks as if there is water nearby.

FIGURE 1: A mirage in the Tanzerouft desert, Algeria. What causes a mirage?

Refraction

Refraction occurs when light passes from one medium (or material) to another. It usually causes a change in the direction of a wave.

When a ray of light passes from air into glass the **angle of incidence**, i, is greater than the **angle of refraction**, r.

Rainbow or spectrum

Sunlight (white light) is made up of a mixture of many different colours. It is refracted when it passes through raindrops and is split up into its different colours. This is called **dispersion**.

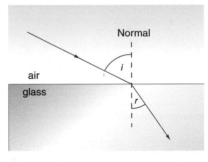

FIGURE 2: Refraction. Why does the ray of light change direction?

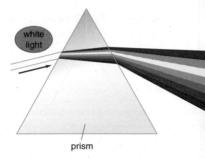

FIGURE 3: Dispersion in a glass prism. What colour light is bent the most?

Remember!

A good way of remembering the order of refraction of white light is: 'ROYGBIV'
red orange yellow green blue indigo violet
Red light has the longest wavelength and violet the shortest wavelength.

Questions

1 Light passes from glass to air. Which is bigger, the angle of incidence or the angle of refraction?

2 Why does light refract when it passes through raindrops?

Refraction and dispersion

When light, and other electromagnetic waves, enter a different medium, the speed of the waves changes.

> When light waves enter an optically denser medium (for example, air to glass) they slow down and the waves bend towards the **normal**; r_1 is less than i_1.

> When light waves enter an optically less dense medium (for example, glass to air) they speed up and the waves bend away from the normal; r_2 is *greater* than i_2.

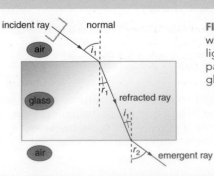

FIGURE 4: Describe what happens to the light waves as they pass through the glass block.

The **refractive index** indicates the amount of bending.

refractive index, $n = \dfrac{\text{speed of light in vacuum}}{\text{speed of light in medium}}$

Example:

> Light travels at a speed of 300 000 000 m/s (3×10^8 m/s) in a vacuum and 200 000 000 m/s (2×10^8 m/s) in a glass block. Calculate the refractive index of glass.
>
> $n = \dfrac{\text{speed of light in vacuum}}{\text{speed of light in medium}} = \dfrac{300\ 000\ 000}{200\ 000\ 000} = 1.5$

Why does dispersion occur?

Dispersion occurs when white light passes through a medium such as glass or water. In a vacuum (or in air) all the spectral colours travel at the same speed. But each spectral colour slows down by a different amount on entering another medium and speeds up by a different amount on leaving. This means the different colours are refracted by different amounts so they separate and spread out.

> Red light has the longest wavelength and is bent least.

> Violet and blue light have the shortest wavelengths and are bent most.

Questions

3 Glass has a greater refractive index than water. Is light bent more or less when entering water than glass? Explain your answer.

4 Copy and complete the two diagrams.

5 Calculate the refractive index of water if the speed of light in water is 225 000 000 m/s and its speed in a vacuum is 300 000 000 m/s.

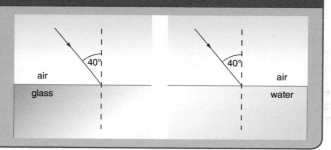

More about refraction

> When light waves enter an optically denser medium they are slowed down; their wavelength decreases. They are refracted *towards* the normal.

> When light waves enter an optically less dense medium they speed up; their wavelength increases. They are refracted *away from* the normal.

How much the light bends depends on the refractive index of the material.

refractive index, $n = \dfrac{\text{speed of light in vacuum}}{\text{speed of light in medium}} = \dfrac{v_o}{v}$

Rearranging: $v_o = n \times v$ $v = v_o / n$

Example:

> How fast does light travel in a diamond of refractive index 2.4 if its speed in a vacuum is 300×10^6 m/s?
>
> $n = \dfrac{v_o}{v}$ $v = \dfrac{v_o}{n} = \dfrac{300 \times 10^6}{2.4} = 125 \times 10^6$ m/s

Air and a vacuum behave in almost the same way, so the refractive index of air is taken to be 1. The denser the medium, the smaller the speed of light in that medium and the greater its refractive index.

medium	speed of light in that medium in m/s	refractive index
air	300×10^6	1.0
glass	200×10^6	1.5
water	225×10^6	1.33

How is the refractive index of a material related to the speed of light in it?

Explaining dispersion

> Red light travels faster than blue light in glass.

> Red light has a smaller refractive index than blue light.

> Red light has a smaller angle of refraction than blue light; it is bent less.

> Each colour has a different angle of refraction, forming a spectrum.

Remember!
Refractive index is a ratio so it doesn't have a unit.

Questions

6 The refractive index of ice is 1.31. How fast does light travel in ice if its speed in a vacuum is 300×10^6 m/s?

7 Explain why blue light has a larger angle of refraction than red light when white light is passed through a glass prism.

Reflection of light at a boundary

When light passes *from* glass or water, to air, most of the light is refracted but there is also a weak reflected ray. If the angle of incidence is big it is possible for all the light to be reflected and none refracted. The surface of the glass acts like a perfect mirror. This is called **total internal reflection**:

> 'total' because all the light is reflected

> 'internal' because it only happens inside a denser medium.

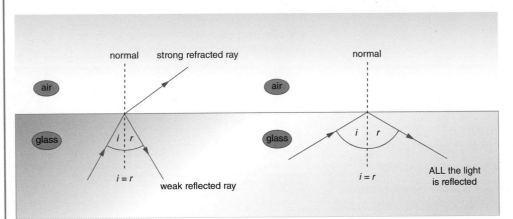

FIGURE 5: Total internal reflection. When total internal reflection happens, is the angle of incidence big or small?

Total internal reflection only occurs when:

> light passes from materials like glass or water, to air

> the angle of incidence is large.

Some uses of total internal reflection:

> optical fibres to transmit pulses of light – for example, in telecommunications and endoscopes

> binoculars to shorten their length

> bicycle reflectors so that the bicycle is seen in a car's headlights

> cat's-eyes to mark the centre of a road in the dark.

 Questions

8 Draw diagrams to show the path of a ray of light when it passes from water to air if the angle of incidence is: **a** small **b** large.

9 Suggest why light that has been totally internally reflected is as bright as the original light source.

Critical angle, c, and total internal reflection

Light from a ray box is aimed at the centre of the flat side of a semicircular glass block. (See Figure 7.) The **critical angle** is the angle of incidence for which the angle of refraction is 90°. The value of the critical angle is different for different media. For glass it is about 42° and for water it is about 49°.

Changing direction

Light can 'bend' round corners along an optical fibre. (See Figures 8 and 9.)

Total internal reflection happens because the fibre is very narrow and the angle of incidence is always large.

Using total internal reflection

> Optical fibres are widely used in telecommunications.

> Binoculars have prisms to turn the image the right way up.

> Cat's-eyes and bicycle reflectors use total internal reflection. (See Figure 10.)

> Paint containing a large number of cylindrical glass beads is used for some road signs.

FIGURE 6: A diamond is cut so that the faces produce total internal reflection. How does this make it sparkle?

🔍 total internal reflection for KS4

FIGURE 7: Critical angle. Why is it called TOTAL internal reflection?

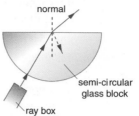

normal

semi-circular glass block

ray box

When the angle of incidence is less than the critical angle, c, most of the light is refracted out of the glass block.

normal

c

i = c and the angle of refraction = 90°

normal

i r

When the angle of incidence is greater than the critical angle, all the light is totally internally reflected.

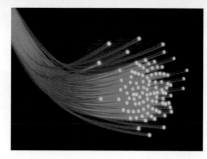

FIGURE 8: Total internal reflection occurs in optical fibres. Give one use of optical fibres.

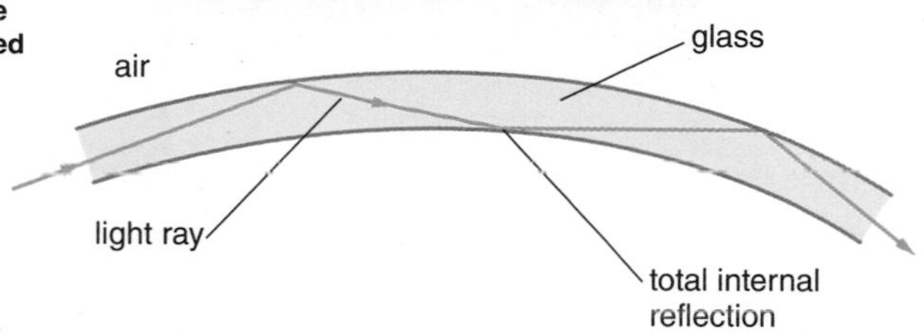

glass

air

light ray

total internal reflection

FIGURE 9: Light travelling along an optical fibre. Why does total internal reflection happen?

FIGURE 10: A bicycle reflector uses total internal reflection. How does this make the cyclist visible to an approaching motorist?

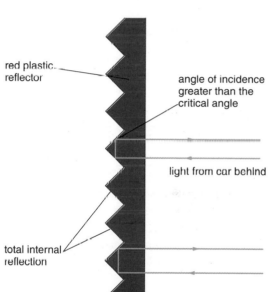

red plastic reflector

angle of incidence greater than the critical angle

light from car behind

total internal reflection

Questions

10 Why is a ray of light not refracted on entering a semicircular glass block as shown in Figure 7?

11 Draw diagrams to show how a ray of light can be turned through **a** 90° **b** 180° inside an isosceles right-angled prism.

Total internal reflection

Total internal reflection only happens when:

> a ray of light travels from one medium towards another with a lower refractive index (for example, glass to air or glass to water) so that the angle of refraction reaches 90° as the angle of incidence is increased.

> the angle of incidence is greater than the critical angle.

The higher the refractive index of a material, the lower its critical angle.

	refractive index	critical angle
glass	1.5	42°
water	1.33	49°

Questions

12 Explain why total internal reflection only happens when a ray of light travels from one medium to another with a lower refractive index.

13 The refractive index of diamond is 2.4. What can you say about its critical angle? Explain how this makes it sparkle.

Q uses of total internal reflection for KS4   how cat's-eyes work

Optics

You will find out:
> about convex lenses
> about the focal length of a convex lens
> how a camera and a projector produce a real image

In the future, will we still go to the 'movies'?

A visit to the cinema to see the latest movie has been a favourite pastime ever since the early days of silent movies over 100 years ago.

Today's cameras use a complex arrangement of lenses to zoom in and focus on the actors. Cinema projectors are also much more sophisticated. Cinema has survived the arrival of television, but will it survive the current rapid developments of new technology?

FIGURE 1: What do you think the future holds for the cinema?

Convex lenses

> A **convex lens**, or **converging lens**, is narrow at each end and 'bulges' in the middle.

> Light travelling parallel to the principal axis of a lens passes through the focal point after refracting through the lens. It is brought to a **focus**.

> The **focal length** is measured from the centre of the lens to the focus (**focal point**).

> Thicker lenses ('fat' lenses) bend light more and are therefore described as more **powerful**. Powerful lenses have short focal lengths.

Projectors and cameras

Both a **projector** and a **camera** use a convex lens to bring light to a focus on a screen or piece of film. This is called a **real image**; light actually reaches the screen or film.

FIGURE 3: Projecting onto a screen during a lecture. What type of image is brought to a focus on the screen?

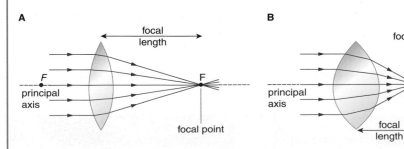

FIGURE 2: The thicker a convex lens, the more it bends light. Which lens has the longer focal length?

Questions

1 The eye can be likened to a camera. What replaces the film?

2 a Which lens shown in Figure 2, A or B, is the more powerful?

b What does this mean?

3 Name **two** things that contain a lens.

Using a convex lens to bend beams of light

A **parallel** beam of light from a distant object can be converged to a focus in the **focal plane**. (See Figure 4.)

> O is the optical centre of the lens.

> F is the focal point (focus) of the lens.

> OF is the focal length of the lens.

If the beam of light is parallel to the **principal axis** it converges to the focal point, F, on the principal axis, as shown in Figure 2.

A **diverging** beam of light from a near object can be converged using a strong convex lens. The light meets at a point beyond the focal plane. (See Figure 5.)

Q convex lenses for KS4

Cameras

> A simple camera is a box with a convex lens that focuses light onto a film. A digital camera focuses light onto a **light sensitive chip** to create a digital image.

> Light from an object passes through the lens and is brought to a focus on the film or light sensitive chip.

> A real image, much smaller than the object, is formed. Light actually reaches it and makes a record of the pattern of light.

Projectors

> In a projector, the slide is placed closer to the lens than the object in a camera.

> A real image, much larger than the slide, is formed on a screen.

> The position and size of the image depends on the distance of the object from the lens and the screen.

> Two examples of ray diagrams used to locate the position of an image are shown in Figure 6.

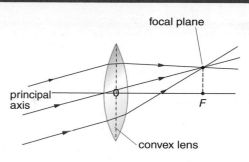

FIGURE 4: Converging a parallel beam of light. Where do the rays come to a focus?

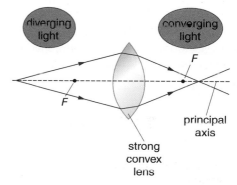

FIGURE 5: Converging a diverging beam of light using a strong convex lens. Where does the light focus?

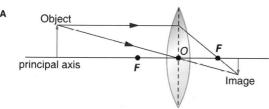

object more than twice the focal length from lens

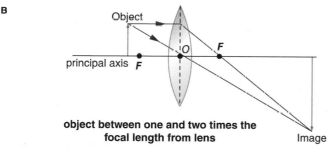

object between one and two times the focal length from lens

FIGURE 6: Which diagram, A or B, shows the action of a convex lens in (i) a camera (ii) a projector?

Questions

4 Explain, in terms of refraction, why light converges after passing through a convex lens.

5 If the rays of light in Figure 5 diverged from a point closer to the convex lens, where would they be brought to a focus?

6 How does the image change when the screen is moved further away from a projector?

Ray diagrams

The position of the image formed by a convex lens can be found by scale drawing. The path of any two rays from the top of an object, O, is known (see Figure 6):

> a ray parallel to the principal axis refracts through the focus, F, of the lens

> also a ray through the optical centre, O, of the lens is not deviated and follows a straight line

> a ray through the focus, F, of the lens refracts parallel to the principal axis

(Note: Light does refract as it passes through a lens but it is shown bending at the central axis for simplicity.) The image of the top of an object is located where the two known rays cross.

Questions

7 Draw a ray diagram to locate the image produced by a convex lens when an object is placed on the principal axis at a distance of more than twice the focal length from the lens. Write down **three** things about the image.

8 Draw a scale diagram to locate the position and size of the image produced by a convex lens of focal length 10 cm when a 4 cm tall object is placed on the principal axis, at a distance of 15 cm from the lens.

You will find out:

> about some uses of convex lenses

> how the images produced by cameras and projectors are focused

> about real and virtual images

> how to calculate magnification

Uses of convex lenses

Magnifying glass

When a convex lens is put *close to* an object it makes it appear bigger.

> In a camera, a convex lens focuses an image onto film or, in the case of a digital camera, on to a light-sensitive chip, storing it in the camera's memory as thousands of minute, coloured, dots called **pixels** (short for 'picture elements').

> In a projector, a convex lens focuses an image onto a screen. There are several types such as a slide, film or data projector.

> Convex lenses are used in some spectacles, for example reading glasses.

FIGURE 7: What does a magnifying glass do?

FIGURE 8: Miniature spy camera used for secret surveillance at home or in the work place. It can be connected to standard video and television equipment. Is this a film or a digital camera?

FIGURE 9: A data projector projects data directly from a computer onto a screen. What sort of lens is used?

Questions

9 Write down **one** similarity and **one** difference between a camera and a projector.

10 Suggest how a telescope helped the astronomer Galileo to discover the planet Jupiter.

11 The image produced by a slide projector is inverted. Why is this not a problem?

Did you know?

The image produced on the retina of your eye is upside down. Your brain has learned how to interpret it.

More on uses of convex lenses

Magnifying glass

A magnifying glass uses a convex lens to produce an enlarged image.

> The lens must be very close to the object.

> The image is the right way up.

> The image cannot be captured on film or screen.

Camera

> A traditional camera uses a convex lens to produce a small, **inverted**, **real** image on a piece of film.

> A digital camera focuses light onto a device that records light electronically, creating a digital image. This is a long series of 1 s and 0 s that represents all the pixels that make up the image. The more pixels the more detail is seen.

> Cheaper cameras have a fixed lens. There is only one object distance that gives a well-focused image. In more expensive cameras the lens is moved in and out, often automatically, to give a sharp image.

> The **shutter** opens and shuts very quickly to let light briefly into the camera. The shutter speed, or exposure time, can be varied in some cameras.

> The **aperture** is the size of the hole through which light enters. In some cameras it can be adjusted depending on the surrounding light level.

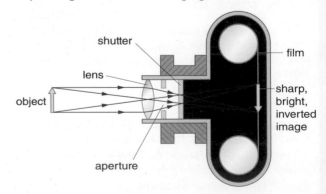

FIGURE 10: How a camera works. In a digital camera the film is replaced with an electronic recording device. What is the aperture?

Q how a camera works   history of telescopes

Projector

A projector uses a convex lens (projection lens) to produce a large, inverted, real image on a screen.

> The projection lens and/or screen can be moved backwards and forwards to give a well-focused image.

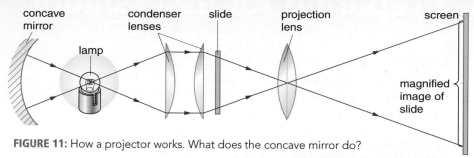

FIGURE 11: How a projector works. What does the concave mirror do?

> The **condenser lenses** concentrate light on the slide so that it is evenly lit and gives a clear, bright image.

> The **concave mirror** reflects light from the lamp back to the condenser lenses.

Digital projectors project a digital image directly from a computer through a projection lens on to a screen. (See Figure 9.)

Magnification

magnification, $m = \dfrac{\text{image size}}{\text{object size}}$

If the image is:

> larger than the object, m is greater than 1

> smaller than the object, m is less than 1.

Example:

> A 3 cm projector slide is 90 cm high when projected onto a screen. Find the magnification produced by the lens.
> $$m = \frac{\text{image size}}{\text{object size}} = \frac{90}{3} = 30$$
> Remember that ratios have no units.

Questions

12 What happens to the image from a slide projector if the screen is moved further away from the lens?

13 Sam is taking Ali's picture. Ali moves closer to the camera. Which way must the camera lens move to keep the image in focus?

14 Abi is 150 cm tall. Joe takes a picture of her. On the film she is 0.5 cm tall. Calculate the magnification.

15 A ladybird is 0.5 cm long. When looked at under a magnifying glass it appears to be 3 cm long. Find the magnification produced.

More on images

Images produced by a projector or camera:

> are real, i.e. can be projected onto a screen or film

> are inverted.

The image produced by a magnifying glass:

> is virtual, i.e. cannot be projected onto a screen or film

> is the right way up.

Example:

> A projector lens has a magnification of 8. How big is the image of a 2 cm high tree on the screen?
> Image size = object size × m = 2 × 8 = 16 cm.

FIGURE 12: A virtual image. When is a virtual image produced?

A virtual image is produced when an object is less than the focal length from the lens. It is called a virtual image because no light passes through it and it cannot be captured on a screen or film.

More about magnification

magnification, $m = \dfrac{\text{image size}}{\text{object size}}$

image size = object size × m object size = m / image size

Questions

16 a What is meant by a virtual image?

b Under what condition does a convex lens form a virtual image?

c What can you say about the magnification of the virtual image produced by a magnifying glass?

17 Nisha uses a microscope with a magnification of 40. She looks at a bacterium 0.3 mm in size under the microscope. How large does it appear to be?

magnification of optical instruments

Preparing for assessment: Analysis and evaluation

To achieve a good grade in science, you not only have to know and understand scientific ideas, but you need to be able to apply them to other situations and investigations. These tasks will support you in developing these skills.

✳ Tasks

> Find out how the position and magnification of the image produced by a lens depends on the distance of the object from the lens.

> Find out the effect, if any, of the focal length of the lens.

✳ Context

Convex lenses are used in a wide range of optical instruments. Cameras and projectors use convex lenses to form a real image of an object.

> In a camera the image is smaller than the object.

> In a projector the image is larger than the object.

This is achieved by changing the distance of the object from the lens. When light strikes a convex lens it is refracted by the lens and brought to a focus. The distance of the focus from the centre of the lens is its focal length. 'Fat' lenses are more powerful and have short focal lengths.

$$\text{Magnification} = \frac{\text{height of image}}{\text{height of object}}$$

✳ Planning your investigation

These are the things you will need to consider when planning your investigation. (You can develop your plan in groups of two or three.)

1 Place a metre rule on the bench to keep your apparatus aligned and to take measurements easily. (You can use a special optical bench if available.)

2 Place a ray box fitted with an object card (e.g. a card with a square hole and cross-wires) so that the 'object' is at the zero mark on the metre rule. Measure the height of the object.

3 Place a convex lens in a holder so that it is about 2 cm more than its focal length from the 'object'.

4 Note the reading of the centre of the lens on the metre rule. Record this distance. It is the distance of the lens from the object (d_o).

5 Move a screen back and forth along the metre rule beyond the lens until a clearly focused image of the object is seen. Note the reading on the metre rule and record the distance of the image from the lens (d_i). Measure the height of the image.

6 Repeat step 5 at least once more and find the mean value of d_i and the height of the image.

7 Repeat the experiment at least five more times increasing the object distance up to about 5 cm more than twice the focal length of the lens.

8 To find out if other lenses behave in the same way, replace the lens with one of different focal length and do the whole experiment again.

Try to write the plan in a logical order and ask yourself if someone can perform the investigation following just your plan.

✳ Results, analysis and evaluation

A group of students found these results when they tested two lenses of different focal lengths.

Lens 1 (focal length = 8 cm) object height = 1.0 cm					
distance of object from lens in cm (d_o)	distance of image from lens in cm (d_i) 1	distance of image from lens in cm (d_i) 2	mean d_i in cm	height of image in cm	magnification
10.0	39.6	40.5		4.0	
12.0	24.2	24.0		2.6	
14.0	18.9	18.3		2.0	
16.0	15.8	16.2		1.1	
18.0	14.1	14.7		0.83	
22.0	12.4	12.8		0.57	
25.0	11.5	12.1		0.47	

Lens 2 (focal length = 5 cm) object height = 1.0 cm					
distance of object from lens in cm (d_o)	distance of image from lens in cm (d_i) 1	distance of image from lens in cm (d_i) 2	mean d_i in cm	height of image in cm	magnification
6.0	31.2	29.0		5.0	
7.0	17.3	19.6		2.6	
8.0	13.3	12.8		1.6	
9.0	10.8	11.1		1.2	
10.0	10.4	9.7		0.97	
12.0	9.2	9.0		0.75	
16.0	7.1	7.5		0.46	

1 Complete the blank columns in each table.

2 Use the same set of axes to plot graphs of mean image distance, d_i, (y-axis) against object distance, d_o, (x-axis).

3 Draw lines of best fit through the points.

4 Is there a simple relationship between the object and image distances? Explain your answer by referring to the shape of the line graph.

5 Do all convex lenses behave in the same way? Explain your answer by referring to the two graphs.

6 It is suggested that magnification is linked to the steepness, or gradient, of the graph. Test this and comment.

7 How could the students improve their experiment?

8 How precise were their results? How do you know?

> Think about the scale for the axes. Use sensible divisions but make the graph as large as possible.

> Should these be straight lines or smooth curves? Never draw dot to dot.

> What shape graph would give a simple relationship?

> What measurement(s) were hard to do?
> How could you do it better?

> Do not confuse reliability with accuracy; precision is about the repeatability of results.

P5 Checklist

To achieve your forecast grade in the exam you'll need to revise

Use this checklist to see what you can do now. It gives you many of the important points you will need to know. Refer back to the relevant pages in this book if you're not sure and to see if there is anything else you need to know. Look across the three columns to see how you can progress.

Remember you'll need to be able to use these ideas in various ways, such as:

> interpreting pictures, diagrams and graphs
> applying ideas to new situations
> explaining ethical implications
> suggesting some benefits and risks to society
> drawing conclusions from evidence you've been given.

Look at pages 272–294 for more information about exams and how you'll be assessed.

To aim for a grade E	To aim for a grade C	To aim for a grade A
recall that gravity is the universal force of attraction between masses **recognise** that a satellite is an object that orbits a larger object in space **describe** how the height above Earth's surface affects the orbit of an artificial satellite and its use **recall** some uses of artificial satellites	**explain** why the Moon remains in orbit around Earth and Earth and other planets around the Sun **describe** the orbit of a geostationary artificial satellite **understand** that gravity provides the centripetal force for orbital motion **explain** why different satellite applications require different orbits	**understand** variation of gravitational force **explain** how the orbital period of a planet depends on its distance from the Sun **understand** that artificial satellites accelerate towards Earth due to gravitational pull but keep moving in an approximately circular orbit **explain** why artificial satellites in low orbit travel faster than those in higher orbits
recall that direction is important when describing motion **understand** how relative speed depends on direction of movement **calculate** distance travelled from distance = average speed × time **use** the equation $v = u + at$ to find final speed	**describe** the difference between scalar and vector quantities **calculate** the vector sum from vector diagrams of parallel vectors **use** and manipulate the equations: $v = u + at$ $s = \dfrac{(u + v) \times t}{2}$	**calculate** the resultant of two vectors that are at right angles to each other **use** and manipulate the equations: $v^2 = u^2 + 2as$ $s = ut + \frac{1}{2}at^2$
recall that the path of an object projected horizontally in Earth's gravitational field is curved and is called the trajectory **recognise** examples of projectile motion **recall** that the range of a projectile depends on the launch angle, with an optimum of 45°	**describe** the trajectory of an object projected in Earth's gravitational field as parabolic **recall** that horizontal and vertical velocities of a projectile are vectors **recall** that a projectile has no horizontal acceleration but the acceleration due to gravity acts vertically **interpret** data on the range of a projectile at varied launch angles	**calculate** the resultant velocity of a projectile **use** the equations of motion for an object projected horizontally above Earth's surface **explain** how the horizontal velocity is constant but the vertical velocity changes due to gravity
recognise that every action has an equal and opposite reaction **describe** the opposite reactions in a parallel collision **explain**, using a particle model, how a gas exerts a pressure **recall** that in a rocket the force on particles backwards equals that pushing the rocket forwards	**know** that when two bodies interact, they exert an equal and opposite force on each other **describe** the opposite reactions in static situations **explain**, using a particle model, how a change in volume or temperature produces a change in pressure **explain** rocket propulsion using simple kinetic theory	**apply** the principle of conservation of momentum **explain** pressure in terms of the rate of change of momentum of particles and frequency of collisions **explain** how sufficient force is created to lift a large rocket to put a satellite into Earth orbit

To aim for a grade E

recall that different frequencies are used for low orbit and geostationary satellites

recall that some radio waves are reflected by part of Earth's upper atmosphere and other radio waves and microwaves pass through

recall that radio waves have a very long wavelength so can spread around large objects or spread out from a gap

---

describe the interference of overlapping waves and its effect in different contexts

recall that light travels in straight lines but under certain circumstances can 'bend'

recall that all electromagnetic waves are transverse

recall that explanations of the nature of light have changed over time

---

recognise that refraction occurs when a wave passes from one medium to another

explain why a ray going from air to glass has angle of incidence greater than angle of refraction

describe that dispersion happens when light is refracted and link the order of spectral colours to orders of wavelengths

recognise that some or all of a light ray can be reflected when it travels from glass or water to air

recall some uses of total internal reflection (TIR)

---

identify the shape of a convex, or converging, lens and link its thickness to its focal length

recall that convex lenses produce real images on a screen

recall that convex lenses are used as magnifying glasses, in cameras, projectors and in some spectacles

To aim for a grade C

describe how information is transmitted to and from artificial satellites using microwaves

describe how electromagnetic waves behave in the atmosphere

recall the wave patterns produced by a plane wave passing through different-sized gaps

---

describe interference in terms of reinforcement and cancellation of two waves

describe diffraction of light for a single slit and double slits, and its evidence for the wave nature of light

explain what is meant by plane polarised light

explain why the particle theory of light is not universally accepted

---

explain why refraction occurs at the boundary between two media and link it to wave speed

use the equation:

$$\text{refractive index} = \frac{\text{speed of light in vacuum}}{\text{speed of light in medium}}$$

explain dispersion in terms of spectral colours having different wave speeds in different media but the same speed in a vacuum

describe what happens to light incident at a glass/air surface at or above the critical angle

describe the optical path in devices using TIR

---

describe the effect of a convex lens on diverging and parallel light beams

describe how a convex lens produces a real image

describe the use of a convex lens as a magnifying glass

use the equation:

$$\text{magnification} = \frac{\text{image size}}{\text{object size}}$$

To aim for a grade A

explain why satellite transmitting and receiving dishes need very careful alignment

describe how electromagnetic waves with different frequencies

describe how the amount of diffraction depends on the size of gap and wavelength of the wave

---

explain interference patterns in terms of constructive and destructive interference

explain a diffraction pattern for light, to include size of gap and interference of diffracted waves

explain how polarisation is used in Polaroid filters and sunglasses

explain how the wave theory supplanted the particle theory as the evidence base changed

---

interpret data on refractive indices and speed of light to predict direction of refraction

use and **manipulate** the equation for refractive index

explain dispersion in terms of spectral colours having a different speed in glass so different refractive indices

explain the conditions under which total internal reflection (TIR) can occur

describe how the refractive index of a medium relates to critical angle

---

explain the refraction of light rays by a convex lens

explain how to find the image formed by a convex lens by drawing ray diagrams

describe the properties of real and virtual images

use and **manipulate** the equation:

$$\text{magnification} = \frac{\text{image size}}{\text{object size}}$$

P5 Exam-style questions

Foundation Tier

1 Two cars are travelling at 20 m/s on a straight road.

AO2 **(a) (i)** What is their relative speed if both are moving in the same direction? [1]

AO2 **(ii)** What is their relative speed if they are moving in opposite directions? [1]

AO1 **(b)** One car accelerates at 3 m/s² for 5 s. How fast
AO2 is it moving after 5 s? [3]
[Total: 5]

2 Joe hits a golf ball on a still day, so there is no air resistance.

AO1 **(a)** Sketch its trajectory. [1]

AO2 **(b) (i)** What is its horizontal acceleration? [1]

AO2 **(ii)** What is its vertical acceleration? [1]

AO2 **(c)** What launch angle should Joe choose for maximum range? [1]
[Total: 4]

3 Pragya has a mass of 50 kg. She runs towards a wall at a speed of 6 m/s.

AO2 **(a)** Calculate Pragya's momentum. [3]

Pragya pushes against a wall with a force of 25 N. This is called an action force.

AO2 **(b) (i)** Describe the reaction force. [1]

AO2 **(ii)** How big is the reaction force? [1]
[Total: 5]

4 (a) The diagram shows light meeting an air–glass boundary.

AO2 **(i)** Complete the path of the ray of light in the glass. [1]

AO2 **(ii)** Mark the angle of refraction, r. [1]

(b) When you see a rainbow, raindrops have refracted the sunlight and split it into a spectrum of colours.

AO1 **(i)** Complete the colours of the rainbow:
red ..... yellow ...... blue ..... violet [1]

AO1 **(ii)** Which colour has the longest wavelength? [1]
[Total: 4]

5 The diagram shows parallel rays of light entering a convex lens.

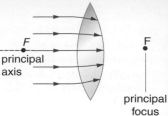

AO2 **(a) (i)** Complete the paths of the rays of light after passing through the lens. [2]

AO2 **(ii)** Label the focal length. [1]

A projector uses a convex lens to produce a real image of a slide.

AO2 **(b) (i)** What is meant by 'a real image'? [1]

AO2 **(ii)** A slide is 25 mm high. When projected, the image on the screen is 1000 mm high. What is the magnification of the lens? [3]
[Total: 7]

AO3 **6** Waves can travel from one side of the Earth to the other in three ways as shown in the diagram.

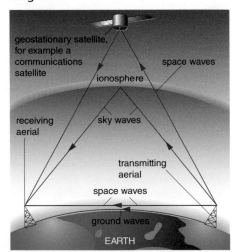

The table shows the frequency and wavelength ranges of some radio waves, **A** to **D**.

wave	frequency	wavelength
A	300 kHz–3 MHz	1 km–100 m
B	3 MHz–30 MHz	100 m–10 m
C	30 MHz–30 GHz	10 m–10 mm
D	above 30 GHz	less than 10 mm

Explain which type of wave is suitable for each type of transmission.

The quality of written communication ✎ will be assessed in your answer to this question.

[6]

AO1 recall the science   AO2 apply your knowledge   AO3 evaluate and analyse the evidence

✱ Worked Example – Foundation Tier

Imran has been looking at the interference of water waves from two point sources in a ripple tank.

He sketches the pattern seen. The red lines show the crests of the waves.

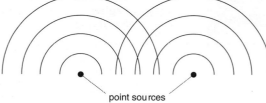

(a) (i) Mark a point on Imran's diagram where the crests of two waves meet. Label it C. [1]

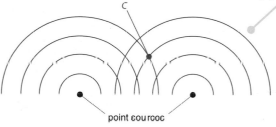

(ii) Mark a point on Imran's diagram where the troughs of two waves meet. Label it T. [1]

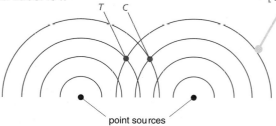

(iii) What did Imran notice at C and T in the ripple tank? [2]

Big waves.

(iv) What happens when a crest and trough meet? [1]

Nothing

(b) Describe the interference of the two sets of water waves. Use words such as crest, trough, reinforcement and cancellation in your answer. [2]

When two crests meet there is reinforcement, making a high wave. When two troughs meet there is cancellation and the water is flat.

How to raise your grade!

Take note of these comments – they will help you to raise your grade.

C has been labelled correctly.
1/1

T is marked incorrectly. Troughs are mid-way between two crests but are not indicated on the diagram. T should be placed in the overlap region so that it is mid-way between two wave crests from both sources. 0/1

The answer needs to be more specific to score both marks. There is a higher crest at C and a deeper trough at T; about twice the amplitude of the wave from a single source. 1/2

'Nothing' is an ambiguous response. There will be calm water as the crest and trough cancel out. 0/1

The first sentence is correct and scores one mark. But two troughs combine to form a very deep trough. Cancellation occurs when a crest and trough meet. 1/2

This student has scored 3 marks out of a possible 7. This is below the standard of Grade C. With more care the student could have achieved a Grade C.

P5 Exam-style questions

Higher Tier

1 The speed of Halley's comet varies a lot during each orbit of the Sun. Explain how and why its speed changes. [4]

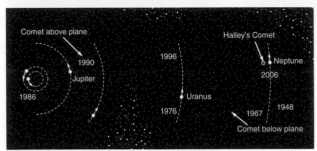

2 An aircraft increases its speed from 150 m/s to 250 m/s with a constant acceleration of 4 m/s².

AO2 **(a)** How far does it travel while accelerating? [3]

The aircraft flies horizontally, maintaining this speed. It starts to descend to drop a food package to a remote village.

AO2 **(b)** Calculate its resultant velocity when it has a vertical velocity of 50 m/s. [4]

The food package is released.

AO2 **(c) (i)** What force(s) act on the food package after it has been released? [1]

AO2 **(ii)** Describe the horizontal and vertical components of the package's motion as it falls. [2]

[Total: 10]

AO1 **3 (a)** Explain what is meant by the statement 'momentum is conserved' when two trolleys collide. [2]

AO2 **(b)** A trolley of mass 2 kg, moving at 4 m/s, collides with and sticks to a stationary trolley of mass 3 kg. With what velocity do the trolleys move after colliding? [3]

[Total: 5]

4 The diagram shows waves diffracting through a gap.

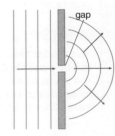

AO1 **(a)** What affects the amount of diffraction? [2]

AO2 **(b) (i)** Why are low-frequency radio waves not suitable for satellite communications? [2]

AO2 **(ii)** What frequency range is used for satellite communication? [2]

AO2 **(iii)** Explain why satellite transmission and receiving dishes need careful alignment. [3]

[Total: 9]

AO3 **5** Sometimes light behaves as if it is a particle and sometimes it behaves as if it is a wave. Explain, using examples, the evidence which has led scientists to decide which theory for the behaviour of light is correct.

The quality of written communication ✏ will be assessed in your answer to this question. [6]

AO2 **6 (a)** Light travels at 300×10^6 m/s in a vacuum or in air. Perspex has a refractive index of 1.5. Calculate the speed of light in Perspex. [3]

AO2 **(b)** When white light enters a triangular glass prism dispersion occurs.

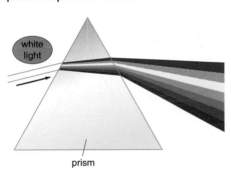

Explain why this happens. Talk about the properties of the spectral colours. [3]

[Total: 6]

7 A convex lens has a focal length of 5 cm. An object 3 cm high is placed on the principle axis, 7.5 cm from the centre of the lens.

AO2 **(a)** Draw an accurate ray diagram to locate the image. [4]

AO2 **(b) (i)** How far is the image from the lens? [1]

AO1 **(ii)** Is the image real or virtual? [1]

AO2 **(iii)** What is the magnification of the lens? [2]

[Total: 8]

AO1 **recall the science** AO2 **apply your knowledge** AO3 **evaluate and analyse the evidence**

Worked Example – Higher Tier

Annie is boating on a river. The boat has pulled into the bank and stopped. Annie steps off the boat onto the bank at 3 m/s. Her mass is 60 kg and the boat has a mass of 500 kg.

(a) Calculate the speed with which the boat moves away from the bank. [3]

There is no momentum before Annie moves so no momentum after she starts to move.

Annie's momentum forwards must equal the boat's momentum backwards.

Momentum = mass × velocity
$60 \times 3 = 500 \times v$
v = speed of boat backwards
$v = \dfrac{180}{500} = 0.36$ m/s

(b) Annie recalls stepping off a stationary skateboard and being surprised at how quickly it moved away. Explain why it moved more quickly than the boat. [2]

The mass of the skateboard is less than the mass of the boat.

(c) Use ideas about momentum to explain how a large force is produced to launch a space rocket. [3]

The momentum of the exhaust gases downwards equals the momentum of the rocket upwards. Since the rocket is very large a lot of exhaust gases are needed to get enough momentum.

P6 Electricity for gadgets

Ideas you've met before

Electric circuits

The components in a circuit affect how current passes through the circuit.

Current is measured with an ammeter.

Voltage is measured with a voltmeter.

Resistance is the opposition to the flow of electricity.

Circuits can be connected in series or in parallel.

 What happens to the brightness of bulbs as more are added into a circuit?

Electromagnetism

The wires carrying an electric current produce a magnetic field.

The magnetic field due to a coil of wire carrying a current is similar to that of a bar magnet.

 Draw the shape of the magnetic field around a bar magnet.

Sources of electricity

Cells are sources of electricity.

A battery is a number of cells joined together.

Energy sources can be used to drive electrical generators to produce electricity.

Electricity is transmitted around the country at high voltages.

What is the critical angle?

Electrostatics

Static electricity is due to the transfer of electrons.

Static electricity can be a nuisance and sometimes dangerous.

 Suggest one example of where electrostatics can be useful.

In P6 you will find out about...

> how an electric current is due to the flow of electrons

> how resistors are used to control the current in circuits

> how to calculate resistance

> electronic devices such as transistors and logic gates

> the uses of motors

> how motors work

> how motors can be made to spin faster

> how and why dynamos work

> generators in power stations

> what transformers do

> why electricity is transmitted at high voltages

> how diodes work

> how diodes can change AC into DC

> charging and discharging capacitors

> why capacitors are useful in smoothing the output from rectifier circuits

Resisting

You will find out:

> how a variable resistor affects a lamp or a motor

> about the relationship between current and voltage

Lighting circuits in theatres

A few minutes before the show the lights in the theatre start to fade.

'Dim the houselights' is the instruction from the lighting director.

FIGURE 1: The lights in the theatre fade as the show is about to start

Dim the lights

Lights at home are usually 'on' or 'off'. The lights are on when a current passes through the circuit. An electric current is a flow of charge carriers called electrons. These electrons normally move at random from one atom to another. Some collide with atoms or other electrons. These collisions cause **resistance**. Different elements have different structures so their resistance is different.

Some lighting circuits have **dimmer switches** that allow you to control the brightness of the bulbs.

FIGURE 2: What does a dimmer switch allow you to do?

Electrical units

> The unit of **voltage** is the **volt**, V.

> The unit of **current** is the ampere, usually abbreviated to **amp**, A.

> The unit of **resistance** is the **ohm**, Ω.

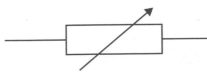

FIGURE 3: Circuit symbol for a variable resistor. Some dimmer switches contain a **variable resistor**. This is sometimes called a rheostat.

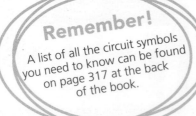

Remember!

A list of all the circuit symbols you need to know can be found on page 317 at the back of the book.

Questions

1 What electrical component is used in a dimmer switch?

2 What does a dimmer switch do to the brightness of a bulb?

3 Look at the equation for resistance on the opposite page. Calculate the resistance of a car headlamp bulb when the supply voltage is 12 V and the current is 2 A.

🔍 resistance in a circuit   DC current in a circuit

Variable resistors

A resistor affects an electric current. The higher the resistance, the lower the current. Resistance is caused by the charge-carrying electrons colliding with the atoms in the conductor and bouncing off in different directions. The current in a bulb affects its brightness. The higher the current, the brighter the bulb. A variable resistor allows the current to be increased from zero to a maximum.

Fast cars

Variable resistors are also used to control the speed of motors. The higher the voltage supplied to the motor, the faster it turns.

Ohm's law

Ohm's law relates voltage, current and resistance in the equation:

$$\text{resistance} = \frac{\text{voltage}}{\text{current}}$$

Questions

4 What happens to the speed of a model train if the resistance in the controller is increased?

5 A wire has a resistance of 25 Ω and a current of 0.2 A passing through it. Calculate the voltage between the ends of the wire.

FIGURE 4: How does a variable resistor control the speed of a model motor car?

Factors affecting resistance

The resistance of a length of wire depends on three factors:

> its length – the longer the wire, the greater the resistance

> its cross-sectional area – the larger the area, the smaller the resistance

> the material the wire is made from.

A variable resistor works by changing the length of the wire. The dimmer switch has a contact that moves around a circular coil of wire.

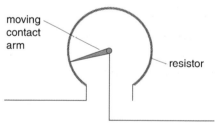

FIGURE 5: How does a dimmer switch work?

Question

6 A voltage of 6 V is connected between the ends of a coil. The coil has a resistance of 200 Ω. Calculate the current passing through the coil.

FIGURE 6: Some variable resistors have a coil of wire with a sliding contact.

You will find out:
> about the heating effect of an electric current
> about the properties of ohmic and non-ohmic conductors

Analysing circuits

> Ammeters are placed in **series** and measure current.

> Voltmeters are placed in **parallel** and measure voltage.

A variable resistor is used to change the current through the fixed resistor in Figure 7. When the current increases, the voltage across the resistor increases as well.

Let there be heat!

When a current passes through a wire, the wire gets hot. When a wire gets hot, its resistance increases.

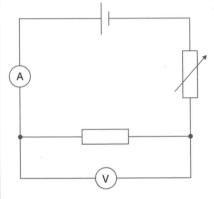

FIGURE 7: Circuit diagram. Is the ammeter placed in series or parallel in the circuit?

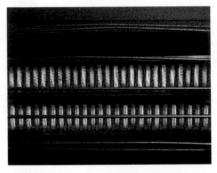

FIGURE 8: Why does an electric fire element glow red?

FIGURE 9: Why does a filament bulb glow white?

Questions

7 Which electrical meter is placed in parallel with a component?

8 The current in a wire is increased. What happens to the:

a voltage across the wire?

b resistance of the wire?

c temperature of the wire?

Measuring resistance

> A power supply, variable resistor, fixed resistor and ammeter are connected in series.

> A voltmeter is connected in parallel across the fixed resistor.

> As the variable resistor is changed, the readings on the ammeter and voltmeter are recorded.

> A graph of voltage against current is plotted.

A straight line passing through the origin shows that voltage is **proportional** to current. Conductors that show this proportional relationship are called **ohmic** conductors.

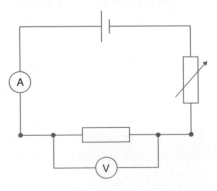

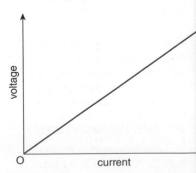

FIGURE 10: Circuit diagram used to measure resistance. What does the straight line through the origin on the graph show?

 measuring resistance GCSE ohmic and non-ohmic conductors

Replacing the fixed resistor with a filament bulb

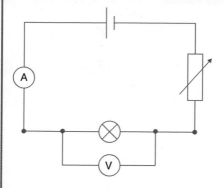

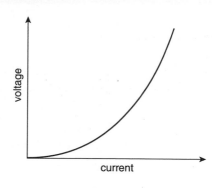

FIGURE 11: Replacing a fixed resistor with a filament bulb. What does the curved line on the graph show?

When the experiment is repeated with a filament bulb in place of the fixed resistor, the graph has a different shape.

It is no longer a straight line. This shows that the resistance is changing.

Resistance of a filament bulb

When charged electrons collide with the atoms in a conductor, this makes the atoms vibrate more.

This increased vibration leads to:

> an increase in the number of collisions so the resistance increases

> an increase in the temperature of the conductor.

Remember!

The resistance of a non-ohmic conductor is found from instantaneous values, not the gradient of the graph.

Questions

9 Sketch a graph to show what happens if current (*y*-axis) is plotted against voltage (*x*-axis) for a filament bulb.

10 Plot a graph to show how the voltage and current for a 10 Ω resistor are related for values of current up to 1 A.

Calculating resistance from a voltage–current graph

For a fixed resistor, the gradient of the graph is equal to the resistance.

The gradient of the voltage–current graph for a filament bulb increases. This means that the resistance increases.

Whilst the gradient of the voltage–current graph for a fixed resistor equals the resistance, the resistance of a filament bulb must be found by taking instantaneous values from the graph and using the resistance formula.

The resistance of a filament bulb cannot be found from the gradient of the graph.

Question

11 The table below shows results from an experiment to investigate the voltage–current relationship for a fixed resistor.

Current in A	Voltage in V
0.0	0.1
0.2	2.3
0.4	5.0
0.6	7.8
0.8	9.2
1.0	11.9

a Plot the points on a grid.

b Draw the best straight line through the origin and use the graph to find the resistance when the current is 0.6 A. (You could use a graph-plotting program on a computer for this question.)

Sharing

You will find out:
> how combinations of fixed and variable resistors are used as potential dividers
> how potential dividers produce the required voltage

Volume control

Music centres are just one household item using variable resistors. They control such things as volume, balance and tone.

Behind each knob is a potential divider which shares out the voltage to give you, the listener, the output you want.

FIGURE 1: How do you control the volume?

Potential divider

If two or more resistors are arranged in series, the total resistance between their ends increases. It is the sum of each resistance.
$R_{total} = R_1 + R_2 + R_3$

Adding a resistor in series increases the resistance.

Imagine a voltage of 10 V across two resistors. The voltage across one of them will be less than 10 V. We can choose the values of the resistors to give any value of voltage up to 10 V. The voltage across a resistor is sometimes called a potential difference. This is why resistors that are used to split a voltage are called potential dividers.

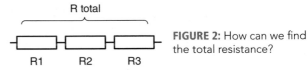

FIGURE 2: How can we find the total resistance?

R total

R1   R2   R3

Questions

1 A 10 Ω and a 33 Ω resistor are connected in series. What is the total resistance?

2 A potential divider can produce an output greater than the input. True or false?

The potential divider circuit

Some circuits use two fixed resistors in series with a supply voltage (V_{in}) between their ends.

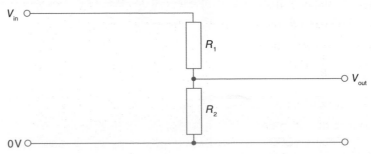

FIGURE 3: How are the fixed resistors arranged in a potential divider circuit?

Q resistors in series and parallel

The voltage between the ends of each resistor is proportional to its resistance because they both have the same current.

The output voltage is between the ends of one of the fixed resistors.

The output voltage depends on the relative values of the resistors.

If one of the resistors is a variable resistor, the output voltage can be altered.

Resistors in parallel

Resistors can be arranged in parallel.

Adding resistors in parallel decreases the total resistance.

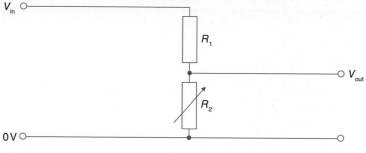

FIGURE 4: Potential divider circuit with a variable resistor. Why is a variable resistor used?

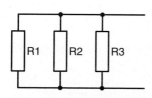

FIGURE 5: What is the effect of adding resistors in parallel?

Remember!

A list of all the circuit symbols you need to know can be found on page 317 at the back of the book

Questions

3 How does the output voltage compare with the input voltage when each resistor in a potential divider has a resistance of 100 W?

4 What happens to the output voltage when the value of the variable resistor shown in Figure 4 is increased?

Output voltage

The output voltage in Figure 3 can be found using the equation:

$$V_{out} = \frac{R_2}{R_1 + R_2} V_{in}$$

When R_2 is very much greater than R_1, the value of V_{out} is approximately V_{in}. When R_2 is very much less than R_1, the value of V_{out} is approximately 0. When one of the pair of resistors in a potential divider circuit is a variable resistor, the value of the output voltage can be altered.

There are some electronic components, such as LDRs and thermistors, whose resistance changes with external conditions. They often need to start working at a particular voltage. This voltage is known as the **threshold voltage**. The variable resistor in the potential divider circuit allows a particular threshold voltage to be set.

Calculating resistance of parallel combination

The total resistance is found using

$$\frac{1}{R_{tot}} = \frac{1}{R_1} + \frac{1}{R_2} + \frac{1}{R_3}$$

Questions

5 A 100 Ω and a 200 Ω resistor are used in a potential divider circuit. Calculate the voltage across the 200 Ω resistor when the supply voltage is 12 V.

6 Three 6 Ω resistors are arranged in parallel. What is the total resistance?

Light-dependent resistors

During the daytime on August 11 1999, a total eclipse of the Sun was seen in the South-West of England. As the people of Plymouth looked out across the city, they saw, 'a street light come on, then another, then a few more, then lots'. **Light-dependent resistors** (LDRs) on top of streetlights had responded to the decrease in light level.

FIGURE 6: Why did the streetlights come on in Plymouth during the day?

Thermistors

The resistance of a thermistor (thermal resistor) changes when the temperature changes.

FIGURE 7: A thermistor. What happens when its temperature changes?

Question

7 What property of a light-dependent resistor changes as the light level changes?

Light-dependent resistor (LDR) characteristics

> The resistance of a light-dependent resistor decreases as the light level increases.

> Its resistance is about 100 Ω in bright sunlight but it can be over 10 MΩ in the dark.

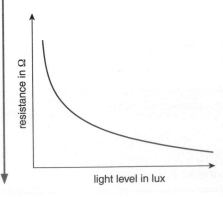

FIGURE 8: Describe how resistance changes with light level in a light-dependent resistor.

Remember!

The resistance of a thermistor decreases as the temperature increases. This is the opposite behaviour to a resistor.

Thermistor characteristics

> The resistance of a thermistor decreases as the temperature increases.

> Each thermistor has its own characteristics but a typical relationship between resistance and temperature is shown in Figure 9.

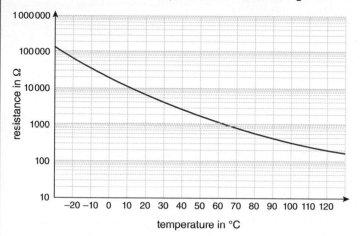

FIGURE 9: Describe how resistance changes with temperature in a typical thermistor.

Questions

8 The resistance of a light-dependent resistor (LDR) is measured as 150 Ω. Suggest where the LDR might be situated.

9 Use the graph in Figure 9 to answer these questions.

a What is the resistance of the thermistor when the temperature is 10 °C?

b Look at the shape of the graph and the values on the y-axis. Does the resistance of the thermistor ever fall to zero?

Light control

Streetlights have potential divider circuits that cause the light to come on automatically at nightfall. When it is dark, the resistance of the light-dependent resistor (LDR) (R_2) is high, which means that the output voltage is high and can be used to switch on the light. If the positions of the LDR and the fixed resistor are reversed, the output voltage decreases as it gets darker.

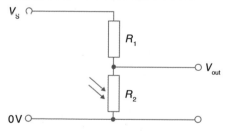

FIGURE 10: Circuit used in streetlights.

This circuit is found in the lightmeter used by umpires to decide whether play should be stopped at a cricket match.

FIGURE 11: How does a lightmeter work?

Temperature control

A thermistor (R_2) can also be used in a potential divider circuit. When the temperature falls, the output voltage rises and can be used to switch on a heater.

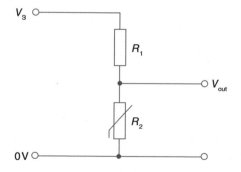

FIGURE 12: How can a thermistor be used to switch on a heater?

Question

10 a Draw a potential divider circuit that can be used to switch on a fan when the temperature rises.

b How can the circuit be changed so that a different threshold temperature can be set for when the fan comes on?

It's logical

You will find out:
> why transistors are important
> how transistors are used to build logic gates

High speed computer access

It is difficult to imagine what life would be like without a PC, laptop or palmtop. Yet it was less than seventy years ago when the first transistor was invented. Transistors were the basic electronic component on which modern technology is based.

The increased miniaturisation of electronic components means that global information and communication is more widely available than ever

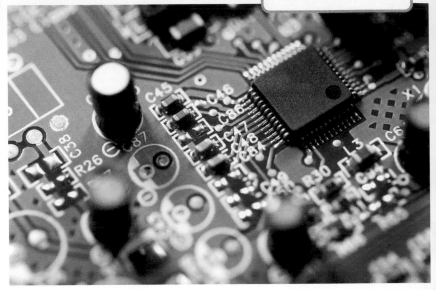

FIGURE 1: A computer chip contains millions, if not billions, of transistors within its circuits.

before. In 1998, less that 10% of households had Internet access. Today, three-quarters of households in the UK own computers and 60% of adults access the Internet every day. The use of Internet-enabled mobile phones is giving everyone greater access. But, not everyone uses the technology wisely. Confidential files are not totally free from hackers; we all know the dangers of cyber bullying and the abuse of social networking sites.

What are transistors?

Transistors are electronic components that have three terminals.

Like all electrical and electronic components, transistors are represented by a special symbol. Each lead from the transistor has a special name.

Transistors form the basic building block for modern electronic circuits.

They behave as a switch, which means they can be connected together to make logic gates. A logic gate controls an output which depends on whether the input(s) to it are switched on or off.

FIGURE 2: A selection of transistors. How many leads are there from a transistor?

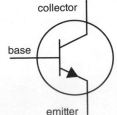

collector

base

emitter

FIGURE 3: The symbol for a transistor.

Questions

1 Why are transistors used as the basic component for logic gates?

2 How might life be different today if the transistor had not been invented?

A transistor works with very low currents

A current passing through the base will then pass into the emitter. If this small base current is large enough, it will switch on a greater current through the collector.

$I_e = I_b + I_c$

Making an AND gate

Two transistors can be combined to make an AND gate. A small input current at A will switch on the upper transistor. A small input current at B will switch on the other transistor. It needs a current at A **and** a current at B to switch on the current from the 6 V supply.

Other logic gates can be made from different combinations of transistors.

> **Remember!**
> At foundation tier, you will be expected to recognise the transistor circuit for an AND gate (Figure 5).

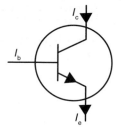

FIGURE 4: What do the symbols I_c, I_b and I_e mean?

Questions

3 A base current of 5 mA switches a collector current of 110 mA. What is the emitter current?

4 Two transistors are joined together to make an AND gate as shown in Figure 5. The transistors will switch on when the base current is I mA. Copy and complete the table to show what the output voltage is.

current at A in mA	current at B in mA	output voltage in V
0	0	
0	1	
1	0	
1	1	

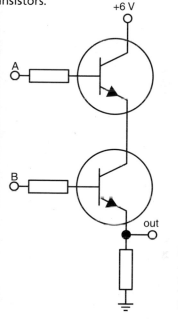

FIGURE 5: How an AND gate is made. Why do there need to be two inputs?

Limiting current

Transistor circuits always have a high value resistor in the base circuit to limit the current.

> **Remember!**
> At higher tier, you will be expected to remember the transistor circuit for an AND gate (Figure 5).

Question

5 Suggest why there is a high value resistor in series with the LED.

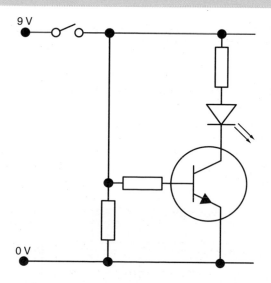

FIGURE 6: A typical circuit for switching a Light Emitting Diode (LED). Why is there a high resistor in the base circuit?

Q transistors and logic gates

Logic signals

Logic gates need a signal to make them work. The signal needs to be at a voltage of about 5 V.

> This signal is described as being 'high' and is sometimes shown by the number '1'.

> If there is no signal (about 0 V) to the logic gate, the signal is described as being 'low' and is sometimes shown by the number '0'.

NOT gate

One of the simplest logic gates is a **NOT** gate. It has one input and one output. The output is the opposite of the input.

A **truth table** shows the way a logic gate behaves. The truth table for a NOT gate is:

Input X	Output Z
0	1
1	0

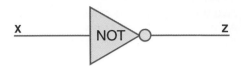

FIGURE 7: NOT gate.

Questions

6 A 5 V signal is applied to the input of a NOT gate. What is the value of the output voltage?

7 The output voltage from a NOT gate is 5 V. What is the value of the input voltage?

Logical output

The logic gate switches an output signal to be high or low. The output signal depends on the logic gate and the input signal or signals.

AND gate

A circuit contains a battery with two switches and a bulb in series. The bulb lights if switch 1 **AND** switch 2 are closed.

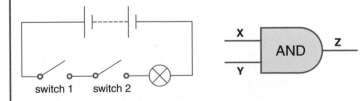

switch 1   switch 2

FIGURE 8: AND gate.

The **AND** gate behaves like two switches in series. The output is high if input X AND input Y are high.

Input X	Input Y	Output Z
0	0	0
0	1	0
1	0	0
1	1	1

OR gate

A circuit contains a battery with two switches and a bulb in parallel. The bulb lights if switch 1 OR switch 2 is closed.

The OR gate behaves like two switches in parallel. The output is high if input X OR input Y is high.

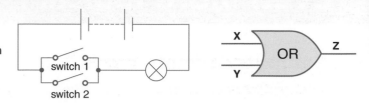

FIGURE 9: OR gate.

Input X	Input Y	Output Z
0	0	0
0	1	1
1	0	1
1	1	1

Questions

8 A lift will not move until the doors are shut and the floor button is pressed. What type of logic gate is used in the lift motor circuit?

9 A car's interior light comes on if any door is opened. What type of logic gate is used in the lighting circuit?

NAND and NOR gates

A **NAND** gate can be thought of as being an AND gate followed by a NOT gate.

Input X	Input Y	Output Z
0	0	1
0	1	1
1	0	1
1	1	0

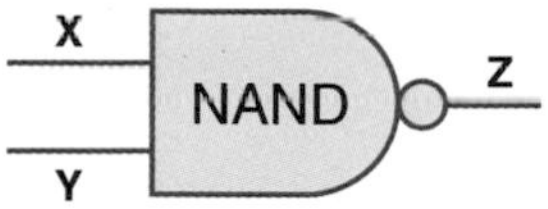

FIGURE 10: NAND gate.

A **NOR** gate can be thought of as being an OR gate followed by a NOT gate.

Input X	Input Y	Output Z
0	0	1
0	1	0
1	0	0
1	1	0

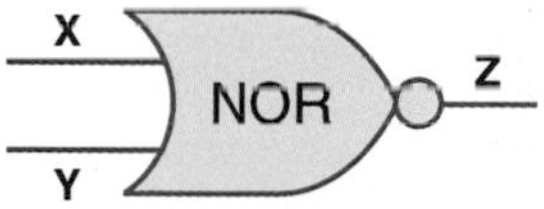

FIGURE 11: NOR gate.

Question

10 Jayne puts a NOT gate before each of the inputs of an AND gate. Copy and complete the truth table below to show that this behaves in the same way as a NOR gate.

System input X	System input Y	AND gate input x	AND gate input y	Output Z
0	0			
0	1			
1	0			
1	1			

Even more logical

You will find out:

> how to work out truth tables for a combination of logic gates

> how switches, LDRs and thermistors are used to provide signals for logic gates

Computer circuit boards

If you could look inside a computer, you would see a lot of electronic components. Many of these contain logic gates that are at the heart of computer circuits. Logic gates are basically electronic switches. A combination of these switches is capable of making thousands of millions of calculations every second.

FIGURE 1: How many calculations can you do in your head every second?

 ## Input and output

Logic gates can be combined to make up a **logic circuit**. The circuit can have many input and output signals. The output signal from one logic gate can be the input signal to another.

In Figure 2, **A**, **B** and **C** are the input signals to the electronic **system**. **E** is the output signal from the system.

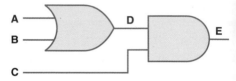

FIGURE 2: In this example, **A** and **B** are the input signals to the OR gate. **D** is the output signal from the OR gate but is an input signal to the AND gate and so is **C**. How would you describe **E** in terms of inputs **A**, **B** and **C**?

 Questions

1 Draw the truth table for the OR gate in Figure 2.

2 Draw the truth table for the AND gate in Figure 2.

 ## More truth tables

When there are several logic gates combined together, truth tables can be used to work out what happens.

A bank vault can be opened when the manager and either the deputy manager or chief cashier key their PINs into an electronic keypad. The logic circuit in Figure 2 shows this. The truth table for the system confirms the fact that there are only certain conditions under which the vault opens. When you work out a truth table for a system, do so one step at a time. Write out all of the possible input values; then work out what happens at each logic gate.

Input signals				Output
A	B	C	D	E
0	0	0	0	0
0	0	1	0	0
0	1	0	1	0
0	1	1	1	1
1	0	0	1	0
1	0	1	1	1
1	1	0	1	0
1	1	1	1	1

 Question

3 Consider the above example of a bank vault.

a Which bank employer is represented by input C?

b What is represented by output E?

More inputs

It does not matter how many inputs there are when a truth table is being worked out.

A homeowner wants an alarm to sound if anyone treads on a pressure pad under the doormat inside the front or back door whilst the house is in darkness.

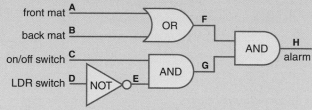

FIGURE 3: Logic system.

The truth table looks like this:

Input							Alarm
A	B	C	D	E	F	G	H
0	0	0	0	1	0	0	0
0	0	0	1	0	0	0	0
0	0	1	0	1	0	1	0
0	0	1	1	0	0	0	0
0	1	0	0	1	1	0	0
0	1	0	1	0	1	0	0
0	1	1	0	1	1	1	1
0	1	1	1	0	1	0	0
1	0	0	0	1	1	0	0
1	0	0	1	0	1	0	0
1	0	1	0	1	1	1	1
1	0	1	1	0	1	0	0
1	1	0	0	1	1	0	0
1	1	0	1	0	1	0	0
1	1	1	0	1	1	1	1
1	1	1	1	0	1	0	0

Question

4 Work out the truth table for this logic circuit.

How could the circuit be simplified to give the same result?

You will find out:

> how to use the output current from a logic circuit

> how LEDs and relays are used in output circuits

> how logic gates can be used to store information

Light-emitting diodes

Light-emitting diodes (LEDs) are often used as indicator lights on such things as laptop computers. They are also used instead of normal filament bulbs in some modern traffic lights. A lot of LEDs are needed to replace a filament bulb. The output current from a logic gate is able to light an LED.

FIGURE 4: The circuit symbol for an LED.

Relays

The output current from a logic gate can also operate a relay.

A relay acts as a switch. It can switch on something that needs a bigger current than the output current from a logic gate.

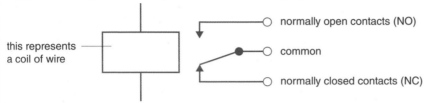

FIGURE 6: The circuit symbol for a relay.

FIGURE 5: What type of light source is used in some modern traffic lights?

Questions

5 A large number of LEDs are used in a traffic light. Suggest why.

6 A coil of wire has a magnetic field around it. Is this statement true or false?

Logical inputs

Switches, light-dependent resistors (LDRs) and thermistors can be used to make the inputs to logic gates high.

> When the switch is open, the input to the logic gate is low.

> When the switch is closed, the input to the logic gate is connected directly to the 5V supply.

> The input to the logic gate is high.

> The resistance of the LDR is high when it is in darkness. The input to the logic gate is low.

> When light shines on the LDR, the resistance is very low. The input to the logic gate is connected almost directly to the 5V supply.

> The input to the logic gate is high.

> The resistance of the thermistor is high when it is cold. The input to the logic gate is low.

> When the thermistor gets hot, the resistance is very low. The input to the logic gate is connected almost directly to the 5V supply.

> The input to the logic gate is high.

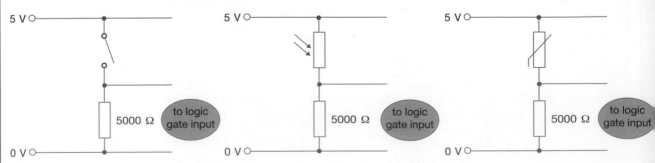

FIGURE 7: What name is given to the arrangement of resistor and thermistor shown above?

Changing the cut-off

If a variable resistor is used instead of the fixed resistor, it is possible to adjust the light level or temperature at which there is a high input (threshold voltage) to a logic gate.

Using logical outputs

If either the passenger's door or the driver's door is open in a car and the ignition is switched on, a warning light appears on the dashboard. An LED provides the warning light. It is connected to the output of the AND gate.

Working relays

A relay is designed to use a low current in the coil to switch a larger current in another circuit.

> When a current passes through the coil, the magnetic field produced attracts the iron armature.

> The armature pivots, pushing the insulating bar against the central contact.

> The central contact moves, opening the normally closed contacts and closing the normally open contacts.

Questions

7 Why is the armature made from iron?

8 Why is the moving bar made from an electrically insulating material?

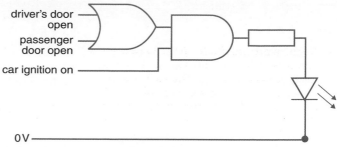

FIGURE 8: How does an LED-containing system provide a warning light in a car?

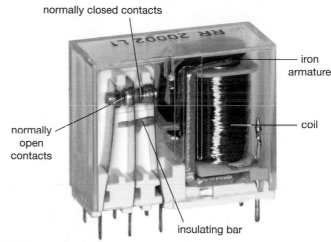

FIGURE 9: What happens when a current passes through the coil in this relay?

Switching large currents

The current output from a logic gate is low. It can be passed through a relay coil to produce the magnetic field necessary to operate the relay.

As well as switching a larger current, the relay isolates the low-current logic circuit from a high-current external circuit. The logic gate would be damaged if the current from the mains passed through it.

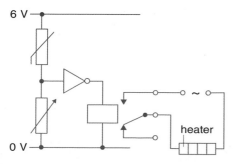

FIGURE 10: An example of use of a relay can be found in an automatic heater.

Questions

9 In the diagram, explain how the input to the logic gate changes as the light level changes when the position of the LDR and resistor are reversed.

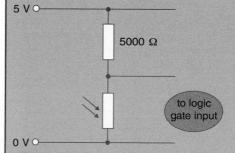

10 Describe how the circuit for an automatic heater works. Explain the purpose of each circuit component.

Preparing for assessment: Applying your knowledge

To achieve a good grade in science, you not only have to know and understand scientific ideas, you also need to be able to apply them to other situations and investigations. These tasks will support you in developing these skills.

✸ An early motor

The year is 1821 and Michael Faraday, the famous experimenter and lecturer, is giving a lecture at the Royal Institution in London. He is explaining his latest scientific discovery to his attentive audience.

He has been working on a new idea and has constructed a piece of equipment to try it out. Faraday is fond of trying out ideas by setting up experiments and he's convinced he's on the edge of something really big.

"Look at this," he says "This is really important … but I'm not sure you're going to understand how important it is. Never mind." The children lean forward in their seats. On the bench is a piece of equipment. It has a wooden base and a metal pole going up the side, with an arm coming out at the top. At the end of the arm is a hook. On the hook hangs a long straight piece of thick bare metal wire. The bottom end of the wire dangles in a dish of mercury and in the centre of the dish is a small cylindrical piece of shiny metal. Wires connect the bottom of the pole to one end of a battery and the mercury to the other end, though a gap has been left in the wires there.

"Watch this," says Faraday and connects the wire to complete the circuit. The wire dangling in the mercury starts to move. It moves around the cylinder in the dish in a circle – and keeps on moving. It goes round and round in circles, twitching slightly but continuing in its path.

"What is it?" asks one of the children, and before Faraday can answer, another asks, "Have you got any sweets?" Faraday's hobby is sweet making and the children are used to new recipes being tried out on them. "Yes, go on then. On the side, in a bowl." Faraday smiles and rolls his eyes as the children rush to the side in search of the sweets.

He stays by the equipment though and continues to smile as the wire continues its little dance.

thick wire — metal support — shiny metal cylinder — mercury in dish — wooden base

key — movement of wire when current connected — battery

✳ Task 1

Faraday had built a simple motor – very important for what it showed, but not a practical application. What would need to be changed to make it of practical use?

✳ Task 2

Faraday's experiment would not be allowed in a school laboratory. Why not?

✳ Task 3

Eventually the motor would stop. Why?

What transfer of energy is taking place during this experiment?

Where does the energy end up?

✳ Task 4

The cylinder of metal in the dish is a magnet; the North pole is sticking out of the dish and the South pole at the bottom. Draw the magnetic field pattern for this magnet.

The dangling wire is connected to the positive terminal of the battery via the pole and to the negative terminal via the mercury. Draw the magnetic field pattern for the wire.

✳ Task 5

Suggest two ways of getting the wire to travel in the opposite direction around the magnet.

✳ Task 6

Faraday also built a version of the motor in which the wire was fixed and the magnet could move. Sketch and label a diagram to show how this might have been set up.

✳ Maximise your grade

Answer includes showing that you can...	
F	Explain that this motor is very weak and is unlikely to be suitable for a serious application.
	Draw the field pattern for the permanent magnet.
	Explain that it is transferring energy from the battery and that eventually all the energy will be transferred and the motor will stop.
	Draw the field pattern for the dangling wire.
C	Explain how each of these fields could be reversed.
	Explain how the fields interact to produce movement.
A	Explain with reference to the magnetic fields how the direction of the rotation of the motor could be reversed.
	Draw and label the diagram of the motor with fixed wire and moveable magnet.
	As above, but with particular clarity and detail.

Motoring

Some motors don't spin

This maglev train is levitated and propelled by linear motors. Maglev trains in Japan have reached 581 km/h in tests and scientists believe they could reach speeds of 650 km/h. The only commercial maglev currently running is in Shanghai where it carries passengers 30 km to the airport in just 7 minutes and 20 seconds. If a proposed maglev system linked Glasgow to London, journey times could be cut from 4.5 hours to just over 2.5 hours.

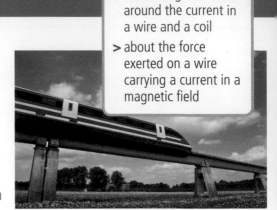

FIGURE 1: A maglev train powered by linear motors. What is its maximum speed?

A current's field

When there is an electric current in a wire, the wire is surrounded by a **magnetic field**. You can show this with the help of a compass.

The magnetic field made by the current in a wire is a circular shape around the wire. The circles are concentric.

If a wire is in a magnetic field and the current is switched on, the two magnetic fields interact and the wire moves.

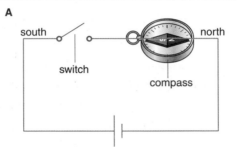

FIGURE 2: If the wire is in a north–south direction, the compass needle is parallel to the wire (A). When the current is switched on, the compass needle rotates until it is at right angles to the wire (B).

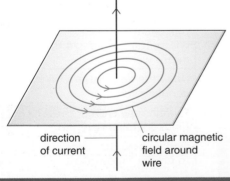

direction of current

circular magnetic field around wire

FIGURE 3: What shape is the magnetic field around a wire when a current passes through it?

Questions

1 What happens when two north poles of a magnet are brought close together?

2 Draw the magnetic field pattern due to a bar magnet.

Field direction

The direction of a magnetic field around a wire can be predicted by using the **right-hand grip rule**.

FIGURE 4: How can you predict the direction of a magnetic field around a wire carrying a current?

Imagine gripping the wire with your right hand, with your thumb pointing in the direction of the current. Your fingers point around the wire in the direction of the field

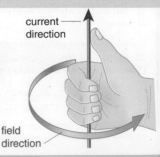

current direction

field direction

Q maglev trains

Different field patterns

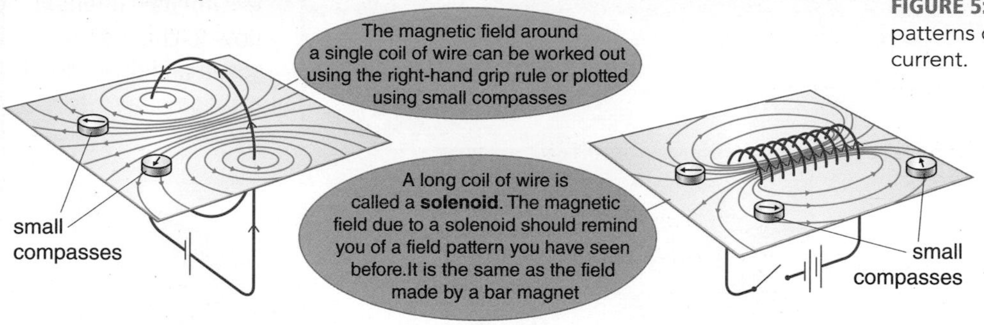

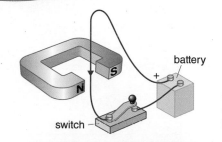

The magnetic field around a single coil of wire can be worked out using the right-hand grip rule or plotted using small compasses

A long coil of wire is called a **solenoid**. The magnetic field due to a solenoid should remind you of a field pattern you have seen before. It is the same as the field made by a bar magnet

small compasses

FIGURE 5: Different magnetic field patterns caused by a wire carrying a current.

small compasses

North and South

The north and south poles of a coil can be worked out by seeing which way the current passes.

north pole

end view

anticlockwise

south pole

clockwise

FIGURE 6: What direction of current produces a north pole at the end of a wire carrying a current?

battery

switch

FIGURE 7: What happens when a wire carrying a current is placed between the poles of a magnet?

Feel the force

If a wire is placed between the poles of a magnet, the wire moves out of the gap when the current is switched on. There is a force on the wire. When the current direction or the poles of the magnet are reversed, the force is in the opposite direction.

Questions

3 A wire is placed between the poles of a magnet. When the current is switched on, the wire moves to the right.

a Which way does the wire move when the poles of the magnet are reversed?

b Which way does the wire move when the poles of the magnet and the direction of the current are reversed?

4 Saleem looks at the end of a solenoid and notes that the current is in a clockwise direction. What is the magnetic pole at that end of the solenoid?

Fleming's left-hand rule

You can use your left hand to work out the relative directions of the current, magnetic field and motion of the wire.

> Your thumb, first and second fingers need to be at right angles to each other.

> Point your first finger in the direction of the magnetic field. (Field direction is from north to south.)

> Point your second finger in the direction of the current. (Current direction is from positive to negative.)

> Your thumb is showing the direction of motion of the wire.

This is called **Fleming's left-hand rule**.

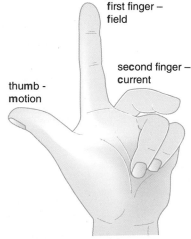

first finger – field

second finger – current

thumb - motion

FIGURE 8: Which hand do you use to predict the direction in which a motor spins?

Question

5 The diagram below shows a wire between the poles of a magnet. There is a current in the wire. In which direction is the force on the wire?

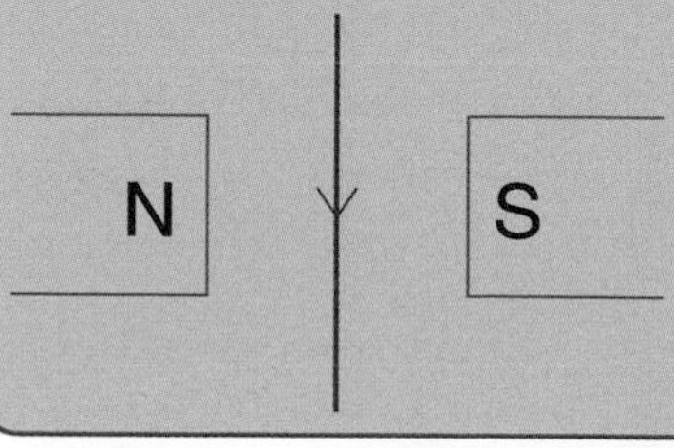

N   S

You will find out:
> how motors are constructed
> how to change the speed and direction of rotation of a motor

Motors at home

The electric motor has many uses around the home.

> In the car, windscreen wipers are just one example of the use of motors.

> In the garden, electric lawnmowers have motors.

> In the kitchen, blenders, whisks and food processors have motors.

> In the workshop, electric drills, sanders and automatic screwdrivers have motors.

> On the move, portable CD players and DVD players need motors to turn the discs.

Motors transfer energy to the load. They do useful work. Some energy is always lost to the surroundings as heat.

FIGURE 9: Lawnmowers and electric drills have motors. What happens to the temperature of a motor as it is working?

Question

6 Which five of the household appliances listed below have a motor in them?

deep-fat fryer   fan   hair dryer   radio   smoke alarm   television   toaster
tumble dryer   video recorder   washing machine

Force on a coil

> When a current passes through a coil placed between the poles of a magnet, there is a force on each side of the coil.

> Fleming's left-hand rule shows that the forces on each side of the coil are in opposite directions. As one side is forced up, the other side is forced down.

> This causes the coil to start to spin.

> This is the basis of the electric motor.

The electric motor

The motor turns faster when:

> the number of turns on the coil is increased

> the size of the current is increased

> the strength of the magnetic field is increased.

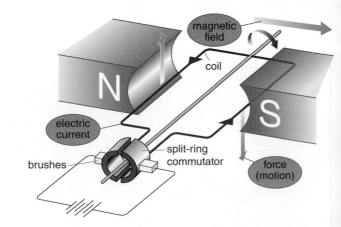

FIGURE 10: A simple electric motor. Why does the coil spin?

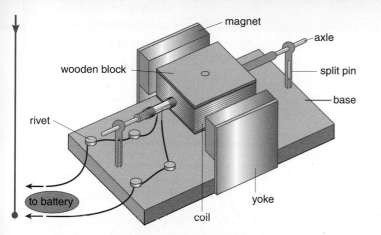

FIGURE 11: Name the parts of a simple electric motor.

Questions

7 What happens to the speed of rotation of a motor if the current is reduced?

8 What happens to the direction of rotation of a motor if the current direction is reversed?

The commutator

A **commutator** allows a motor to continue to spin. Look at Figure 10. The force on the left side of the coil is upwards. Without a commutator, the coil would just turn until it was vertical.

A radial field

Practical motors have pole pieces that are curved. This produces a **radial field**. The coil is always at right angles to the magnetic field. This increases the force and keeps it constant as the coil turns.

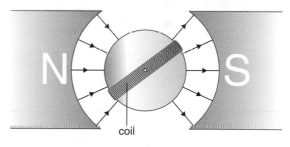

FIGURE 13: Why is a radial field used in practical motors?

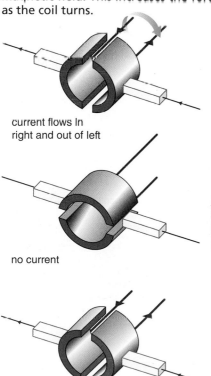

current flows in right and out of left

no current

current flows in right and out of left

- The force on the 'red' side if the coil is upwards
- The motor spins clockwise
- Although there is no current, the coil has enough **inertia** to keep spinning past the vertical
- The current in the coil is now reversed, and the force on the 'blue' side is now upwards, so the motor continues to spin clockwise

FIGURE 12: How a commutator works.

Question

9 What would happen to the coil of a motor if the coil was connected directly to a battery without a commutator?

Generating

You will find out:

> how a dynamo generates electricity

> about induced voltages

> how a magnetic field affects induced voltage

The clockwork radio

This clever 'wind-up radio' was developed for use in Third World countries where affordable energy is scarce or non-existent. It uses no batteries and does not need electricity to run it. Instead, it is powered by an internal clockwork generator. As the spring unwinds it turns the shaft that it is wound around. The shaft is connected to a small generator that works like an electric motor in reverse – we turn a handle and it produces electrical power.

The radio has been personally endorsed by Nelson Mandela.

FIGURE 1: Clockwork radio, created by Trevor Baylis.

The dynamo effect

Moving a magnet near a wire or a wire near a magnet causes a current in the wire.

When a coil is rotated between the poles of a magnet, a current is produced in the coil. The direct current (DC) generator (dynamo) is a motor working in reverse. In the UK, electricity is produced using alternating current (AC) **generators** that supply electricity with a **frequency** of 50 Hz. AC generators are constructed differently to DC generators.

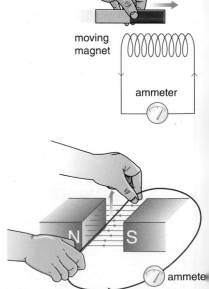

FIGURE 2: Making a current in a wire.

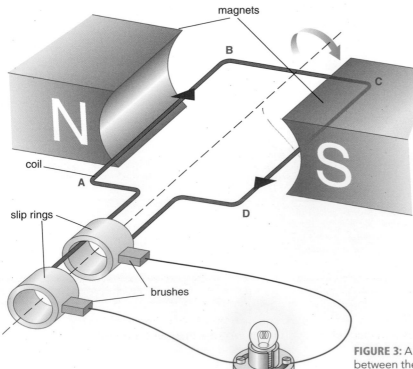

FIGURE 3: A model generator. What is the difference between the construction of a DC motor and an AC generator?

Electricity is very useful. It allows energy to be:

> transmitted over long distances through the National Grid

> stored in special hydro-electric power stations for use later.

Questions

1 Write down one difference between the output from a dynamo and the output from a power station generator.

2 What is the frequency of mains voltage in the UK?

Induced voltage

> Reversing the direction of a magnetic field also reverses the induced voltage and current direction.

> A voltage is induced whenever a magnetic field changes.

> The same effect is noticed when a magnetic field inside a coil changes. A voltage is induced across the coil and a current passes through the coil.

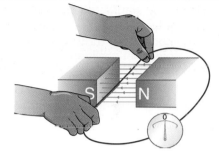

When a wire is held stationary between the poles of a magnet, there is no current in the wire

FIGURE 4: How can a voltage be induced in a wire in a magnetic field?

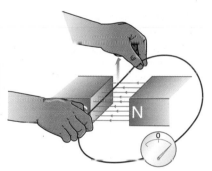

If the wire is moved upwards, a voltage is **induced** in the wire and a current passes. The same thing happens if the magnet is moved downwards. It is the *relative movement* of the wire and the field that is important

Question

3 An ammeter needle moves to the right when a wire is moved down between the poles of a magnet.

What **three** things can be done to make the ammeter needle move to the left?

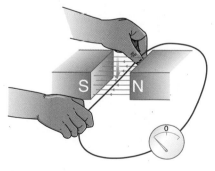

When the wire moves downwards or the magnet moves up the induced voltage is reversed and the current is in the *opposite direction*

Changing the induced voltage

The size of the induced voltage depends on the rate at which the magnetic field changes. If the wire is moved slowly between the poles of a magnet, the voltage is low. If a magnet is pushed quickly into a coil, the induced voltage is high.

🔍 bicycle dynamo animation   induced voltage

Generating an alternating current

You will find out:

> how alternating current is generated

> how the size and frequency of induced current is affected by the speed of rotation of a magnet

> how slip rings work

> Remember that it is the change in a magnetic field that causes a current to be induced.

> Most practical generators have a changing magnetic field inside a coil. In the case of the bicycle dynamo, it is a **permanent magnet** that is rotated within a coil.

> As the magnetic field changes, so the current changes.

> An alternating current is produced in the coil.

Electromagnets

In a power station, the magnetic field inside the coil is produced by an **electromagnet**. An electromagnet has coils of wire with large numbers of turns to produce a strong magnetic field. The electromagnet's coils are the rotor coils and are turned by the turbine. Relatively low-voltage direct current is supplied to the coils by a commutator.

The coils surrounding the rotor are called stator coils. An alternating voltage is generated in these coils.

> Increasing the speed of rotation of the electromagnet increases both the current induced in the stator coils and the frequency of the voltage generated.

> Increasing the number of turns on the electromagnet increases the magnetic field and therefore a larger voltage is induced in the **stator** coils.

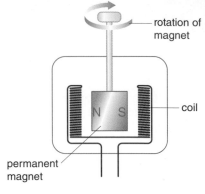

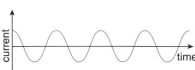

FIGURE 5: How is the magnetic field changed in a dynamo?

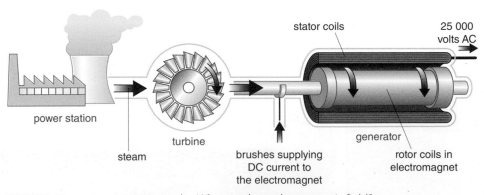

FIGURE 6: How a power station works. What produces the magnetic field?

Questions

4 The pressure of the steam from the power station is reduced. Describe the effect on the electricity generated in the stator coils.

5 The current supplied to the electromagnet has to be DC. Suggest why.

Q AC generator

AC generators

Some AC generators do have a rotating coil between the poles of a magnet.

> As the coil rotates, the direction of the induced current will always be as shown next to each pole.

> In Figure 7 the current is from A to B and from C to D. After the coil has rotated half a turn, the current is from B to A and from D to C. An alternating current has been induced.

> Slip rings are connected to the ends of the coil to allow the coil to spin without winding the wire around itself.

> The brushes are contacts that touch the slip rings and complete the circuit.

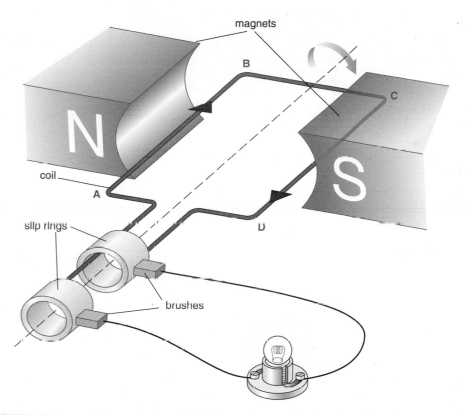

FIGURE 7: How an AC generator works. Why are slip rings connected to the ends of the coil?

Question

6 Describe, using diagrams, how an AC generator works differently from a DC generator.

Experimental research in electricity

The Royal Institution Christmas Lectures, which you may have watched on television, were first presented to young people by Michael Faraday in 1826.

But it was on 29 August 1831 that Faraday made one of his greatest discoveries. Using an induction ring, he generated electricity in a wire by means of the electromagnetic effect of a current in another wire. This was the first electric transformer.

FIGURE 1: Who started Christmas lectures at the Royal Institution?

FIGURE 2: Faraday's induction ring.

What is a transformer?

Whether it is Faraday's induction ring or the local electricity sub-station, **transformers** all do one thing. They change the size of an AC voltage.

> A transformer does not work from a DC supply.

> A transformer does not change AC into DC. The controller for a model train set is sometimes called a transformer. This is not a correct scientific name. It does contain a transformer that changes mains voltage (230 V AC) to 12 V AC; but there are other electrical components that change the AC into DC. DC is needed for the train's motor.

> Your mobile phone charger contains a transformer and so does your radio and laptop. They are **step-down** transformers because the output voltage is less than the input voltage.

> Some transformers increase the voltage. These are **step-up** transformers.

> A bathroom shaver socket contains an isolating transformer. The socket provides an output voltage that is the same as the input voltage, 230 V. Shower sockets usually have an alternative output voltage, 115 V. This means that people who travel around the world can use their electric shavers in their hotel rooms. Not all countries use 230 V; some use 115 V.

FIGURE 4: An electricity sub-station. What does a transformer do?

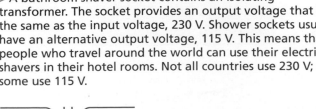

FIGURE 3: The symbol for a transformer

Questions

1 What type of transformer is used in the controller for a train set?

2 What are the **two** most common mains voltages used around the world?

How does a transformer change the voltage?

A transformer consists of two coils of wire wound onto an iron **core**. The input AC voltage is applied to the **primary** coil and the output AC voltage is obtained from the **secondary** coil.

> A step-down transformer has more turns of wire on the primary coil than on the secondary coil.

> A step-up transformer has fewer turns of wire on the primary coil than on the secondary coil.

An isolating transformer is used for safety. There is a lot of water and steam in a bathroom. There is a risk of electrocution or damage to the house wiring system if a normal 13 A socket is used.

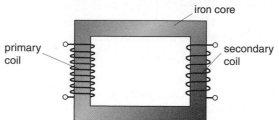

iron core

primary coil

secondary coil

FIGURE 5: A simple transformer. Where is the output obtained from?

Questions

3 A transformer has 300 turns on the primary coil and 3600 turns on the secondary coil. What type of transformer is it?

4 A step-down transformer has 10 000 turns on its primary coil and 200 turns on its secondary coil. 500 V are applied to the primary coil. What is the output voltage?

5 A transformer has the same number of turns on its primary coil as on its secondary coil. How does the secondary voltage compare to the primary voltage?

Induced voltage

The primary coil has an alternating current passing through it. The alternating current produces an alternating (changing) magnetic field in the iron core. Remember that a dynamo works because a changing magnetic field (caused by rotation) induces a changing voltage. In a similar way, a transformer works because a changing magnetic field (caused by the changing current in the primary coil) induces a changing voltage in the secondary coil.

The amount by which the voltage increases or decreases is dependent on the number of turns on each coil.

$$\frac{V_p}{V_s} = \frac{N_p}{N_s}$$

where:
V_p = voltage across primary coil
V_s = voltage across secondary coil
N_p = number of turns on primary coil
N_s = number of turns on secondary coil

The isolating transformer

The output voltage from an isolating transformer is the same as the input voltage. The number of turns on the primary coil is the same as the number of turns on the secondary coil. So why use a transformer? Why not just have normal sockets?

The mains supply is hidden behind the socket face. The only 'bare' terminals exposed are the two holes that the shaver plugs into. These are not 'live' and therefore there is no risk of electrocution if you touch the socket with damp hands and complete the circuit to earth.

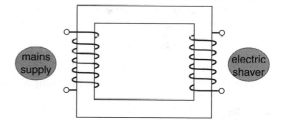

mains supply

electric shaver

FIGURE 6: Is it possible for a person to receive an electric shock when they plug a shaver into a shaving socket?

Question

6 The diagram shows an iron bar inside a coil of wire. An aluminium ring sits on top of the coil. When an AC supply is connected to the coil, the ring jumps in the air and floats. Use your ideas about transformers to explain why.

ring

iron bar

Transformers and the National Grid

A power station generates electricity at 25 000 V.

> A step-up transformer increases the voltage to 400 000 V for transmission around the country through the **National Grid**.

> Step-down transformers reduce the voltage near to where the electricity is to be used.

Homes, shops and offices use 230 V; but large factories may receive 33 000 V whilst small factories, hospitals and schools receive 11 000 V. There will be an electricity sub-station near to where you live. This reduces the voltage from 11 000 V to 230 V.

Questions

7 David's football goes into an electricity sub-station. What should he do to get his ball back?

8 A remote farm has its own transformer. It is mounted at the top of a tall pole in the farmyard. What type of transformer is it?

FIGURE 7: Sub-stations are very dangerous places and you must never go past the danger signs. If you do go too close to the wires or terminals, electricity can jump through the air and pass through your body causing severe burns or death.

Transmission loss

When a current passes through a wire, the wire gets hot. This is why we can use electric fires to heat our homes.

When electricity is passing through the National Grid, the cables heat up. They do not get red hot, like the bars of an electric fire, but the temperature of the wires does increase.

The National Grid contains hundreds of kilometres of cable, all of which are getting warm and transferring **energy** to the surroundings. The transformers also get warm. The energy is lost to the surroundings because it is not being transmitted through the Grid.

FIGURE 8: How does an electric fire work?

Questions

9 Birds often sit on overhead power lines. Suggest why.

10 Why is it better to transmit electricity at low current?

transformers   National Grid

Reducing transmission loss

Power is a measure of how fast energy is transferred. Electrical power loss depends on the current in a wire, but it is not a direct relationship.

power loss = (current² × resistance)

> So, if the current is doubled, the power loss is increased by a factor of four.

> If the current is reduced to a tenth of its value, the power loss is reduced by a factor of 100.

Electrical power depends on voltage and current.

power = voltage × current

The electrical power supplied to the primary coil of a transformer depends on the input voltage and current.

$P_p = V_p I_p$

Similarly, the output power of the secondary coil is given by:

$P_s = V_s I_s$

Assuming the transformer is 100% efficient, no energy is lost to the surroundings.

$P_p = P_s$

$V_p I_p = V_s I_s$

> This means that if the voltage is increased by a step-up transformer, the current is reduced by an equivalent amount.

> Electricity is generated at 25 kV and transmitted at 400 kV.

The voltage is increased by a factor of 16. The current in the secondary coil is only $\frac{1}{16}$ of the current in the primary coil.

> Since energy transfer depends on the square of the current, this means power loss is reduced by a factor of 256.

> The power loss by transmitting electricity at 400 kV is $\frac{1}{256}$ of that which would be lost if the electricity were transmitted at 25 kV.

Remember!

At higher tier, you will have to rearrange the transformer and power equations.

Question

11 Chris has run a cable from his house to a store in a field 200 m away. When he boils a 2 kW kettle in the store, he finds it takes much longer than when he boils it in the house. It does not take any longer to boil the kettle in his house than it does at the power station 25 miles away. Explain these observations.

Charging

You will find out:
> how a diode allows a current to pass in one direction only
> about half-wave rectification

Keeping the animals in and the people out

Anyone who lives in the countryside or who keeps livestock will have seen or used an electric fence. A 12 V battery energises an electronic circuit that can deliver a very high voltage for a fraction of a second. This is usually enough to keep animals where they should be.

FIGURE 1: A horse kept safe behind an electric fence. What voltage battery is needed to work the fence?

Diodes

A diode is an electronic component that only allows a current to pass in one direction. A single diode is used in a circuit to change alternating current (AC) into direct current (DC). The DC is not a constant value, but the current is always in the same direction.

This is called **half-wave rectification** because half of the AC input wave is allowed to pass through the diode as DC. For half of each input cycle, there is no output current.

diode circuit symbol

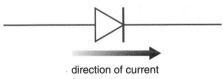

direction of current

FIGURE 2: A diode is shown in an electric circuit by a circuit symbol. What does the arrow tell you?

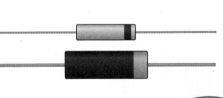

FIGURE 3: On diodes, the direction of the current is shown by a band. The current flows towards the end with the band. In which direction is the current flowing in the two diodes?

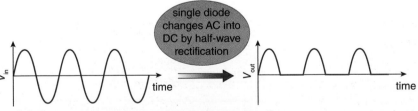

single diode changes AC into DC by half-wave rectification

FIGURE 4: Half-wave rectification. What does it mean?

Questions

1 Draw a circuit diagram to show a cell, resistor, diode and ammeter in series so that there is a reading on the ammeter.

2 Draw a circuit diagram to show a cell, resistor, diode and ammeter in series so that there is no reading on the ammeter.

Diode characteristics

The current–voltage characteristics for a diode can be found using a circuit similar to the one used to find the characteristics of a resistor or a bulb.

> Most diodes start to conduct when there is a voltage of about 0.6 V across them.

> The steep slope of the graph in Figure 5 shows that when conducting, current passes easily through the diode. When the diode is connected so that a current passes it is **forward biased**.

> There is virtually no current in the circuit if the battery is reversed. The diode is **reverse biased**.

Half-wave rectifier circuit

An AC supply to a single diode produces a half-wave rectified output.

The resistor is in the circuit to protect the diode. If unprotected, a very large current would cause it to burn out.

Q half-wave rectifier   how diodes work

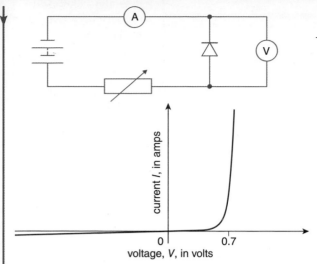

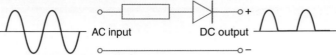

FIGURE 6: Half-wave rectifier circuit.
Why is there a resistor in the circuit?

FIGURE 5: Current–voltage characteristics for a diode.
What does the shape of the trace on the graph show?

Questions

3 A transformer changes 230 V AC from the mains supply to 12 V AC. How can the 12 V AC be used to power the DC motor of a model train?

4 What is the difference between the behaviour of a diode that is forward biased and one that is reverse biased?

Diode resistance

> The steep **gradient** of the current–voltage characteristic graph in Figure 5, when the diode is forward biased, shows that the resistance of the diode is very low.

> The shallow gradient and very low current when the diode is reverse biased shows that the resistance of the diode is very high.

Electrons and holes

A diode consists of a piece of n-type and a piece of p-type semi-conductor joined together to form a junction.

Electrons in the n-type half of the diode are repelled from the junction by the negative ions in the p-type region. Holes in the p-type half are repelled by the positive ions in the n-type region. This is known as the **depletion layer**.

There is no current in the depletion layer because of the absence of electrons and holes. The depletion layer behaves as an **insulator**.

> If a small voltage is connected across the diode, with the positive terminal of the power supply connected to the n-type end of the diode, the current carriers are attracted away from the junction. The depletion layer gets wider. This is reverse bias.

> If the power supply is connected the other way round, the carriers are repelled and driven towards the junction. This makes the depletion layer narrower. This is forward bias.

> Increasing the voltage with the diode forward biased makes the depletion layer disappear and a current passes.

When electrons and holes meet at the junction, they combine. The power supply continues to provide a source of electrons.

key:
(+) positive ion
(−) negative ion
● electron
✳ 'hole'

FIGURE 7: What is the difference between the number of electrons in n-type and p-type semi-conductors?

the space on either side of the junction is left without any electrons or holes – it is called the depletion layer

FIGURE 8: The depletion layer in a diode. What is a depletion layer?

diode conducting

FIGURE 9: Explain how the diode conducts.

Question

5 Explain why a diode starts to work with a voltage of about 0.6 V across it.

Full-wave rectification

An electronic circuit containing four diodes produces an output where current passes all the time, instead of just for half each input cycle. This is called **full-wave rectification**.

four diodes make an output where current passes all the time, this is called full-wave rectification

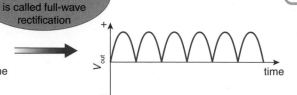

FIGURE 10: Full-wave rectification. What does it mean?

Capacitors

A **capacitor** is an electronic component that stores charge that can be **discharged** later.

> Some capacitors have positive and negative ends. These must be connected into a circuit in the right way.

> Capacitors used in electronic circuits are often coloured blue. They are clearly marked to show the positive and the negative connections.

When a capacitor is placed across the output of a rectifier circuit, the output is smoother. Electronic circuits, such as logic gates, will not work unless they have a fairly constant voltage supply.

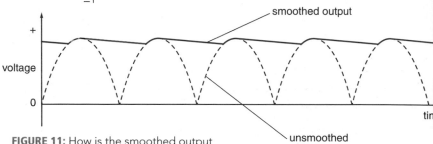

smoothed output

voltage

unsmoothed output

FIGURE 11: How is the smoothed output different from the unsmoothed output?

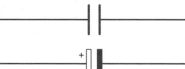

FIGURE 12: A capacitor is shown in circuits by one of two circuit symbols.

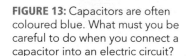

220 µF
25 V

FIGURE 13: Capacitors are often coloured blue. What must you be careful to do when you connect a capacitor into an electric circuit?

Questions

6 Draw a circuit symbol of a capacitor that can be connected into a circuit any way around.

7 The diagram on the right shows part of a circuit diagram. What is wrong with the way the circuit has been drawn?

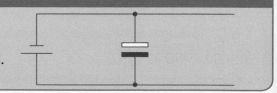

Rectifier circuit

Four diodes arranged into a rectifier circuit make up a **bridge circuit** to obtain full-wave rectification. An AC input becomes a DC output.

Action of a capacitor

> When a DC supply is connected to a capacitor, the capacitor becomes charged. The charge is stored and the voltage across the capacitor increases to a maximum. The maximum voltage is the voltage of the DC supply.

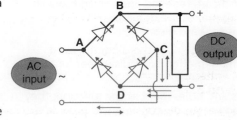

AC input

DC output

FIGURE 14: A bridge circuit. What does this circuit produce?

> When the capacitor is discharged, the voltage decreases over a period of time.
The current in the circuit decreases over a period of time also. The time it takes to charge
and discharge the capacitor depends on the sizes of the circuit resistance and the capacitor.

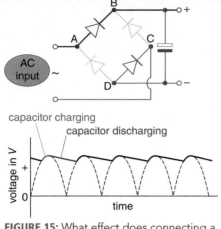

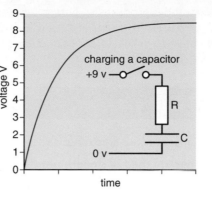

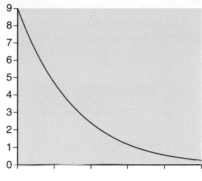

FIGURE 15: What effect does connecting a large capacitor across the output have?

FIGURE 16: Graph to show voltage against time when a capacitor is charged.

FIGURE 17: Graph to show voltage against time when a capacitor is discharged.

Question

8 The boxes in the diagram represent three components used when changing a 230 V mains supply into a 12 V DC supply.

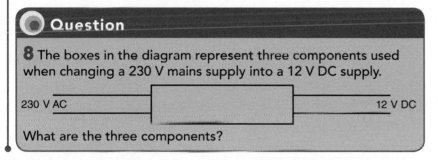

230 V AC   12 V DC

What are the three components?

Why rectify?

In Figure 14:

> During the positive half cycle: current passes from the AC input to **A**, then **B**, from the + DC output to the circuit, back to **D**, then **C** and finally back to the AC supply.

> During the negative half cycle: current passes from the AC input to **C**, then **B**, from the + DC output to the circuit, back to **D**, then **A** and finally back to the AC supply.

> The DC output is therefore always positive at **B** and negative at **D**.

Storing charge

> A battery stores energy. Chemicals in the battery provide the energy source and when they are used up, the battery is no longer useful. Energy is released from the battery as a flow of electrons called a current.

> A capacitor stores *electrical* energy. There is no continual energy source. As the electrons flow, the number of electrons stored decreases. This is why the voltage decreases.

How smooth can you get?

> Smoothing needs a large-value capacitor connected across the DC supply.

The capacitor acts as a reservoir for electrons.

> When the DC voltage from the rectifier falls, the capacitor supplies current to the output.

> The capacitor charges quickly near the peak of the varying DC and then discharges as it supplies current to the output.

Question

9 Sketch a graph to show how the current in a circuit changes as a capacitor is being charged from a DC supply.

Q how capacitors work

Preparing for assessment: Planning and collecting primary data

To achieve a good grade in science, you not only have to know and understand scientific ideas, but you need to be able to apply them to other situations and investigations. These tasks will support you in developing these skills.

✹ Tasks

> Plan an investigation to see how a capacitor discharges and compare the capacitors you have been given.

> Once your plan has been approved, perform the investigation, record your results and write a simple conclusion.

✹ Context

Capacitors are electronic devices that store charge.

They are used to smooth out the voltage from a rectifier circuit.

The longer it takes for a capacitor to discharge, the smoother the output.

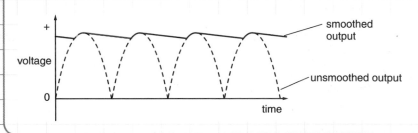

✹ Planning your investigation

You can measure the voltage across a capacitor using a digital voltmeter.

This is the circuit you will need to use for charging and discharging the capacitor.

These are the things you will need to consider when planning your investigation. (You can develop your plan in groups of two or three.)

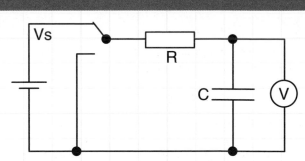

1 How will you time the voltage across the capacitor as it discharges?

2 What do you need to keep the same to make it a fair test?

3 Will it be easier to use voltage or time as the independent variable? The independent variable is usually plotted on the x-axis but not always.

4 How many different capacitors will you need to be able to identify a trend?

5 Will you need to repeat your readings? If so, how many times? •————— Why do scientists repeat readings?

6 You should carry out a risk assessment before you start the investigation. What precautions should you take?

7 Write the plan for the investigation.

Try to write the plan in a logical order and ask yourself if someone can perform the investigation following just your plan.

✸ Results, analysis and evaluation

Once your plan has been approved you can perform the investigation.

1 You will need to record discharge tables for each of the capacitors you are given.

2 If you repeated any readings, all of these will need to be recorded as well as the average result.

3 Record your results in a table like this. You may need to add extra rows or columns.

voltage across capacitor in V	time in s	

4 In order to complete this as a controlled assessment you need to plot a graph and evaluate the investigation.

a What graph(s) would you draw?

b What would the labels be on the axes?

c Would you plot different graphs for each capacitor or use the same set of axes?

d How would you use the graphs to decide on the answer to the task?

> Think about accuracy and precision. How are they different?

5 Is there any way in which you could have improved on how you performed the investigation?

6 What have you found out about how capacitors discharge?

7 What have you found out about how the value of a capacitor affects the time it takes to discharge?

P6 Checklist

To achieve your forecast grade in the exam you'll need to revise

Use this checklist to see what you can do now. It gives you many of the important points you will need to know. Refer back to the relevant pages in this book if you're not sure and to see if there is anything else you need to know. Look across the three columns to see how you can progress.

Remember you'll need to be able to use these ideas in various ways, such as:

> interpreting pictures, diagrams and graphs
> applying ideas to new situations
> explaining ethical implications
> suggesting some benefits and risks to society
> drawing conclusions from evidence you've been given.

Look at pages 272–294 for more information about exams and how you'll be assessed.

To aim for a grade E

recognise and draw common circuit symbols

describe the effect of a variable resistor on a lamp

use the equation:
resistance = voltage ÷ current

understand that current in a wire is a flow of charge

use models of atomic structure to explain electrical resistance

recognise how the resistance changes with temperature

---

recall the use of a potential divider circuit

calculate the total resistance for resistors in series

recognise and draw symbols for an LDR and a thermistor

recall that LDRs respond to light and thermistors respond to temperature

---

recall that the transistor is an electronic switch used in computers

recognise and draw the symbol for a transistor

recall that transistors can be connected to make logic gates

recall that the input for logic gates is either high or low

describe the truth table for a NOT gate

To aim for a grade C

explain the effect of a variable resistor in a circuit

compare resistance qualitatively from a graph

use the kinetic theory to explain resistance and temperature

describe how a V–I graph shows the resistance change of a bulb

---

explain how two resistors can be used as a potential divider

describe how the resistance of an LDR and a thermistor can vary

understand that having resistors in parallel reduces the resistance

---

describe benefits and drawbacks of miniaturisation of electronic components

understand the relationship between currents in base, emitter and collector

recall that logic gates are made from a combination of transistors

recognise the circuit diagram for an AND gate as two transistors

describe the truth tables for AND and OR gates

To aim for a grade A

explain the effect on resistance of wire length

calculate resistance from a voltage–current (V–I) graph

explain the shape of the V–I graph for a non-ohmic conductor

---

calculate the value of V_{out} when R_1 and R_2 are in a simple ratio

explain how LDRs and thermistors are used in potential divider circuits

calculate the total resistance for resistors in parallel

---

explain the impact on society of increased use of computers

explain why a high resistor is placed in the base circuit

draw a circuit diagram to show a transistor used as a switch

show how an AND gate is made from two transistors

describe the truth table for NAND and NOR gates

To aim for a grade E

identify the input and output signals in a system of logic gates
recognise and draw the symbols for an LED and a relay
recognise that the current from a logic gate is able to light an LED
recall that a relay can be used as a switch

describe the magnetic field around a current-carrying wire
explain why a wire placed in a magnetic field can move
recall that motors are found in a variety of everyday applications
recall that motors transfer energy to the load and surroundings

describe and recognise the dynamo effect
label a diagram of an AC generator
describe a generator as a motor working in reverse
explain why electricity is useful
recall that in the UK, mains electricity is supplied at 50 Hz

recall that transformers are devices that work with AC
understand the terms step-up and step-down transformer
recall everyday uses of step-down transformers
recall where an isolating transformer is used
recall where transformers are used in the National Grid

recognise that a diode only allows a current to pass in one direction
recognise half-wave rectification from a voltage–time graph
recognise full-wave rectification from a voltage–time graph
recognise the circuit symbols for a diode and a capacitor
describe the function of a capacitor to smooth output

To aim for a grade C

complete a truth table of a logic system with up to three inputs
describe how to use switches, LDRs, thermistors and resistors to provide input signals for logic gates
explain how an LED and series resistor can be used to indicate the output of a logic gate
describe how a relay works

describe the magnetic field around a coil, and a solenoid
understand that a current-carrying wire in a magnetic field experiences a force
explain how the forces on a coil in a magnetic field produce a turning effect and how this is used in a simple DC motor
describe the effect of changing: current, turns and magnetic field

understand two ways that a voltage is induced
describe the effect of changing the direction of the magnetic field
explain what happens when a magnet is rotated inside a coil
describe how electricity is generated and what happens when the speed of rotation of the coils is changed
describe how the number of turns on the coils affects the voltage

describe the construction of a transformer
describe the difference between step-up and step-down transformers
explain why an isolating transformer is used
recall that power is lost during the distribution of electricity

recognise the current–voltage characteristics for a silicon diode
recall that a single diode produces half-wave rectification
recall how four diodes can be used for full-wave rectification
describe the charging and discharging of a capacitor

To aim for a grade A

complete a truth table of a logic system with up to four inputs
explain how a thermistor or an LDR can be used to generate a signal for a logic gate, which may have a threshold voltage
explain why a relay is needed for a logic gate to switch a mains current

explain how Fleming's left-hand rule is used to predict the force on a wire
describe the use of a splitting-ring commutator and how the direction of the force on the motor coil is maintained
explain why practical motors have curved pole pieces

explain what affects the size of the induced voltage
explain how an AC generator works

explain how transformers work with alternating current
use the equation: $\dfrac{V_p}{V_s} = \dfrac{N_p}{N_s}$
explain the construction of isolating transformers and how they improve safety
understand why electricity is transmitted at high voltages

explain the current–voltage graph for a diode in terms of resistance
describe the action of a diode in terms of holes and electrons
explain how four diodes can produce full-wave rectification
explain the action of a capacitor in a smoothing circuit

267

Foundation Tier

AO1 **1** Physicists use symbols to represent electrical components. Copy and finish the table by naming each component and stating what the component does. [3]

symbol	component	job
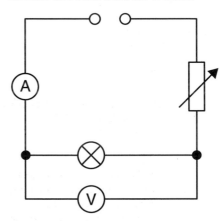		

2 Michael builds a circuit to find the voltage–current characteristics for a bulb.

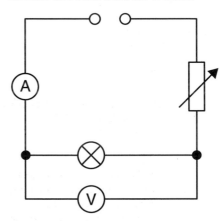

He records these results.

current in A	voltage in V
0.0	0.0
0.5	1.1
1.0	4.7
1.5	6.9
2.0	12.0

AO2 **(a)** Plot a graph of Michael's results, drawing a line of best fit. [3]

AO3 **(b)** Sandy says that Michael could improve the precision of his results. Why are Michael's results not precise? [1]

AO1 **(c)** Calculate the resistance of the bulb when the current passing is 2.0 A. [2]
AO2

AO1 **(d)** How does the graph show that the resistance of the bulb is changing as the current in the bulb is increased? [1]

[Total: 7]

AO1 **3 (a)** Draw the circuit symbol for a transistor. [1]

AO1 **(b)** Label the base, collector and emitter. [2]

AO2 **(c)** When the value of the base current is 1 mA, the emitter current is measured as 110 mA. What is the value of the collector current? [1]

AO2 **(d) (i)** What would be the value of the collector current if the base current was 0 A? [1]

AO1 **(ii)** What is the job of the base current? [1]

[Total: 6]

AO1 **4** The diagram represents a step-up transformer.
AO2

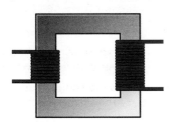

A mobile phone charger has an output of 6 V. It is plugged into the 230 V mains supply. Some power is lost in the transformer. Explain why the step-up transformer is not suitable and describe how a suitable transformer would be constructed. How is power lost in the transformer? [6]

The quality of written communication ✐ will be assessed in your answer to this question.

AO1 recall the science   AO2 apply your knowledge   AO3 evaluate and analyse the evidence

⊛ Worked Example – Foundation Tier

A wind turbine is used to generate electricity.

The turbine causes coils of wire to spin. These coils carry an electric current so they are electromagnets.

They spin inside another coil of wire.

An alternating current is induced in this outer coil.

Helena builds a model wind turbine. The voltage produced is not very large. She wants to increase the voltage produced.

Explain two ways she can increase the voltage output from her generator and how one of these ways will also affect the frequency of the output voltage. [6]

The quality of written communication ✎ will be assessed in your answer to this question.

she can

- *make the coils spin faster*

- *increase number of turns*

spinning faster will also make frequency bigger

> These longer 6 mark answers usually have marks awarded for the Quality of Written Communication shown by this symbol ✎ so answers need planning, and care is needed with spelling, punctuation and grammar. (See the example banded mark scheme on page 291.)
>
> This answer is simplistic. There is limited use of specialist terms. Errors of grammar, punctuation and spelling prevent communication of the science.
>
> This student has scored 1 or 2 marks out of a possible 6. This is below the standard of Grade C. With more care the student could have achieved a Grade C.

How to raise your grade!
Take note of these comments – they will help you to raise your grade.

> On this occasion, it is not clear which of the coils is being referred to.

> The student has not stated which of the coils is spinning faster but in context the electromagnetic coils can be assumed. Always be safe and make sure your answers cannot be misunderstood.

> Increasing the speed of rotation of the electromagnet coils will increase the frequency.

Higher Tier

1   Michael builds a circuit to find the resistance of a resistor.

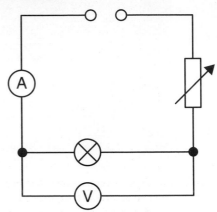

He records these results.

current in mA	average voltage in V
0	0.0
200	4.0
400	7.9
600	12.1
800	16.1

AO2 **(a)** Plot a graph of Michael's results, drawing a line of best fit.   [3]

AO2 **(b)** What has Michael done to improve the precision of his results?   [1]

AO2 **(c)** Use the graph to calculate the resistance of the resistor.   [2]

AO1 **(d)** Michael continues to increase the current through the resistor. Explain how and why the shape of the graph changes.   [2]

[Total: 8]

2   Janet is investigating a NAND gate.

AO1 **(a)** Copy and complete the truth table for a NAND gate.   [1]

input 1	input 2	output
0	0	
0	1	
1	0	
1	1	

AO1 **(b)** AO2   Janet wants the output from the NAND gate to switch on a 24 V electric motor. Explain why she cannot do this directly and how the relay allows this to be achieved.

The quality of written communication ✐ will be assessed in your answer to this question.   [6]

[Total: 7]

3   The diagram represents a step-down transformer.

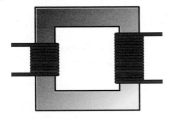

(a) A mobile phone charger has an output of 6 V. It is plugged into the 230 V mains supply.

AO1 **(i)** AO2   There are 4000 turns on the primary coil. Calculate the number of turns on the secondary coil.   [2]

AO3 **(ii)** In practice, there may be more turns on the secondary coil than calculated. Suggest why.   [1]

AO1 **(b)** AO2   Step-down transformers are also found at the end of transmission lines before the electricity is supplied to homes. Explain why it is necessary to use both step-up and step-down transformers when electricity is distributed around the country.   [3]

[Total: 6]

❊ Worked Example – Higher Tier

The diagram shows a rectifier circuit.

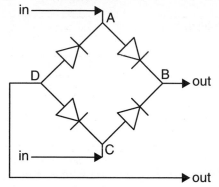

How to raise your grade!

Take note of these comments – they will help you to raise your grade.

↓

(a) The rectifier has a 12 V AC input. Explain how the arrangement of four diodes produces a DC output. [3]

During the positive cycle, current flows from A to B, through the other circuit then back to D and A. During the negative cycle, it flows from C to B through the circuit to D and A. This means current is always flowing in the same direction in the circuit.

This is quite a good answer. The student has correctly identified the current direction in both half-cycles. However, has mentioned cycles instead of half-cycles and could have been clearer by mentioning external circuit rather than a vague other circuit. **2/3**

(b) A capacitor has been connected across the output terminals.

(i) Sketch a graph to show how the voltage across a capacitor changes when it is being discharged. [2]

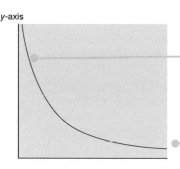

When a capacitor discharges, it starts with a particular voltage. The graph should touch the *y*-axis.

The student has correctly labelled the *y*-axis but forgotten to label the *x*-axis. **0/2**

(ii) How does the capacitor help to maintain a smooth output from the rectifier circuit? [2]

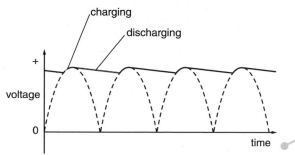

The high value capacitor charges quickly while the current is flowing from the rectifier and then discharges as the output from the rectifier circuit falls.

This is a good answer. The diagram helps to make clear what is being written. **2/2**

This student has scored 4 marks out of a possible 7. This is below the standard of Grade A. With more care the student could have achieved a Grade A.

Carrying out controlled assessments in GCSE Separate Sciences

Introduction

As part of your GCSE Biology, Chemistry and Physics courses, you will have to carry out a controlled assessment. This will be divided into three parts:

> Research and collecting secondary data

> Planning and collecting primary data

> Analysis and evaluation.

The tasks will be set by OCR, the awarding body, and marked by the teachers in school. The marking will be checked to make sure standards are the same in every school.

Some of the work you do must be supervised and you will have to work on your own under examination conditions.

Some experimental work may be performed in groups and the results shared.

Some research work may be done as part of a homework exercise.

As well as your scientific skills, the quality of written communication will also be assessed as part of the controlled assessment.

Controlled assessment is worth 25% of the marks for your GCSE. It's worth doing it well.

 ### Part 1: Research and collecting secondary data

You will have to plan and carry out research. The task will be given to you in the form of a handout. You will be allowed to research in class and/or as a homework activity.

Secondary data needs to be appropriate and you will have to select the information from a variety of sources to answer the questions you have been set.

Secondary data can be collected by a variety of different methods:

> survey

> questionnaire

> interview

> textbook

> newspapers and magazines

> internet search.

Always make sure that you reference the material you use from any secondary source.

The work you do for your GCSE examination will need to be handwritten or typed and printed out. You will take the work to a supervised lesson where you will use the information you have collected to answer specific questions in an answer booklet. Your research will be retained. It will be needed when you complete the analysis and evaluation part of your controlled assessment.

Your research may also be needed when you are planning your experiment.

Assessment tip

Make sure that the data you look at is relevant and appropriate. The data is more likely to be reliable if the same results and conclusions are obtained by a number of different researchers.

Assessment tip

As part of your research, you may need to design your own survey or questionnaire as well as collecting the data.

Choose the search terms you use on the internet carefully.

Assessment tip

When referencing, make sure you record:
> book title, author and page
> newspaper or magazine title, date, author and page
> full website address, author (if possible).

Definition

Secondary data are measurements/observations made by anyone other than you.

Part 2: Planning and collecting primary data

As part of your GCSE Separate Sciences course, you will develop practical skills which will help you to plan and collect primary data from a science experiment. Your experimental work will be divided into several parts.

A scientific investigation usually begins with a scientist testing an idea, answering a question, or trying to solve a problem.

You first have to plan how you will carry out the investigation.

Your planning will involve producing and testing a **hypothesis**. For example, you might observe that plants grow faster in a heated greenhouse than an unheated one.

So your hypothesis might be 'the rate of photosynthesis increases because the temperature increases'.

To formulate a hypothesis you may have to research some of the background science.

First of all, use your lesson notes and your textbook. The topic you've been given to investigate will relate to the science you've learned in class.

Also make use of the Internet, but make sure that your Internet search is closely focused on the topic you're investigating.

✔ The search terms you use on the Internet are very important. 'Investigating photosynthesis' is a better search term than just 'photosynthesis', as it's more likely to provide links to websites that are more relevant to your investigation.

✔ The information on websites also varies in its reliability. Free encyclopaedias often contain information that hasn't been written by experts. Some question-and-answer websites might appear to give you the exact answer to your question, but be aware that they may sometimes be incorrect.

✔ Most GCSE Science websites are more reliable but, if in doubt, use other information sources to verify the information.

When you produce a hypothesis you can use your lesson notes, the research you have already done and textbooks.

> **Definition**
>
> A **hypothesis** is a possible explanation that someone suggests to explain some scientific observations.

> **Assessment tip**
>
> When you're formulating a hypothesis, it's important that it's testable. In other words, you must be able to test the hypothesis in the school lab.

> **Assessment tip**
>
> In the planning stage, scientific research is important if you are going to obtain higher marks.

Example 1

Investigation: Plan and research an investigation into the effect of pH on how well yeast catalyses the fermentation of sugars.

Your hypothesis might be 'I think that the yeast will produce the most carbon dioxide per minute at pH7'. You should be able to justify your hypothesis by some facts that you find. For example 'yeast contains enzymes and I know that most enzymes work best in neutral pH solutions'.

✳ Choosing a method and suitable apparatus

As part of your planning, you must choose a suitable way of carrying out the investigation.

You will have to choose suitable techniques, equipment and technology, if this is appropriate. How do you make this choice?

You will have already carried out the techniques you need to use during the course of practical work in class (although you may need to modify these to fit in with the context of your investigation). For most of the experimental work you do, there will be a choice of techniques available. You must select the technique:

✔ that is most appropriate to the context of your investigation, and

✔ that will enable you to collect valid data, for example if you are measuring the effects of light intensity on photosynthesis, you may decide to use an LED (light-emitting diode) at different distances from the plant, rather than a light bulb. The light bulb produces more heat, and temperature is another independent variable in photosynthesis.

Your choice of equipment, too, will be influenced by measurements you need to make. For example:

✔ you might use a one-mark or graduated pipette to measure out the volume of liquid for a titration, but

✔ you may use a measuring cylinder or beaker when adding a volume of acid to a reaction mixture, so that the volume of acid is in excess of that required to dissolve, for example, the calcium carbonate.

In science, the measurements you make as part of your investigation should be as precise as you can, or need to, make them. To achieve this, you should use:

✔ the most appropriate measuring instrument

✔ the measuring instrument with the most appropriate size of divisions.

The smaller the divisions you work with, the more precise your measurements. For example:

✔ in an investigation on how your heart rate is affected by exercise, you might decide to investigate this after a 100 m run. You might measure out the 100 m distance using a trundle wheel, which is sufficiently precise for your investigation

✔ in an investigation on how light intensity is affected by distance, you would make your measurements of distance using a metre rule with millimetre divisions; clearly a trundle wheel would be too imprecise

Assessment tip

Technology, such as data-logging and other measuring and monitoring techniques, for example heart sensors, may help you to carry out your experiment.

Definition

The **resolution** of the equipment refers to the smallest change in a value that can be detected using a particular technique.

Assessment tip

Carrying out a preliminary investigation, along with the necessary research, may help you to select the appropriate technique to use.

✔ in an investigation on plant growth, in which you measure the thickness of a plant stem, you would use a micrometer or Vernier callipers. In this instance, a metre rule would be too imprecise.

 Variables

In your investigation, you will work with independent and dependent variables.

The factors you choose, or are given, to investigate the effect of are called **independent variables**.

What you choose to measure, as affected by the independent variable, is called the **dependent variable**.

 Independent variables

In your practical work, you will be provided with an independent variable to test, or will have to choose one – or more – of these to test. Some examples are given in the table.

Investigation	Possible independent variables to test
activity of amylase enzyme	> temperature > sugar concentration
rate of a chemical reaction	> temperature > concentration of reactants
stopping distance of a moving object	> speed of the object > the surface on which it's moving

Independent variables can be **discrete** or **continuous**.

> When you are testing the effect of different disinfectants on bacteria, you are looking at discrete variables.

> When you are testing the effect of a range of concentrations of the same disinfectant on the growth of bacteria, you are looking at continuous variables.

Range

When working with an independent variable, you need to choose an appropriate **range** over which to investigate the variable.

You need to decide:

✔ which treatments you will test, and/or

✔ the upper and lower limits of the independent variables to investigate, if the variable is continuous.

Once you have defined the range to be tested, you also need to decide the appropriate intervals at which you will make measurements.

Definition

Variables that fall into a range of separate types are called **discrete variables**.

Definition

Variables that have a continuous range are called **continuous variables**.

Definition

The **range** defines the extent of the independent variables being tested.

The range you would test depends on:

✔ the nature of the test
✔ the context in which it is given
✔ practical considerations, and
✔ common sense.

Example 2

1 Investigation: Investigating the energy changes that occur when different fuels are burned.

You may decide which fuels to test based on a range that you have been provided with. You may wish to test only a liquid or you may choose to test both solid and liquid fuels. You will not test certain fuels, such as pressurised gas, because of safety reasons.

2 Investigation: Comparing the focal lengths of a number of lenses.

The range of lenses you can choose will depend on the availability of the resources at your school. As a minimum though you would choose two different lenses to compare, such as convex and plane or concave and convex.

Temperature

You might be trying to find out the best temperature to grow tomatoes.

The 'best' temperature is dependent on a number of variables that, taken together, would produce tomatoes as fast as possible whilst not being too costly.

You should limit your investigation to just one variable, temperature, and then consider other variables such as fuel cost later.

> **Assessment tip**
>
> Again, it's often best to carry out a trial run or preliminary investigation, or carry out research, to determine the range to be investigated.

✹ Dependent variables

The dependent variable may be clear from the problem you're investigating, for example the stopping distance of moving objects. But you may have to make a choice.

Example 3

1 Investigation: Investigating the amount of sodium hydroxide needed to neutralise 25 cm³ of hydrochloric acid.

There are several ways that you could establish the neutralisation point in this investigation. These include:

> measuring the volume of sodium hydroxide solution being added by counting the number of drops or, more precisely, by using a pipette

> identifying the neutralisation point using an indicator such as methyl orange or a pH probe to measure pH.

> **Assessment tip**
>
> The value of the *depend*ent variable is likely to *depend* on the value of the independent variable. This is a good way of remembering the definition of a dependent variable.

2 Investigation: Measuring the rate of a chemical reaction.

You could measure the rate of a chemical reaction in the following ways:

> the rate of formation of a product

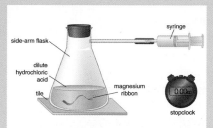

> the rate at which the reactant disappears

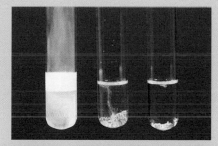

> a colour change

> a pH change.

Control variables

The validity of your measurements depend on you measuring what you're supposed to be measuring.

Some of these variables may be difficult to control. For example, in an ecology investigation in the field, factors such as varying weather conditions are impossible to control.

Experimental controls

Experimental controls are often very important, particularly in biological investigations where you're testing the effect of a treatment.

Definition

Other variables that you're not investigating may also have an influence on your measurements. In most investigations, it's important that you investigate just one variable at a time. So other variables, apart from the one you're testing at the time, must be controlled, and kept constant, and not allowed to vary. These are called **control variables**.

Definition

An **experimental control** is used to find out whether the effect you obtain is from the treatment, or whether you get the same result in the absence of the treatment.

Example 4

Investigation: Comparing the rate of reaction between ethanoic acid and hydrochloric acid with marble chips.

Many factors affect the rate of a chemical reaction, so you need to control those factors that you are not investigating.

You are comparing ethanoic acid and hydrochloric acid. Therefore you need to make sure the volumes and concentrations are the same (for example, 25 cm³ of 1 mol/dm³ acid), the quantity of marble chips and their sizes are the same and the temperature is the same.

Assessing and managing risk

Before you begin any practical work, you must assess and minimise the possible risks involved.

Before you carry out an investigation, you must identify the possible hazards. These can be grouped into biological hazards, chemical hazards and physical hazards.

Biological hazards include:

> microorganisms
> body fluids
> animals and plants.

Chemical hazards can be grouped into:

> irritant and harmful substances
> toxic
> oxidising agents
> corrosive
> harmful to the environment.

Physical hazards include:

> equipment
> objects
> radiation.

Scientists use an international series of symbols so that investigators can identify hazards.

Hazards pose risks to the person carrying out the investigation.

A risk posed by chlorine gas produced in the electrolysis of sodium chloride will be reduced if you increase the ventilation of the process, or devise a method to remove the gas so that workers cannot inhale it.

When you use hazardous materials, chemicals or equipment in the laboratory, you must use them in such a way as to keep the risks to an absolute minimum. For example, one way is to wear eye protection when using hydrochloric acid.

Definition

A **hazard** is something that has the potential to cause harm. Even substances, organisms and equipment that we think of as being harmless may, if used in the wrong way, be hazardous.

Hazard symbols are used so that hazards can be identified

Definition

The **risk** is the likelihood of a hazard to cause harm in the circumstances it's being used in.

Assessment tip

When assessing risk, and suggesting control measures these should be specific to the hazard and risk, and not general. Hydrochloric acid is dangerous as it is 'corrosive, and skin and eye contact should be avoided' will be given credit but wear 'eye protection' is too vague.

Risk assessment

Before you begin an investigation, you must carry out a risk assessment. Your risk assessment must include:

✔ all relevant hazards (use the correct terms to describe each hazard, and make sure you include them all, even if you think they will pose minimal risk)

✔ risks associated with these hazards

✔ ways in which the risks can be minimised

✔ results of research into emergency procedures that you may have to take if something goes wrong.

You should also consider what to do at the end of the practical. For example, used agar plates should be left for a technician to sterilise; solutions of heavy metals should be collected in a bottle and disposed of safely.

Assessment tip

To make sure that your risk assessment is full and appropriate:

> remember that for a risk assessment for a chemical reaction, the risk assessment should be carried out for the products and the reactants

> when using chemicals, make sure the hazard and ways of minimising risk match the concentration of the chemical you're using; many acids, for instance, while being corrosive in higher concentrations, are harmful or irritant at low concentrations.

Collecting primary data

✔ You should make sure that observations are recorded in detail, if appropriate. For example, it's worth recording the colour of your precipitate when making an insoluble salt, in addition to any other measurements you make.

✔ Measurements should be recorded in tables. Have one ready so that you can record your readings as you carry out the practical work.

✔ Think about the dependent variable and define this carefully in your column headings.

✔ You should make sure that the table headings describe properly the type of measurements you've made, for example 'time taken for magnesium ribbon to dissolve'.

✔ It's also essential that you include units – your results are meaningless without these.

✔ The units should appear in the column head, and not be repeated in each row of the table.

Definition

When you carry out an investigation, the data you collect are called **primary data.** The term 'data' is normally used to include your observations as well as measurements you might make.

Repeatability and reproducibility of results

When making measurements, in most instances it's essential that you carry out repeats.

value 1	value 2	value 3	average value

These repeats are one way of checking your results.

Definition

One set of results from your investigation may not reflect what truly happens. Carrying out repeats enables you to identify any results that don't fit. These are called **outliers** or **anomalous results**.

Results will not be repeatable, of course, if you allow the conditions that the investigation is carried out in to change.

You need to make sure that you carry out sufficient repeats, but not too many. In a titration, for example, if you obtain two values that are within 0.1 cm³ of each other, carrying out any more will not improve the precision of your results.

This is particularly important when scientists are carrying out scientific research and make new discoveries.

Once you have planned your experiment and collected your primary data, your work will be retained. It will be needed when you complete the analysis and evaluation part of your controlled assessment.

Definition
If, when you carry out the same experiment several times, and get the same, or very similar results, we say the results are **repeatable**.

Definition
Taking more than one set of results will help to make sure your data is **precise**.

Part 3: Analysis and evaluation

Calculating the mean

Using your repeat measurements you can calculate the arithmetical mean (or just 'mean') of these data. Often, the mean is called the 'average'.

Here are the results of an investigation into the energy requirements of three different MP3 players. The students measured the energy using a joulemeter for ten seconds.

Definition
The **reproducibility** of data is the ability of the results of an investigation to be reproduced by someone else, who may be in a different lab, carrying out the same work.

MP3 player	energy used in joules (J)			
	Trial 1	Trial 2	Trial 3	Mean
viking	5.5	5.3	5.7	5.5
anglo	4.5	4.6	4.9	4.7
saxon	3.2	4.5	4.7	4.6

You may also be required to use formulae when processing data. Sometimes, these will need rearranging to be able to make the calculation you need. Practise using and rearranging formulae as part of your preparation for assessment.

Definition
The **mean** is calculated by adding together all the measurements, and dividing by the number of measurements.

Significant figures

When calculating the mean, you should be aware of significant figures.

For example, for the set of data below:

18	13	17	15	14	16	15	14	13	18

The total for the data set is 153, and ten measurements have been made. The mean is 15, and not 15.3.

This is because each of the recorded values has two significant figures. The answer must therefore have two significant figures. An answer cannot have more significant figures than the number being multiplied or divided.

Definition
Significant figures are the number of digits in a number based on the precision of your measurements.

Using your data

When calculating means (and displaying data), you should be careful to look out for any data that don't fit in with the general pattern.

It might be the consequence of an error made in measurement. But sometimes outliers are genuine results. If you think an outlier has been introduced by careless practical work, you should ignore it when calculating the mean. But you should examine possible reasons carefully before just leaving it out.

Definition
An **outlier** (or **anomalous result**) is a reading that is very different from the rest.

Displaying your data

Displaying your data – usually the means – makes it easy to pick out and show any patterns. And it also helps you to pick out any anomalous data.

It is likely that you will have recorded your results in tables, and you could also use additional tables to summarise your results. The most usual way of displaying data is to use graphs. The table will help you to decide which type to use.

Type of graph	When you would use the graph	Example
bar charts or bar graph	where one of the variables is discrete	'The energy requirements of different MP3 players'
line graph	where independent and dependent variables are both continuous	'The volume of carbon dioxide produced by a range of different concentrations of hydrochloric acid'
scatter graph	to show an association between two (or more) variables	'The association between length and breadth of a number of privet leaves' In scattergraphs, the points are plotted, but not usually joined

If it's possible from the data, join the points of a line graph using a straight line, or in some instances, a curve. In this way graphs can also help us to process data.

☀ Conclusions from differences in data sets

When comparing two (or more) sets of data, we often compare the values of two sets of means.

Example 5

Investigation: Two groups of students measured the amount of sweat they produced after exercising for one hour. Their results are shown in the table below. Group 1 rarely exercise, group 2 exercise regularly.

Student	Amount of sweat produced (cm³)					
	1	2	3	4	5	Mean
Group 1	15	12	17	20	12	15.2
Group 2	10	8	11	12	9	10.0

When the means are compared, it appears that Group 1 produced more sweat than Group 2. The difference may be due to the amount and type of exercise each group did, or it may be purely by chance.

Scientists use statistics to find the probability of any differences having occurred by chance. The lower this probability is, which is found out by statistical calculations, the more likely it is that students who exercise regularly sweat less than those who do not exercise regularly.

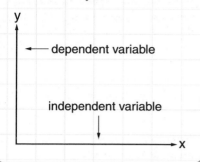
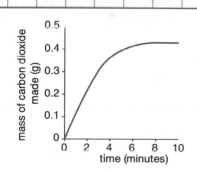

We can calculate the rate of production of carbon dioxide from the gradient of the graph

Drawing conclusions

Observing trends in data or graphs will help you to draw conclusions. You may obtain a linear relationship between two sets of variables, or the relationship might be more complex.

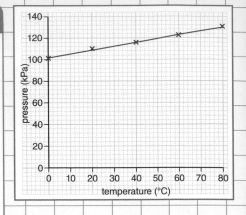

Example 6

Conclusion: As the temperature of the gas increased its pressure also increased.

Conclusion: Increasing the temperature increased the energy of the gas particles, causing them to move around faster. This means that there were more collisions between the gas particles and the sides of the container so therefore the pressure increased.

When drawing conclusions, you should try to relate your findings to the science involved.

> In the first point in Example 6, your discussion should focus on describing what your results show, including any patterns or trends between them.

> In the second point in Example 6, there is a clear scientific mechanism to link the increase in temperature to an increase in gas pressure.

> Sometimes, you can see correlations between data which are coincidental, where the independent variable is not the cause of the trend in the data.

Evaluating your investigation

Your conclusion will be based on your findings, but must take into consideration any uncertainty in these introduced by any possible sources of error. You should discuss where these have come from in your evaluation.

The two types of errors are:

✔ random error
✔ systematic error.

Random errors can occur when the instrument you're using to measure lacks sufficient sensitivity to indicate differences in readings. It can also occur when it's difficult to make a measurement. If two investigators measure the height of a plant, for example, they might choose different points on the compost, and the tip of the growing point, to make their measurements.

Systematic errors are either consistently too high or too low. One reason could be down to the way you are making a reading, for example taking a burette reading at the wrong point on the meniscus. Another could be the result of an instrument being incorrectly calibrated, or not being calibrated.

Definition

Error is a difference between a measurement you make, and its true value.

Definition

With **random error**, measurements vary in an unpredictable way.

Definition

With **systematic error**, readings vary in a controlled way.

Assessment tip

A pH meter must be calibrated before use using buffers of known pH.

Accuracy and precision

When evaluating your investigation, you should mention accuracy and precision. But if you use these terms, it's important that you understand what they mean, and that you use them correctly.

The terms accuracy and precision can be illustrated using shots at a target.

The shots are precise but not accurate.

The shots are precise and accurate.

The shots are not precise and not accurate.

Definition

When making measurements:

> the **accuracy** of the measurement is how close it is to the true value

> **precision** is how closely a series of measurements agree with each other.

Improving your investigation

When evaluating your investigation, you should discuss how your investigation could be improved. This could be by:

✔ confirming your data. For example, you could make more repeats, or more frequent readings, or 'fine-tune' the range you chose to investigate, or refine your technique in some other way

✔ improving the accuracy and precision of your data, by using more precise measuring equipment.

Using secondary data

As part of controlled assessment, you will be expected to compare your data – primary data – with **secondary data** you have collected.

The secondary data you collected earlier and the primary data you collected from your experiment will now be returned to you. You will be provided with an answer booklet to complete. This will help you to organise the data you have collected and answer specific questions about the topic you have been investigating.

You should review secondary data and evaluate it. Scientific studies are sometimes influenced by the **bias** of the experimenter.

✔ One kind of bias is having a strong opinion related to the investigation, and perhaps selecting only the results that fit with a hypothesis or prediction.

✔ Or the bias could be unintentional. In fields of science that are not yet fully understood, experimenters may try to fit their findings to current knowledge and thinking.

There have been other instances where the 'findings' of experimenters have been influenced by organisations that supplied the funding for the research.

You must fully reference any secondary data you have used, using one of the accepted referencing methods.

Assessment tip

Make sure you relate your conclusions to the hypothesis you are investigating. Do the results confirm or reject the hypothesis? Quote some results to back up your statement, for example: 'My results at 35°C and 65°C show that over a 30°C change in temperature, the time taken to produce 50 cm³ of carbon dioxide halved.'

Do the data support the hypothesis?

You need to discuss in detail whether all, or which, of the primary and secondary data you have collected, support your original hypothesis. They may, or may not.

You should communicate your points clearly, using the appropriate scientific terms, and checking carefully your use of spelling, punctuation and grammar. You will be assessed on this written communication as well as your science.

If your data do not completely match the hypothesis, it may be possible to modify the hypothesis or suggest an alternative one. You should suggest any further investigations that can be carried out to support your original hypothesis or the modified version.

It is important to remember, however, that if your investigation does support the hypothesis, it can improve the confidence you have in your conclusions and scientific explanations, but it can't prove your explanations are correct.

Referencing methods

The two main conventions for writing a reference are the:

✔ Harvard system
✔ Vancouver system.

In your text, the Harvard system refers to the authors of the reference, for example 'Smith and Jones (1978)'.

The Vancouver system refers to the number of the numbered reference in your text, for example '... the reason for this hypothesis is unknown.[5]'

Though the Harvard system is usually preferred by scientists, it is more straightforward for you to use the Vancouver system.

Harvard system

In your references list, a book reference should be written:

> Author(s) (year of publication). *Title of Book*, publisher, publisher location.

The references are listed in alphabetical order according to the authors.

Vancouver system

In your references list, a book reference should be written:

> 1 Author(s). *Title of Book.* Publisher, publisher location: year of publication.

The references are numbered in the order in which they are cited in the text.

Assessment tip

Remember to write out the URL of a website in full. You should also quote the date when you looked at the website.

How to be successful in your GCSE Separate Sciences written examinations

Introduction

OCR uses assessments to test how good your understanding of scientific ideas is, how well you can apply your understanding to new situations and how well you can analyse and interpret information you've been given. The assessments are opportunities to show how well you can do these.

To be successful in exams you need to:

✔ have a good knowledge and understanding of science

✔ be able to apply this knowledge and understanding to familiar and new situations

✔ be able to interpret and evaluate evidence that you've just been given.

You need to be able to do these things under exam conditions.

✸ The language of the external assessment

When working through an assessment paper, make sure that you:

✔ re-read a question enough times until you understand exactly what the examiner is looking for

✔ highlight key words in a question. In some instances, you will be given key words to include in your answer

✔ look at how many marks are allocated for each part of a question. In general, you need to write at least as many separate points in your answer as there are marks.

✸ What verbs are used in the question?

A good technique is to see which verbs are used in the wording of the question and to use these to gauge the type of response you need to give. The table lists some of the common verbs found in questions, the types of responses expected and then gives an example.

Verb used in question	Response expected in answer	Example question
write down state give identify	these are usually more straightforward types of question in which you're asked to give a definition, make a list of examples, or choose the best answer from a series of options	'Write down three types of microorganism that cause disease' 'State one difference and one similarity between radio waves and gamma rays'
calculate	use maths to solve a numerical problem	'Calculate the percentage of carbon in copper carbonate $(CuCO_3)$'

estimate	use maths to solve a numerical problem, but you do not have to work out the exact answer	'Estimate from the graph the speed of the vehicle after 3 minutes'
describe	use words (or diagrams) to show the characteristics, properties or features of, or build an image of, something	'Describe how meiosis halves the number of chromosomes in a cell to make egg or sperm cells'
suggest	come up with an idea to explain information you're given, usually in a new or unfamiliar context	'Suggest why tyres with different tread patterns will have different braking distances'
demonstrate show how	use words to make something evident, using reasoning	'Show how enzyme activity changes with temperature'
compare	look for similarities and differences	'Compare aerobic and anaerobic respiration'
explain	to offer a reason for, or make understandable, information you're given	'Explain how carbon-14 dating could be used to estimate the age of the remains of an animal'
evaluate	to examine and make a judgement about an investigation or information you're given	'Evaluate the benefits of using a circuit breaker instead of a fuse in an electrical circuit'

What is the style of the question?

Try to get used to answering questions that have been written in lots of different styles, before you sit the exam. Work through past papers, or specimen papers, to get a feel for these. The types of questions in your assessment fit the three assessment objectives shown in the table.

Assessment objective	Your answer should show that you can...
AO1 Recall the science	Recall, select and communicate your knowledge and understanding of science
AO2 Apply your knowledge	Apply skills, knowledge and understanding of science in practical and other contexts
AO3 Evaluate and analyse the evidence	Analyse and evaluate evidence, make reasoned judgements and draw conclusions based on evidence

Assessment tip

Of course you must revise the subject material adequately. But it's important that you are familiar with the different question styles used in the exam paper, as well as the question content.

How to answer questions on: AO1 Recall the science

These questions, or parts of questions, test your ability to recall your knowledge of a topic. There are several types of this style of question:

✔ Describe a process
✔ Explain a concept
✔ Complete sentences, tables or diagrams
✔ Tick the correct statements
✔ Use lines to link a term with its definition or correct statement.

Example 7

a Which is the correct equation to calculate the pressure that an object exerts?

Tick (✓) **one** box.

☐ $P = \dfrac{F}{A}$

☐ $P = \dfrac{A}{F}$

☐ $P = F \times A$.

How to answer questions on: AO1 Recall the science in practical techniques

You may be asked to recall how to carry out certain practical techniques; either ones that you have carried out before, or techniques that scientists use.

To revise for these types of questions, make sure that you have learnt definitions and scientific terms. Produce a glossary of these, or key facts cards, to make them easier to remember. Make sure your key facts cards also cover important practical techniques, including equipment, where appropriate.

Example 8

Describe two factors that scientists can change to affect the amount of product produced in equilibrium reactions such as the Haber process or Contact process.

Assessment tip

Don't forget that mind maps – either drawn by you or by using a computer program – are very helpful when revising key points.

How to answer questions on: AO2 Apply skills, knowledge and understanding

Some questions require you to apply basic knowledge and understanding in your answers.

You may be presented with a topic that's familiar to you, but you should also expect questions in your Science exam to be set in an unfamiliar context.

Questions may be presented as:

✔ experimental investigations

✔ data for you to interpret

✔ a short paragraph or article.

The information required for you to answer the question might be in the question itself; but for later stages of the question, you may be asked to draw on your knowledge and understanding of the subject material in the question.

Practice will help you to become familiar with contexts that examiners use, and their question styles. But you will not be able to predict many of the contexts used. This is deliberate; being able to apply your knowledge and understanding to different and unfamiliar situations is a skill the examiner tests.

Practise doing questions where you are tested on being able to apply your scientific knowledge and your ability to understand new situations that may not be familiar. In this way, when this type of question comes up in your exam, you will be able to tackle it successfully.

Assessment tip

Work through the Preparing for Assessment: Applying your knowledge tasks in this book as practice.

Example 9

When light enters the eye from the air it first passes through the cornea – a jelly-like substance – before entering the pupil. Use your knowledge of refraction to explain what happens to the light as it passes through the cornea.

How to answer questions on: AO2 Apply skills, knowledge and understanding in practical investigations

Some opportunities to demonstrate your application of skills, knowledge and understanding will be based on practical investigations. You may have carried out some of these investigations, but others will be new to you, and based on data obtained by scientists. You will be expected to describe patterns in data from graphs you are given or that you will have to draw from given data.

Again, you will have to apply your scientific knowledge and understanding to answer the question.

Example 10

A student measured the pH of two different brands of beer on six days to find out which one oxidised from ethanol into ethanoic acid first. His results are shown in the table on the right.

a Calculate the total pH decrease for each beer during the time.

b Plot the results as a line graph.

c Ethanoic acid has a pH of 2.4. Use the graph to work out which beer oxidises the ethanol to ethanoic acid the faster.

Beer	pH on each day					
	0	1	2	3	4	5
Old Brew	6.6	5.8	4.0	3.0	2.1	2.1
Eagle lager	6.3	5.9	4.1	3.5	2.2	2.1

 How to answer questions on: AO3 Analysing and evaluating evidence

For these types of questions, you will analyse and evaluate scientific evidence or data given to you in the question. It's likely that you won't be familiar with the material.

When describing patterns and trends in the data, make sure you:

✔ explain a pattern or trend in as much detail as you can

✔ mention anomalies where appropriate.

You must also be able to evaluate the information you're given. This is one of the hardest skills. Think about the validity of the scientific data: did the technique(s) used in any practical investigation allow the collection of accurate and precise data?

Your critical evaluation of scientific data in class, along with the practical work and controlled assessment work, will help you to develop the evaluation skills required for these types of questions.

Assessment tip

Remember, when carrying out any calculations, you should include your working at each stage. You may get credit for getting the process correct, even if your final answer is wrong.

Example 11

The table shows the properties of some metals.

metal	melting point in °C	density in g/cm³	relative electrical conductivity	cost per tonne in £
aluminium	660	2.7	38	2491
copper	1083	8.9	60	9048
silver	962	10.5	63	1 125 276
zinc	420	7.1	17	2260

Pylon wires are made from metal.

Which metal would be most suitable for using to make pylon wires?

Use information about each of the metals in the table to explain your answer.

Example 12

When investigating the activity of an enzyme, why is it hard to achieve repeatable results when the enzyme is exposed to temperatures above 40 °C?

✸ The quality of your written communication

Scientists need good communication skills to present and discuss their findings. You will be expected to demonstrate these skills in the exam. Questions will [end] with the sentence: The quality of your written communication will be assessed in your answer to this question.

✔ You must also try to make sure that your spelling, punctuation and grammar are accurate, so that it's clear what you mean in your answer. Again, examiners can't award marks for answers where the meaning isn't clear.

✔ Make sure your language is concise. When describing and explaining science, use correct scientific vocabulary.

Practise answering some longer, 6 mark questions. These will examine your quality of written communication as well as your knowledge and understanding of science. Look at how marks are awarded in mark schemes provided by the awarding body. You'll find these in the specimen question papers, and past papers.

You will also need to remember the writing and communication skills you've developed in English lessons. For example, make sure that you understand how to construct a good sentence using connectives.

Assessment tip

You will be assessed on the way in which you communicate science ideas.

Assessment tip

When answering questions, you must make sure that your writing is legible. An examiner can't award marks for answers that he or she can't read.

All long answer, 6 mark, questions will have the same general mark scheme which will be amplified and made specific to the question by examiners. Three levels of answer will gain credit.

Level 3

All information in answer is relevant, clear, organised and presented in a structured and coherent format. Specialist terms are used appropriately. Few, if any, errors in grammar, punctuation and spelling. (5–6 marks)

Level 2

For the most part the information is relevant and presented in a structured and coherent format. Specialist terms are used for the most part appropriately. There are occasional errors in grammar, punctuation and spelling. (3–4 marks)

Level 1

Answer may be simplistic. There may be limited use of specialist terms. Errors of grammar, punctuation and spelling prevent communication of the science. (1–2 marks)

Level 0

Insufficient or irrelevant science. Answer not worthy of credit. (0 marks)

✺ Revising for your Science exam

You should revise in the way that suits you best. But it's important that you plan your revision carefully, and it's best to start well before the date of the exams. Take the time to prepare a revision timetable and try to stick to it. Use this during the lead-up to the exams and between each exam.

When revising:

✔ find a quiet and comfortable space in the house where you won't be disturbed. It's best if it's well ventilated and has plenty of light

✔ take regular breaks. Some evidence suggests that revision is most effective when you revise in 30 to 40 minute slots. If you get bogged down at any point, take a break and go back to it later when you're feeling fresh. Try not to revise when you are feeling tired. If you do feel tired, take a break

✔ use your school notes, textbook and possibly a revision guide. But also make sure that you spend some time using past papers to familiarise yourself with the exam format

✔ produce summaries of each topic or module

✔ draw mind maps covering the key information on a topic or module

Assessment tip

Try to make your revision timetable as specific as possible – don't just say 'science on Monday, and Thursday', but list the [modules] that you'll cover on those days.

✔ set up revision cards containing condensed versions of your notes

✔ ask yourself questions, and try to predict questions, as you're revising topics or modules

✔ test yourself as you're going along. Try to draw key labelled diagrams, and try some questions under timed conditions

✔ prioritise your revision of topics. You might want to allocate more time to revising the topics you find most difficult.

Assessment tip

Start your revision well before the date of the exams, produce a revision timetable, and use the revision strategies that suit your style of learning. Above all, revision should be an active process.

How do I use my time effectively in the exam?

Timing is important when you sit an exam. Don't spend so long on some questions that you leave insufficient time to answer others. For example, in a 75-mark question paper, lasting 75 minutes, you will have, on average, one minute per mark.

If you're unsure about certain questions, complete the ones you're able to do first, then go back to the ones you're less sure of.

If you have time, go back and check your answers at the end of the exam.

What will my exam look like?

Your Biology, Chemistry and Physics exams each consist of two papers.

Paper 1 contains three sections and lasts 1 hour 15 minutes.
There are 25 marks for each section.
You should spend about 25 minutes answering each section.
The questions in each section will test objectives AO1, AO2 and AO3 using structured questions.

Paper 2 contains four sections and lasts 1 hour 30 minutes.
The first three sections will be similar to the Paper 1 sections.
Section D contains a ten mark data response question which primarily assesses objective AO3. You will be required to analyse and evaluate evidence, make reasoned judgements and draw conclusions based on evidence.

You should spend about 15 minutes answering this section.

On exam day

A little bit of nervousness before your exam can be a good thing, but try not to let it affect your performance. When you turn over the exam paper keep calm. Look at the paper and get it clear in your head exactly what is required from each question. Read each question carefully. Don't rush.

If you read a question and think that you have not covered the topic, keep calm – it could be that the information needed to answer the question is in the question itself or the examiner may be asking you to apply your knowledge to a new situation.

Finally, good luck!

✳ Mathematical skills

You will be allowed to use a calculator in all assessments.

These are the maths skills that you need to complete all the assessments successfully.

You should understand:

✔ the relationship between units, for example, between a gram, kilogram and tonne

✔ compound measures such as speed

✔ when and how to use estimation

✔ the symbols = < > ~

✔ direct proportion and simple ratios

✔ the idea of probability.

You should be able to:

✔ give answers to an appropriate number of significant figures

✔ substitute values into formulae and equations using appropriate units

✔ select suitable scales for the axes of graphs

✔ plot and draw line graphs, bar charts, pie charts, scattergraphs and histograms

✔ extract and interpret information from charts, graphs and tables.

You should be able to calculate.

✔ using decimals, fractions, percentages and number powers, such as 10^3

✔ arithmetic means

✔ areas, perimeters and volumes of simple shapes.

In addition, if you are a higher tier candidate, you should be able to:

✔ **change the subject of an equation**

and should be able to use:

✔ **numbers written in standard form**

✔ **calculations involving negative powers, such as 10^{-1}**

✔ **inverse proportion**

✔ **percentiles and deciles.**

Some key physics equations

With the written papers, there will be an equation sheet. Below are some of the key equations found on the sheet; it will help if you practise using them.

$$s = \frac{(u + v)t}{2}$$

$$v = u + at$$

$$v^2 = u^2 + 2as$$

$$s = ut + \tfrac{1}{2}at^2$$

$$m_1 u_1 + m_2 u_2 = (m_1 + m_2)v$$

$$\text{refractive index} = \frac{\text{speed of light in vacuum}}{\text{speed of light in medium}}$$

$$\text{magnification} = \frac{\text{image size}}{\text{object size}}$$

$$\text{resistance} = \frac{\text{voltage}}{\text{current}}$$

$$R_T = R_1 + R_2 + R_3$$

$$\frac{1}{R_T} = \frac{1}{R_1} + \frac{1}{R_2} + \frac{1}{R_3}$$

$$I_e = I_b = I_c$$

$$\frac{\text{voltage across primary coil}}{\text{voltage across secondary coil}} = \frac{\text{no. primary turns}}{\text{no. secondary turns}}$$

$$\text{power loss} = \text{current}^2 \times \text{resistance}$$

$$V_p I_p = V_s I_s$$

Glossary

A

absorbed   taken in

accelerate   an object accelerates if it speeds up

acid strength   a measure of the ability of an acid to ionise

action (force)   applying a force on an object

addition reaction   reactions across a carbon-to-carbon double bond

adrenaline   hormone produced by adrenal gland

aerating   increasing the amount of air in soil

aerial   a device for receiving or transmitting radio signals

aerobic respiration   respiration that involves oxygen

agglutination   when red blood cells group together

alginate   chemical used to immobilise enzymes

amniocentesis   test during pregnancy for foetal abnormalities

amp   unit used to measure electrical current

amplifies   increases the size, e.g. of a radio signal

amylase   enzyme that digests starch

anaerobic respiration   respiration without using oxygen

AND gate   a logic gate which only delivers an output if both input terminals are on

(in) antiphase   when two waves are 'out of step' with each other; crests coincide with troughs

angle of incidence   the angle between the incident ray of light and the normal at a given point

angle of refraction   the angle between the refracted ray of light and the normal at a given point

anion   ion with a negative charge

anode   positive electrode in an electrolysis cell

antagonistic muscles   a pair of muscles working together, as one contracts the other relaxes

antibiotic   therapeutic drug acting to kill bacteria which is taken into the body

anti-coagulant   chemical that stops blood clotting

anti-diuretic hormone (ADH)   hormone which controls re-absorption of water in kidneys (controls water levels in the blood)

antiseptic   used to kill microorganisms in wounds

aperture   the size of the hole through which light enters a camera

aseptic technique   precautions taken to ensure there is no contamination when growing bacteria

asexual reproduction   reproduction involving only one parent

assaying   technique used in genetic engineering to find out if bacteria have taken up the genes

asthma   condition where airways inflamed and constricted

attenuate   lose energy

B

bacteria   single-celled microorganisms which can either be free-living organisms or parasites (they sometimes invade the body and cause disease)

barium chloride   chemical used to test for sulfate ions

binary fission   reproduction in bacteria

biofuels   fuels made from plants – these can be burned in power stations

biogas   biofuel containing methane

biological catalyst   molecules in the body that speed up chemical reactions

biological indicators   organisms which live in water – their presence or absence tells scientists how polluted water is

biomass   waste wood and other natural materials which are burned in power stations

blood transfusion   when blood from one person is put into another person (the two people must have compatible blood groups)

bone marrow   centre of long bones

budding   reproduction in yeast

burette   a graduated tube with a tap for accurately adding a liquid, showing the amount added

C

camera an optical instrument that produces a reduced image on a piece of film (film camera) or light sensitive chip (digital camera)

cancellation when two waves cancel to give reduced amplitude; destructive interference

capacitor an electronic device for storing electric charge

carbohydrates chemicals found in all living things. They contain the elements carbon, hydrogen and oxygen. Sugars and starches are carbohydrates

carnivore animal that eats other animals

cartilage softer and more flexible than bone, found in internal skeletons

catalyst a substance that speeds up a particular reaction without being used up

cathode negative electrode in an electrolysis cell

cation ion with a positive charge

centripetal force force towards the centre of a circle essential for circular motion

chemical digestion enzymes carry out chemical digestion when they break down large food molecules into smaller ones

chlorofluorocarbons inert molecules that used to be used in aerosols. Now banned in the EU as they damage the ozone layer

chromosomes thread-like structures in the cell nucleus that carry genetic information

clone genetically identical copy

closed circulatory system blood carried in blood vessels

coalesce join together / move as one

coherent waves having the same frequency, amplitude and phase (or constant phase difference)

colloids small particles dispersed through a liquid that do not settle

comets lumps of rock and ice found in space – some orbit the Sun

commutator part of an electric motor that reverses the current direction every half turn

compound fracture where a broken bone pierces skin

concave curving inwards

concentration the amount of solute dissolved in 1 dm³ of solution

condenser lens used in a projector to concentrate light on a slide

conservation of mass a law stating that in any reaction the mass of reactants equals the mass of products, as matter cannot be created or destroyed, just rearranged

constructive (interference) when two waves combine to give increased amplitude

Contact Process how sulfuric acid is manufactured

contractile vacuole structure in amoeba used to remove excess water

converging coming towards a point

convex curving outwards

coronary artery blood vessel that supplies the heart

corpuscles light particles as suggested by Newton in his particle theory of light

covalent bonds bonds between atoms where some of the electrons are shared

crests peaks of a wave

critical angle the angle of incidence for which the angle of refraction is 90°; larger angles of incidence result in total internal reflection

current flow of electrons in an electric circuit

D

decelerate an object decelerates if it slows down

decomposer organisms that break down dead animals and plants

denatured an enzyme is denatured if its shape changes so that the substrate cannot fit into the active site

dermis layer of skin

destructive (interference) when two waves cancel to give reduced amplitude

detergents soapless cleaning agents

detritivores organisms that feed on decaying matter

dibromo compound compound made when bromine reacts across a carbon-to-carbon double bond

diffraction the spreading out of a wave when it passes through a gap or around an edge

digestive system the 9-metre long system that handles and digests food (starts at mouth, ends at anus)

digital (signal) signal which is either 'on' or 'off'

dimmer switches switches that contain a variable resistor which can dim lights

diode an electronic component that only lets current pass through it in one direction

dispersed particles spread though a colloid

dispersion the splitting of light into its different wavelengths

displacement reaction chemical reaction where one element displaces or 'pushes out' another element from a compound

distillation using evaporation and condensation to increase the alcohol content of drinks

diverging spreading out / moving away from a point

DNA molecule found in all body cells in the nucleus – its sequence determines how our bodies are made (e.g. whether we have straight or curly hair), and gives each one of us a unique genetic code

DNA fingerprint identification obtained by examining a person's unique sequence of DNA

double circulatory system blood system of two circuits, found in mammals

dry cleaning cleaning without water using an organic solvent

E

egestion expulsion of solid waste

electrode terminal that conducts electricity put into a cell to perform electrolysis

electrolysis breaking up a dissolved ionic compound using an electric current

electrolyte solution or molten liquid through which electricity is passed

electromagnet a magnet which is magnetic only when a current is switched on

electromagnetic waves a group of waves that carry different amounts of energy – they range from low-frequency radio waves to high-frequency gamma rays

electron small negatively-charged particle

empirical formula the simplest way of writing a whole number ratio of the atoms in a compound

emulsion a mixture, often of a solid in liquid, which is held but not fully dissolved

endocrine system the body system that is made up of endocrine glands (these secrete hormones)

endothermic reaction chemical reaction which takes in heat

energy the ability of a system to do something (work), we detect energy by the effect it has on the things around us, e.g. heating them up

energy profile diagram diagram showing energy taken in or given out during a chemical reaction

enzymes biological catalysts that increase the speed of a chemical reaction

epidermis the outer layer of the skin

equilibrium when the forward and backward reactions are happening at the same rate

eutrophication when waterways become too rich with nutrients (from fertilisers) which allows algae to grow wildly and then die, oxygen is used up by the bacteria that decompose dead algae

excretion the process of getting rid of waste from the body

exothermic reaction chemical reaction in which heat is given out

external skeleton skeleton on outside of body, usually of chitin

F

fermentation the process of using yeast to break down sugars to alcohol

finite resource resource that will run out and there will be no more made

Fleming's left-hand rule if the fingers of the left hand are placed around a wire so that the thumb points in the direction of electron flow, the fingers point in the direction of the magnetic field produced by the conductor

focal length the distance from the optical centre of a lens to its focus

focal plane the plane that includes the focus (focal point)

focal point   focus (of a lens)

focus   (of lens or mirror) the point to which rays of light converge or from which they diverge

follicle   a ball of cells in the ovary containing an 'unripened' egg cell

follicle-stimulating hormone (FSH)   the hormone in females (made in the pituitary) which stimulates a follicle in an ovary to develop into a mature egg

fractional distillation   separation of a mixture by boiling followed by condensation, which is successful because each component has a different boiling point

fracture   break in bone

frequency   the number of vibrations per second, frequency is measured in hertz

fringes   light and dark bands of light produced by two-slit interference of monochromatic (single wavelength) light

fuel cell   uses the heat energy generated by the reaction between hydrogen and oxygen to convert to electrical energy

fungi   living organisms which can break down complex organic substances (some are pathogens and harm the body)

G

galvanising   coating another metal with zinc

gaseous exchange   the movement of gases across an exchange membrane, e.g. in the lungs of mammals – gaseous exchange usually involves carbon dioxide and oxygen moving in opposite directions

gasohol   biofuel that contains alcohol and petrol

generator   a device for converting energy of movement (kinetic energy) into electrical energy (current flow)

genetic engineering   transfer of genes from one organism to another

geostationary satellite   a satellite in orbit above the equator taking 24 hours for one orbit

gill filaments   part of the gills, through which oxygen is taken into the blood from the water passing through the gills

Global Positioning System (GPS)   a satellite navigation system that involves many satellites orbiting Earth

glomerulus   a tiny ball-shaped clump of blood vessels that filters substances from blood (found in kidney tubules)

glycerol   together with fatty acids, these make up fats

gravitational attraction   force of attraction between two bodies due to their mass

gravitational field   a region in which a mass experiences a force

gravity   an attractive force between objects (dependent on their mass)

greenstick fracture   bone not completely broken

growth hormone   a hormone produced by the pituitary gland that stimulates growth

guideline daily allowances   recommended values for safe amounts of fats, saturates, sugar and salt

H

heart attack   damage to heart muscle, can be fatal

humus   organic matter in soil

hydrogenated   reacted with hydrogen over a catalyst to saturate a compound with double carbon-to-carbon bonds

hydrolysis   reaction with OH^- ions of water or alkali

hydrophilic   water loving

hydrophobic   water hating

I

immobilised enzyme   placing enzymes inside gel beads

incubation period   time between infection by a pathogen and when the first symptoms appear

indicators   chemicals which change colour according to the pH (indicators show how acid or alkali a substance is)

insulin   hormone made by the pancreas which controls the level of glucose in the blood

interference   the formation of points of reinforcement and cancellation when two sets of waves overlap

intermolecular force   force between molecules

internal skeleton   skeleton inside body, made of cartilage or bone

inverse square law   when one variable is inversely proportional to the square of another

inverted   upside down

ion exchange resin   resin used in water softeners to exchange Na^+ ions on resin for Ca^{2+} ions in hard water

ionic lattice   regular arrangement of charged ions held together by ionic bonds

ionisation   (chemistry) – breaking up into ions in solution

ionosphere   a region of Earth's atmosphere where ionisation caused by incoming solar radiation affects the transmission of radio waves; it extends from 70 km to 400 km above Earth

isolating transformer   a transformer whose output voltage is the same as its input voltage

J

jet engine   a vehicle, similar to a rocket, that does not travel into space so does not need to carry an oxygen supply

L

laser   source of intense, narrow beam of light 'Light Amplification by Stimulated Emission of Radiation'

launch angle   the angle at which a projectile is thrown

lead nitrate   chemical used to test for halide ions

lens   a piece of transparent material, often glass, that is fatter in the middle than at the ends (convex) or thinner in the middle than at the ends (concave)

lever   system of gaining mechanical advantage

ligament   tissue joining bone to bone

ligase   enzyme used to stick DNA together

light-emitting diode (LED)   a very small light in electric circuits that uses very little energy

light-sensitive chip   surface in a digital camera that records light electronically, producing a digital image

light-dependent resistor (LDR)   device in an electric circuit whose resistance falls as the light falling on it increases

lightmeter   an electronic device for measuring the intensity of light

limescale   hard white substance found inside 'furred up' kettles (mostly calcium carbonate)

limiting reactant   the reactant that gets used up

line-of-sight   in direct line with no obstructions

lipases   enzymes that break down fats into fatty acids and glycerol

logic circuits   circuits composed of a series of logic gates

logic gates   electronic components that respond to signals by following preset logical rules

luteinising hormone (LH)   hormone (made in the pituitary) which, together with FSH, controls the release of an egg from the ovary

M

magnetic field   an area where a magnetic force can be felt

magnification   the ratio of the height of the image to the height of the object

magnitude   size of something

marine snow   organic matter that falls from the surface of oceans to the depths where it is used for food

menstrual cycle   monthly hormonal cycle which starts at puberty in human females

microvilli   microscopic projections from cells lining small intestine

microwaves   non-ionising waves used in satellite and mobile phone networks – also in microwave ovens

molar mass   the relative formula mass in grams – unit g/mol

mole   a number of particles (Avogadro's number) the same as in 12 g of carbon

molecular formula   the formula of a chemical using symbols in the periodic table, e.g. methane has a molecular formula of CH_4

molten   melted (e.g. rock or salt)

momentum   the product of mass and velocity of an object. Unit: kgm/s or Ns

mutation   where the DNA within cells have been altered (this happens in cancer)

N

NAND gate   a combination of an AND gate followed by a NOT gate

National Grid   network that carries electricity from power stations across the country (it uses cables, transformers and pylons)

negative electrode   electrode attached to negative pole of DC source which attracts positive ions in electrolysis

negative ions   ions formed from atoms that have accepted extra electrons to make a full outer electron shell

neutralisation   reaction between H⁺ ions and OH⁻ ions (acid and base react to make a salt and water)

noise   unwanted signals

non-renewable   something which is used up at a faster rate than it can be replaced, e.g. fossil fuels

NOR gate   a combination of an OR gate followed by a NOT gate

normal   a line perpendicular to a surface

NOT gate   a logic gate whose output is opposite to its input

O

oestrogen   female hormone secreted by the ovary and involved in the menstrual cycle

ohm   the unit used to measure electrical resistance

open circulatory system   blood system where blood is not contained in blood vessels

optical brightener   ingredient added to washing powder to make clothes 'whiter than white'

optical fibres   very thin glass fibres that light travels along by total internal reflection

optimum temperature   the best temperature for a reaction or process to occur, which may not be the lowest or highest

OR gate   a logic gate which delivers an output if any input terminal is on

orbit   the path taken by a satellite

organic solvent   solvent containing carbon and hydrogen bonded together or with other elements

osmosis   net movement of water from an area of high water concentration to an area of low water concentration

ossification   the formation of bone from cartilage

osteoporosis   disease of the bones in which bones become very brittle

oxidation   process of electron loss

P

pacemaker   cells in heart generate nerve impulses to stimulate muscle contraction

parabolic   shaped like a parabola, which looks a bit like an opened umbrella

parallelogram of forces   a method of finding the resultant of two forces (or other vectors)

particulates   particles such as soot released when fossil fuels burn

pasteurisation   heating of a liquid to 72 °C for 15 minutes to kill microorganisms

pathogen   harmful organism which invades the body and causes disease

permanent hardness   hardness of water which is removed not by boiling, containing calcium sulfate

pH meter   a device which measures the pH of a substance accurately

pH scale   scale running from 0 to 14 which shows how acid or alkali a substance is

(in) phase   when two waves are 'in step' with each other; crests coincide and troughs coincide

physical digestion   breaking down of food particles by teeth or muscles

phytoplankton   microscopic organisms in water that can photosynthesise

pipette   apparatus to transfer a measured amount of liquid

pixels   short for picture elements; stores data in the light-sensitive chip of a digital camera

plankton   microscopic organisms in water

plasmid   circular DNA in bacteria

platelets   cell fragments which help in blood clotting

polar orbit   a satellite orbit that passes over Earth's North and South poles

polarised (light)   light in which the oscillations are confined to one plane only

Polaroid   a material that absorbs light except that polarised in one particular plane, producing polarised light

pollution   contaminating or destroying the environment as a result of human activities

polyunsaturated fat   fat with many carbon-to-carbon double bonds

positive electrode   electrode attached to positive pole of DC source which attracts negative ions in electrolysis

positive ions   ions formed from atoms that have lost any extra electrons in their outside shell to make a stable electronic structure

potential divider   a combination of two resistors which allows an output voltage that is a fraction of the input voltage

power   the rate that a system transfers energy, power is usually measured in watts (W)

precipitate   solid formed in a solution during a chemical reaction

pressure   the force acting normally per unit area of a surface. Pressure – force / area. Unit: pascals (Pa) or N/m²

primary coil   the input coil of a transformer

principal axis   the axis, perpendicular to the face of a lens, that passes through the optical centre

products   molecules produced at the end of a chemical reaction

progesterone   hormone, produced by the ovary, which prepares the uterus for pregnancy

projectile   any object thrown in Earth's gravitational field

projector   an optical instrument that produces an enlarged image on a screen

proteases   enzymes which break down proteins into amino acids

R

radical   a very reactive species with one spare electron

radio waves   non-ionising waves used to broadcast radio and TV programmes

randomly   in no set pattern; haphazardly

range   the horizontal distance covered by an object

reactants   chemicals which are reacting together in a chemical reaction

reaction force   when an object feels a force it pushes back with an equal reaction force in an opposite direction

real (image)   an image that can be projected onto a screen; light actually passes through it

receiver   device which receives waves, e.g. a mobile phone

recoils   rebounds

red blood cells   blood cells which are adapted to carry oxygen

redox   simultaneous process of **red**uction and **oxi**dation

reduction   process of electron gain

refraction   a change in speed, and usually direction, when light passes from one medium to another, e.g. from air to glass or water

refractive index   the ratio of the speed of light in a vacuum (or air) to the speed of light in a medium

reinforcement   when two waves combine to give increased amplitude; constructive interference

relative speed   the speed of a moving object with respect to another

relay   an electronic switch which allows a very small current in one circuit to switch a larger current in another circuit

resistance   measurement of how hard it is for an electric current to flow through a material

resistor   a conductor that reduces the flow of electric current

respiration   process occurring in living things where oxygen is used to release the energy in foods

restriction enzyme   enzyme used to cut DNA

re-transmits   sends out a signal again (often after amplification)

reversible reaction   a reaction where products are made, which break back down into the reactants

rhesus   blood can be grouped into rhesus-positive and rhesus-negative groups

ripple tank   equipment containing a water surface to observe wave motion

rocket a vehicle that travels into space carrying its own oxygen supply

rusting process of forming hydrated iron(III) oxide from the reaction of iron with water and air

S

sacrificial protection using another, more reactive metal to protect a less reactive metal

saponification process of making soap from fat and sodium hydroxide

satellite a body orbiting a larger body, e.g. communications satellites orbit Earth

saturated fat fats, most often of animal origin, which are solid at room temperature

scalar a quantity having magnitude but no direction

scattered moved in random directions

secondary coil the output coil of a transformer

shutter in a camera, it opens and closes very quickly to let light into the camera

simple fracture clean break in bone

single circulatory system blood system with only one circuit, e.g. in fish

slip rings allow the transfer of current from a rotating coil in an AC generator

solutes substances which dissolve in a liquid to form a solution

solvents liquids in which solutes dissolve to form a solution

spectator ions ions that do not directly take part in a reaction

state symbols symbols used in equations to show whether something is solid, liquid, gas or in solution in water

stroke sudden change in blood flow to the brain – can be fatal

strong acid has a pH between 1 and 3 and completely ionises

sucrase enzyme that breaks down sucrose (sugar)

suspension liquid with very small solid particles suspended throughout

sustainable resources or processes which renew and are not wasteful

sweat liquid produced by the skin; it cools you down when it evaporates

synovial joint joint containing synovial fluid

T

temporary hardness hardness of water which is removed by boiling, containing calcium hydrogen carbonate

tendon tissue joining muscle to bone

terrestrial Earth-based

thermal decomposition breaking down a compound by heating it

thermistor an electronic device whose resistance changes with temperature

thinning agent agent added to detergent to allow it to flow more freely

threshold voltage the voltage needed to switch on a circuit

titration an accurate method for neutralisation

total internal reflection complete reflection of a light ray within glass when the ray hits the glass/air boundary at an angle which is greater than the critical angle

trajectory the path of a projectile

transformers devices which can change the voltage and current of electricity

transgenic organism organism that contains DNA from another organism

transistor an electronic component used to amplify or switch electronic signals

transmitter device which transmits waves, e.g. a mobile-phone mast

troughs lowest points of a wave

truth table a mathematical way of describing the behaviour of logic gates

U

ultraviolet light light from the Sun associated with sun-tanning. The light in the electromagnetic spectrum that has higher frequency than blue light

urea nitrogen-containing substance cleared from the blood by kidneys and excreted in urine

uric acid excretory product found in urine

V

variable resistor a device whose resistance can change, often used in volume controls and dimmer switches

vector a carrier, e.g. DNA plasmid that carries human DNA into the bacteria

villi 'finger-like' structures on the surface of the small intestine which give it a greater surface area for absorption

virtual image image formed on the same side of the lens as the object; a virtual image formed by reflection can be seen but cannot be projected onto a screen

viruses very small infectious organisms that reproduce within the cells of living organisms and often cause disease

volt unit used to measure voltage

voltage the potential difference across a component or circuit

W

washing soda sodium carbonate used to soften water

wavelength distance between two wave peaks

weak acid acid that has a pH between 4 and 6 and only partly ionises

weight force on an object due to gravitational attraction (= mg). Unit: N

white blood cells blood cells which defend against disease

Y

yeast single-celled fungus used in making beer and bread

Index

M

N

O

Y

Z

Internet research

The Internet is a great resource to use when you are working through your GCSE Science course.

Below are some tips to make the most of it.

1   Make sure that you get information at the right level for you by typing in the following words and phrases after your search: 'GCSE', 'KS4', 'KS3', 'for kids', 'easy', or 'simple'.

2   Use OR, AND, NOT and NEAR to narrow down your search.

> Use the word OR between two words to search for one or the other word.

> Use the word AND between two words to search for both words.

> Use the word NOT, for example, 'York NOT New York' to make sure that you do not get unwanted results (hits).

> Use the word NEAR, for example, 'London NEAR Art' to bring up pages where the two words appear very close to each other.

3   Be careful when you search for phrases. If you search for a whole phrase, for example, A Room with a View, you may get a lot of search results matching some or all of the words. If you put the phrase in quote marks, 'A Room with a View', it will only bring search results that have that whole phrase and so bring you more pages about the book or film and less about flats to rent!

4   For keyword searches, use several words and try to be specific. A search for 'asthma' will bring up thousands of results. But a search for 'causes of asthma' or 'treatment of asthma' will bring more specific and fewer returns. Similarly, if you are looking for information on cats, for example, be as specific as you can by using the breed name.

5   Most search engines list their hits in a ranked order so that results that contain all your listed words (and so most closely match your request) will appear first. This means the first few pages of results will always be the most relevant.

6   Avoid using lots of smaller words such as A or THE unless it is particularly relevant to your search. Choose your words carefully and leave out any unnecessary extras.

7   If your request is country-specific, you can narrow your search by adding the country. For example, if you want to visit some historic houses and you live in the UK, search 'historic houses UK' otherwise it will search the world. With some search engines you can click on a 'web' or 'pages from the UK only' option.

8   Use a plus sign (+) before a word to force it into the search. That way only hits with that word will come up.

Modern periodic table

Group 1	Group 2												Group 3	Group 4	Group 5	Group 6	Group 7	Group 0
							1 1 **H** hydrogen											4 2 **He** helium
7 3 **Li** lithium	9 4 **Be** beryllium												11 5 **B** boron	12 6 **C** carbon	14 7 **N** nitrogen	16 8 **O** oxygen	19 9 **F** fluorine	20 10 **Ne** neon
23 11 **Na** sodium	24 12 **Mg** magnesium												27 13 **Al** aluminium	28 14 **Si** silicon	31 15 **P** phosphorus	32 16 **S** sulfur	35 17 **Cl** chlorine	40 18 **Ar** argon
39 19 **K** potassium	40 20 **Ca** calcium	45 21 **Sc** scandium	48 22 **Ti** titanium	51 23 **V** vanadium	52 24 **Cr** chromium	55 25 **Mn** manganese	56 26 **Fe** iron	59 27 **Co** cobalt	59 28 **Ni** nickel	64 29 **Cu** copper	65 30 **Zn** zinc		70 31 **Ga** gallium	73 32 **Ge** germanium	75 33 **As** arsenic	79 34 **Se** selenium	80 35 **Br** bromine	84 36 **Kr** krypton
85 37 **Rb** rubidium	88 38 **Sr** strontium	89 39 **Y** yttrium	91 40 **Zr** zirconium	93 41 **Nb** niobium	96 42 **Mo** molybdenum	99 43 **Tc** technetium	101 44 **Ru** ruthenium	103 45 **Rh** rhodium	106 46 **Pd** palladium	108 47 **Ag** silver	112 48 **Cd** cadmium		115 49 **In** indium	119 50 **Sn** tin	122 51 **Sb** antimony	128 52 **Te** tellurium	127 53 **I** iodine	131 54 **Xe** xenon
133 55 **Cs** caesium	137 56 **Ba** barium	139 57 **La** lanthanum	178 72 **Hf** hafnium	181 73 **Ta** tantalum	184 74 **W** tungsten	186 75 **Re** rhenium	190 76 **Os** osmium	192 77 **Ir** iridium	195 78 **Pt** platinum	197 79 **Au** gold	201 80 **Hg** mercury		204 81 **Tl** thallium	207 82 **Pb** lead	209 83 **Bi** bismuth	210 84 **Po** polonium	210 85 **At** astatine	222 86 **Rn** radon
223 87 **Fr** francium	226 88 **Ra** radium	227 89 **Ac** actinium																

Modern periodic table. You need to remember the symbols for the highlighted elements.

junction of conductors		ammeter	
switch		voltmeter	
primary or secondary cell		indicator or light switch	
battery of cells		or	
power supply		motor	
fuse		generator	
fixed resistor		variable resistor	
diode		capacitor	
electrolytic capacitor		relay	NO COM NC
LDR		LED	
thermistor		NOT gate	NOT
AND gate	AND	OR gate	OR
NOR gate	NOR	NAND gate	NAND
NPN transistor	Collector Base Emitter		

Acknowledgements

The publishers gratefully acknowledge the following for permission to reproduce images. Every effort has been made to trace copyright holders. However, any cases where this has not been possible, or for any inadvertent omission, the publishers will gladly rectify at the first opportunity.

Cover & p.1 Sovereign, ISM/Science Photo Library, p.8t Reshavskyi/Shutterstock, p.8c Francis Leroy, Biocosmos/Science Photo Library, p.8b Monkey Business Images/Shutterstock, p.9t Dept. Of Clinical Radiology, Salisbury District Hospital/Science Photo Library, p.9cl Alexonline/Shutterstock, p.9cr Picsfive/Shutterstock, p.9l CC Studio/Science Photo Library, p.9b Jellyfish Pictures/Science Photo Library, p.10t Colin Bell, p.10cr Caitlin Mirra/Shutterstock, p.10bl Linda Bucklin/Shutterstock, p.10br D. Roberts/Science Photo Library, p.11 Mircea Bezergheanu/Shutterstock, p.12 Dept. Of Clinical Radiology, Salisbury District Hospital/Science Photo Library, p.14t Jean-Loup Charmet/Science Photo Library, p.14c Astris & Hanns-Frieder Michler/Science Photo Library, p.14b Phototake Inc./Alamy, p.16 DocCheck Medical Services GmbH/Alamy, p.17 Alexonline/Shutterstock, p.18 Scott Camazine/Alamy, p.20t Gina Sanders/Shutterstock, p.20b Antonia Reeve/Science Photo Library, p.22 Buzz Pictures/Alamy, p.24 Peter Elvidge/Shutterstock, p.25 Hattie Young/Science Photo Library, p.27 BSIP, Laurent/Science Photo Library, p.28t Bettmann/Corbis, p.28b pnicoledolin/Shutterstock, p.29 Martin/Custom Medical Stock Photo/Science Photo Library, p.32 Stephen J. Krasemann/Science Photo Library, p.36 Jordan Tan/Shutterstock, p.38tl Motta & Familiari/Anatomy Dept./University 'La Sapienza', Rome/Science Photo Library, p.38tc Peter Arnold, Inc./Alamy, p.38tr Francis Leroy, Biocosmos/Science Photo Library, p.38c Monkey Business Images/Shutterstock, p.38b J.C. Revy, ISM/Science Photo Library, p.39 CC Studio/Science Photo Library, p.40 Khoroshunova Olga/Shutterstock, p.44 Kacso Sandor/Shutterstock, p.45 Universal Images Group Limited/Alamy, p.52t Sebastian Kaulitzki/Shutterstock, p.52c Richard Griffin/Shutterstock, p.52b neelsky/Shutterstock, p.53t Manfred Kage/Science Photo Library, p.53u Power and Syred/Science Photo Library, p.53l Laguna Design/Science Photo Library, p.53b Pborowka/Shutterstock, p.54t Charles B. Ming Onn/Shutterstock, p.54b Eye of Science/Science Photo Library, p.55l magmarcz/Shutterstock, p.55r BSIP, Chassenet/Science Photo Library, p.56 Jacopin/Science Photo Library, p.59t Hilary Brodey/iStockphoto, p.59b CDC/Science Photo Library, p.60t Darrenp/Shutterstock, p.60b Keichi Nakane/AP/Press Association Images, p.62t Joy Brown/Shutterstock, p.62b Roman Ivaschenko/iStockphoto, p.64t robert paul van beets/Shutterstock, p.64b Robert Asento/Shutterstock, p.66t Anna Lurye/Shutterstock, p.66c Alexey Gorichenskiy/iStockphoto, p.66b LianeM/iStockphoto, p.68t dbimages/Alamy, p.68b Caro/Alamy, p.69t Pichugin Dmitry/Shutterstock, p.69b Alleyn Plowright/iStockphoto, p.70 Willie Nelson in Blonde Ambition courtesy of AF archive/Alamy, p.72t NASA, p.72c Julija Sapic/Shutterstock, p.72b malcolm romain/Shutterstock, p.73 Bart Coenders/iStockphoto, p.75 Vinicius Tupinamba/Shutterstock, p.76t European Space Agency, p.76b Jan Hinsch/Science Photo Library, p.78 wim claes/Shutterstock, p.80t Kristina Pchelintseva/Shutterstock, p.80b Sinisa Botas/Shutterstock, p.81 Andy Hooper/Daily Mail/Rex Features, p.82 Cordelia Molloy/Science Photo Library, p.84 Alexey Stiop/Shutterstock, p.85 James King-Holmes/Science Photo Library, p.86t Alexander Raths/Shutterstock, p.86b TEK Image/Science Photo Library, p.88 AM29/iStockphoto, p.96t haveseen/Shutterstock, p.96u Charles D. Winters/Science Photo Library, p.96l RabidBadger/Shutterstock, p.96b Gustoimages/Science Photo Library, p.97t Seleznev Valery/Shutterstock, p.97u alxpin/iStockphoto, p.97l NASA, p.97b Andrew Lambert Photography/Science Photo Library, p.98t ErickN/Shutterstock, p.98b Andrew Lambert Photography/Science Photo Library, p.100 Andrew Lambert Photography/Science Photo Library, p.102t Andrew Lambert Photography/Science Photo Library, p.102b Adam Hart-Davis/Science Photo Library, p.103 Andrew Lambert Photography/Science Photo Library, p.104l Rainer Walter Schmied/iStockphoto, p.104r Charles D. Winters/Science Photo Library, p.106t AJ Photo/Science Photo Library, p.106b Prill Mediendesign & Fotografie/iStockphoto, p.110t Laurence Gough/Shutterstock, p.110b alina_hart/iStockphoto, p.112 Andrew Lambert Photography/Science Photo Library, p.116t Ramon Berk/Shutterstock, p.116b Charles D. Winters/Science Photo Library, p.120 Charles Brutlag/iStockphoto, p.121 NASA, p.122tl MaxFX/Shutterstock, p.122tr Linda Hides/iStockphoto, p.122bl marema/Shutterstock, p.122br Rob Bouwman/Shutterstock, p.123 Bram van Broekhoven/Shutterstock, p.124t Sabine Kappel/Shutterstock, p.124c Andrew Lambert Photography/Science Photo Library, p.124b Andrew Lambert Photography/Science Photo Library, p.125 Andrew Lambert Photography/Science Photo Library, p.126 Charles D. Winters/Science Photo Library, p.128t Javier Trueba/MSF/Science Photo Library, p.128b Andrew Lambert Photography/Science Photo Library, p.129 David Taylor/Science Photo Library, p.140t tunart/iStockphoto, p.140u Richard Megna/Fundamental Photos/Science Photo Library, p.140l Natalija Brenca/Shutterstock, p.140b Luis Santos/Shutterstock, p.141t chiara levi/iStockphoto, p.141u craftvision/iStockphoto, p.141l fdimeo/Shutterstock, p.141b Anna Hoychuk/Shutterstock, p.144 Charles D. Winters/Science Photo Library, p.146t Alan Freed/Shutterstock, p.146b Philippe Psaila/Science Photo Library, p.147 Nick Fraser/Alamy, p.148 Volker Steger/Science Photo Library, p.150t Lucy Baldwin/Shutterstock, p.150c W. Disney/Everett/Rex Features, p.150b Andrew Dorey/Shutterstock, p.152l Andrew Lambert Photography/Science Photo Library, p.152r Andrew Lambert Photography/Science Photo Library, p.154t Art Directors & TRIP/Alamy, p.154c ARENA Creative/Shutterstock, p.154l Bloomberg/Getty Images, p.154r svetlana foote/iStockphoto, p.156 Khram/Shutterstock, p.158 Simon Owler/iStockphoto, p.160 haveseen/Shutterstock, p.162l Andrew Lambert Photography/Science Photo Library, p.162r i love images/Alamy, p.162b Knud Nielsen/Shutterstock, p.163 NASA/GSFC, p.164 Sheila Terry/Science Photo Library, p.166t Sheila Terry/Science Photo Library, p.166b David J. Green/Alamy, p.166r mediablitzimages (uk) Limited/Alamy, p.167 Andrew Lambert Photography/Science Photo Library, p.168tl aleks.k/Shutterstock, p.168tr szefei/Shutterstock, p.168bl russ witherington/Shutterstock, p.168br Natalia Klenova/Shutterstock, p.169 Andrew Lambert Photography/Science Photo Library, p.170l Stephen Aaron Rees/Shutterstock, p.170r Robyn Mackenzie/Shutterstock, p.176 Nomad_Soul/Shutterstock, p.184t ollirg/Shutterstock, p.184u Michael Krinke/iStockphoto, p.184l Lion_&_Croc/Shutterstock, p.184b Vadim Ponomarenko/Shutterstock, p.185t Snaprender/Shutterstock, p.185u Scott Andrews/NASA, p.185l Sybille Yates/Shutterstock, p.185b yakup yücel/Shutterstock,